ECONOMIC HISTORY

THE ECONOMIC HISTORY OF THE BRITISH IRON AND STEEL INDUSTRY, 1784–1879

THE UK

THE ECONOMIC HISTORY OF THE BRITISH IRON AND STEEL INDUSTRY, 1784–1879

Essays in industrial and economic history with special reference to the development of technology

ALAN BIRCH

Routledge
Taylor & Francis Group
LONDON AND NEW YORK

First published in 1967

Published in 2006 by
Routledge
2 Park Square, Milton Park, Abingdon, Oxfordshire OX14 4RN
711 Third Avenue, New York, NY 10017

First issued in paperback 2014

Routledge is an imprint of the Taylor and Francis Group, an informa business

British Library Cataloguing in Publication Data
A CIP catalogue record for this book
is available from the British Library

The Economic History of the British Iron and Steel Industry, 1784-1879
ISBN 0-415-38248-3 (volume)
ISBN 0-415-37841-9 (subset)
ISBN 0-415-28619-0 (set)

ISBN 13: 978-1-138-86532-7 (pbk)
ISBN 13: 978-0-415-38248-9 (hbk)

Routledge Library Editions: Economic History

THE ECONOMIC HISTORY OF THE BRITISH IRON AND STEEL INDUSTRY 1784-1879

THE ECONOMIC HISTORY OF THE

BRITISH IRON AND STEEL INDUSTRY

1784-1879

Essays in industrial and economic history with special reference to the development of technology

ALAN BIRCH

FRANK CASS & CO. LTD.
1967

First published in 1967 by
FRANK CASS AND COMPANY LIMITED
67 Great Russell Street, London WC1

CONTENTS

PART III

APPENDICES

ILLUSTRATIONS

ACKNOWLEDGEMENTS

IT will be quite evident from the sources cited in this study that I have been given boundless access to records and much help by archivists, librarians and fellow research workers; throughout the industry, too, I have received every courtesy and kind attention. To add to my sins of omission in not being able to express my gratitude to these many friends individually, I must also run the risk of apparently and invidiously distinguishing a number of people and institutions without whose help this study would not have been written.

First, I must thank the University of Manchester for encouraging and financing the initial work as a research student; later, during the years 1962–3 for the award of a Simon Fellowship for me to write this present book. (I should also acknowledge the earlier assistance of the Bank of England Houblon-Norman Foundation.) In 1963, I was the honoured guest—at least, that is how my good friend Herr Reiffer, the Librarian, made me feel—at the Eisenbibliothek, Paradies, Switzerland.

Apart from his many kindnesses, I have to thank Herr Reiffer as Editor of the Eisenbibliothek's *Nachrichten* for permission to reprint a section of this book on Bessemer and the Steel Revolution which was first published there; I have to make similar acknowledgements in respect of an article appearing in the *Bulletin* of the John Rylands Library; for a contribution of which I was joint author with my fellow-student and friend, Michael Flinn, on the early Yorkshire steel industry in the *Yorkshire Bulletin of Economic and Social Research*. Mr. R. B. Pugh, General Editor of *The Victoria History of the Counties of England*, has also most kindly put aside any possible difficulties about the use of a piece written for a volume of the *V.C.H. Staffordshire*, which will also appear this year.

Mrs. Sybil Jack, Senior Tutor in Economic History at the University of Sydney, very kindly read the galley proofs and helped remove several awkward and ambiguous excrescences of language—she is, of course, not responsible for the final product. My wife, too, like all wives a stern critic (as Professor Ashton once wisely remarked, an under-estimated force behind the decisions which have made up economic history), has influenced the work throughout in various subtle, but potent ways.

Finally, and here most certainly, I am aware of my inability to express adequately my gratitude for the pervasive, practical help and friendship of my teacher, Dr. W. H. Chaloner, without whom, quite literally, this book would never have been published. Indeed, almost every economic historian working upon the Industrial Revolution must have received the benefit of Dr. Chaloner's vast store of learning and out-of-the-way fact to help them on their way; so much so that one is likely to take his generosity for granted. To him, to his colleagues at Manchester, and to the late Professor Arthur Redford, as an imperfect economic historian trying to live up to the Manchester tradition, I owe all.

ALAN BIRCH.

SYDNEY, N.S.W.
NOVEMBER, 1966.

PREFACE

Innovation and Enterprise

The Rise of the Modern British Iron and Steel Industry

As I explain later, a considerable time has elapsed between the first draft of this book and its final form. During that time the study of economic history in Britain has changed to new directions and perspectives and I, too, have had time enough to consider what contribution I can make to this subject. By way of introduction to this book I would like to bring together a brief discussion of some of the problems which have been perplexing me, particularly as I grappled to forge my history of the iron industry into something new. First of all, a brief statement of an attitude to the study of economic history in general, since, if I am not mistaken, English historians, more than most, are particularly concerned with this problem—the beginning and end of writing history—that is how to put flesh on to the economic skeleton of quantities and statistics. In what follows, I am in fact all too aware I have not succeeded as well as I might have done.

There is a danger of economic history becoming too abstract. Where would the economic historian be without his terms 'development', 'growth', 'trends', etc.? Indeed, much as one feels sympathy for Professor Hexter's stricture that 'these simple words have enabled historians to conceal their ignorance from others and themselves', the nature of the subject seems to make the use of such concepts imperative. After all, the study of an economy is the study of an abstracted level of one aspect of the activities of man in society, and since the economic historian studies the changes in the size and structure of that abstraction it is understandable if he speaks of some organic process such as growth or evolution. Professor Hexter is talking about that seamless web of history, and, as he says, these seemingly more precise descriptions of change are like 'sealed junction boxes', needing to be prised open, to reveal the connections which have induced the currents of history to change direction. Nevertheless, it is true that our understanding of critical phases of economic history such as the Industrial Revolution has been sharpened by using the statistical and analytical terms which have enabled us to state more precisely those changes in the population, in agriculture, industry, trade and so on. At the very least, these 'trends' reveal to us the contours of the changing economic landscape clothed now in the newly-hedged enclosed fields and the smoking chimney stacks of the town's power-driven manufactories.

To return to my first point, however. When we speak of 'the development of enterprise' or of 'innovation' we should not lose sight of the concrete realities of men shaping their historical contexts. We are far more likely to understand the extent of the achievements and problems of the pioneers of industrialisation by studying in detail, for instance, the lives and business careers of such men as Boulton and Watt, for example, than by using these abstract terms.

I was brought back to this conclusion very forcibly quite recently when looking at the history of one of Britain's largest steel firms. Here is this firm with a vast works on a site extending over several square miles, employing something like 17,000 men and women, to say nothing of representing a capital expenditure of over £60 millions. All this had started from the launching out by a Lancashire clog maker into nail making with a new machine costing £40 just over 110 years ago. There followed the train of rolling mills, puddling furnaces, steel mills, blast furnaces, coke ovens and the great automotive strip mill, the cost of which is to be reckoned in terms of millions of pounds. Here, I think, is a challenge to the economic historian: to explain a transformation which would have been beyond the realms of possibility to the man who set this chain of circumstances in motion three generations ago. And here one would agree with Hexter, it is not much use talking of 'development' or any other such term, no matter how dramatic. Nor for that matter is it particularly enlightening to refer to this 'explosion' of enterprise and organisation as an example of the process of 'built-in compound interest'. This does show, of course, the measure of 'automatic' growth, just as a sum of money deposited in the bank, merely by lying in an account, will double or treble itself over a period of time.

But if we discount the exponential version, it must be kept in mind that, as industrial historians, we are faced with a challenge to explain the unforeseeable, this miracle of growth. Can we be sure when we have looked carefully at the characters of the businessmen who took the important decisions along the way, and when we have taken into account the conjuncture of a firm's progress—the elements of forecasting and chance bearing upon the efficacy of those decisions—that we as historians have reduced the course of even a narrow sector of human history to a clear and all-embracing formula? I do not like to think so.

From such generalities I turn to certain thoughts prompted again by the question of what the historian is endeavouring to say in the particular context in which he writes. The writer of a volume of economic history is much like one of his entrepreneurs whose beginnings in business he tries to describe. The emergent captain of industry is seldom launching a completely new unique product to be bought up rapidly by eager customers—each economic good is in some way a substitute for another, to be fitted into an adjusted pattern of expenditure. This train of thought does not have to be elaborated. However, the economic historian, too, is bringing his product—his picture of the past—to the academic market place, here to establish its place among the standard works on the book-shelves. That picture or interpretation is seldom revolutionary in breaking away completely from its predecessors; it has evolved out of them and commends itself to the attention of readers by offering a changing perspective to the study of the solution of the economic problem, the allocation of scarce resources. If the 'new' interpretation is convincing it succeeds by changing the pattern of economic development already in scholars' minds.

Now, in the world of commerce, the evolution of a product generally follows an appraisal or estimate of consumers' demands. Some innovations mould consumers' tastes, others are created after a shrewd diagnosis of what the public is likely to demand. Yet, of course, the suggestion that pieces of academic research are stimulated by an awareness of the demand for what is likely to be accepted may appear shocking to scholars. The traditional picture of the scholar pursuing a line of enquiry arising purely out of his own disinterested interests persists. Nevertheless, one's own academic interests are influenced in some degree by the kind of problems which are being talked about around one. Often the kind of book one would like to write is that very book which would definitively work out the implications of approaches to the subject which are being currently discussed. This then is the demand for the academic product.

All this is, perhaps, a clumsy preamble to a statement how this present work has evolved. It had its beginning in the stimulus of Professor Ashton's brilliant analytical and descriptive book: *Iron and Steel in the Industrial Revolution.* This has, of course, long been regarded as a classic and its influence is such that it has created for us an accurate and memorable map to the historical territory of the Industrial Revolution. The idea of my own research was to supplement the details given by Professor Ashton in the light of records of ironmasters and ironworks made available to the historian after his pioneering survey. On due consideration I now regard any attempt to rewrite *Iron and Steel* as an impertinence. I shall not attempt to do this. I had also intended to extend the account of the industry's history beyond the traditional boundary set to the Industrial Revolution in Britain (*c.* 1830) and to take up the story to 'The Age of Steel', when Bessemer and Gilchrist Thomas laid the foundations to the modern iron and steel industry, with a new technology and a new set of factors of localisation.

This I did with what I believe was some measure of success in a thesis written in 1951–53. Since then that work had achieved the accolade of academic anonymity by figuring as a reference in several detailed studies of the industry.

In that long interval the study of economic history has developed a great deal; in particular there has been a thorough sifting and shaping of the statistical analyses of the growth of the British economy at that period; there has been achieved a deeper understanding of what were the economic and social changes of the Industrial Revolution. Economic historians have benefited from an increasing awareness of the many studies of economic growth carried out by economists, sociologists and others. In short, though no one could ever accuse Ashton of ignoring economic analysis in his economic history, it is not so easy for the economic historian working today to concentrate, for example, on technological progress to the exclusion of a consideration of problems of capital and enterprise; nor, for that matter, the end point of production—the market.

In the introductory essay to this volume I try to work out some of the implications of this changed situation in terms of both technology and management in the iron and steel industry and significantly, I hope, I reach out to Rostow's conception of industrial history treated as an aspect of the study of economic growth. In the rest of this volume I aim to forge a structure, not too 'cold-short', that will measure up to Rostow's specification.

In this task I am now helped by a number of works which have been published since my initial researches and which I now include in the amended bibliography. There has been a study of the changing location of the industry in Britain; there have been a number of studies examining the individual histories of old-established concerns, such as Carron, in considerable detail. There has recently appeared a history of the industry setting out its salient features. Of course, I cannot pretend that this book presents the results of researches warm from the excitement of working through hitherto neglected and newly-discovered manuscripts—though I once experienced this to set me on the historian's course.

The themes of this reappraisal, as I have rather loosely set them out, are Innovation and Enterprise. By the latter term I mean not only the personal qualities of those industrial pioneers such as Darby, Crowley, Guest, Bessemer, but also the forms of business enterprise which they shaped. The process of industrialisation has often been analysed in terms of capital investment; the creative drive of the entrepreneur, as I have suggested above, is no less important. But the one cannot work without the other. My term Enterprise embraces both these critical factors. Finally, I shall also have to give a brief account of the changes wrought in the industry—in its great expansion of plant and product and in its regional setting. Out of all this will emerge, I hope, a picture of an ancient industry being transformed geographically and technologically until, by 1880, the British iron and steel industry stands on the threshold of modern developments.

Part I

B

CHAPTER I

The British Iron and Steel Industry 1760–1879

A straightforward economic history of the iron industry—and here one is thinking of the coke-iron industry during the period from 1760 (when Dowlais and Carron, the two giants of the modern industry, came into production) to 1870 when the age of steel comes into being—would treat the following subjects:

(i) Technology.
(ii) Growth—output, location, markets.
(iii) Organisation and integration, large firms, associations.
(iv) Labour in the industry.

This approach is compartmental and, of course, is open to the objection that, since the economic historian is as much concerned with the process of change as with the results, the task of the historian is to trace and show the interaction, the movements between the factors of production. The springs of this movement have traditionally been sought in technological advances—coke smelting, the puddling and steam-driven rolling processes and the successive improvement on these. Conversely, although this has not so clearly been shown, the motive force of supersession could well be demonstrated as forging the structure and location of the industry.[1] It is true that this appears to take us back to Schumpeter's concept of innovation (involving creative destruction by the new leaders of industry), particularly as a production function.[2] As Schumpeter sets it out:

[1] As long ago as the 1920s, Professor J. H. Jones referred to 'obsolescence', this constant factor of industrial development, 'as an economic factor . . . one of the neglected subjects in the scientific study of industry'.

[2] V. W. Ruttan, 'Usher and Schumpeter on invention, innovation and technological change', *Quarterly Journal of Economics*, Vol. 73, 1959, p. 599.

> 'This function describes the way in which quantity of products varies if quantity of factors vary. If, instead of quantities of factors, we vary the form of function, we have an innovation.'[3]

However, as has been pointed out, Schumpeter has given explicit attention to the process by which innovation—technological and organisational change—is generated. The problem of invention, which may be said to spark off the process of innovation, is a fascinating one and it would be tempting to dwell on A. P. Usher's theories about this. The implications of accepting Usher's 'accumulative synthesis approach',[4] which is not incompatible with Schumpeter's work on the effects of innovation on economic development, reinforces Schumpeter's emphasis upon the practical application of innovations in technology and in economic innovation.

It is this practical application of technological advances which Professor Habakkuk examines in his recent book and which bears directly on the British experience in the iron and steel industry. The iron industry affords, in the case of the long time-interval between the early experiments in coke smelting in the early part of the seventeenth century and Abraham Darby's success in Coalbrookdale—and then the still slow adoption of the new process through the eighteenth century, a classic instance of the process of innovation. It is essentially this problem of determining the factors retarding or hastening innovation in its full sense which Habakkuk analyses.[5] One of his starting-points is the statement:

> 'Technical progress has a logic of its own; one invention raises problems intellectual and mechanical, which require time and effort for their solution and which, being solved, suggest new lines of advance. At any moment, only certain lines of advance are technically possible, and with the best will in the world one can only go a certain distance. At some periods, for purely autonomous reasons, the technical knowledge available to be incorporated into practice is richer than at others.'

[3] Schumpeter, *Business Cycles*, Vol. I, pp. 87–8, cited in Ruttan, *loc. cit.*, p. 598.

[4] As Ruttan explains it, Usher's theory of invention is defined in terms of the emergence of 'new things' which require an 'act of insight' going beyond the normal exercise of technical or professional skill. They are 'unlearned activities that result in new organisations of prior knowledge and experience' (Usher, *Technical Change and Capital Formation*, p. 526). For an account of Usher's 'cumulative synthesis approach', see Ruttan, pp. 601–2. It is significant that it removes the distinction between the processes of invention and innovation since it includes the final stage of critical revision where the newly perceived relations become fully understood and effectively worked into the entire context to which they belong.

[5] H. J. Habakkuk, *American and British Technology in the Nineteenth Century* (Cambridge, 1962), pp. 181–2.

Another feature of technical progress, a fact of common experience to industrial researchers but now only being recognised by historians of technology, is the need for the researcher to go back to former experiments written off as failures or inconclusive. This applies to theoretical and applied knowledge; industry aided with new fundamental techniques of analysis—I am thinking of the statistical approach—and new means of control and investigation, must learn from the past in this way.

Habakkuk incidentally illustrates this line of development:

> 'Compared with the last forty years of the eighteenth century', he says, 'the first half of the nineteenth was not a period of major technological breakthroughs in English industry. Down to the mid-century and indeed well beyond, the principal advances sprang from remedying imperfections in the basic inventions of the industrial revolution and applying the principles involved in them to new fields. It was not until the 1870s or 1880s that additions to fundamental knowledge assumed technological significance.'[6]

Apart from the stream of know-how available to the industrialist, he must also take into account the economics of supplying capital for the machine or even research and the economics of labour, which the innovation may utilise more efficiently, giving greater productivity. However, the ironmaster has to give priority to the problem of fuel and power. We would agree with Habakkuk that a shortage of both was an incentive to invention in England.[7] The problem of timber supplies for the forges and furnaces of the eighteenth century, we have come to learn with the guidance of Michael Flinn, and other scholars, must not be considered in absolute terms of scarcity. And though this is not the place to discuss at length this much-vexed fuel problem, we recognise that taking into account the possibility of securing crops of wood at increasing costs; of tapping new virgin areas of woodland in Furness and Scotland, and *per contra* the increased demand for iron from the growing Black Country metal industries; there was pressure on the ironmaster to seek a substitute for charcoal.[8] Further, as we shall see, when the use of coal had been proved to be a technical success,[9] and even when it

[6] Habakkuk, *op. cit.*, p. 182.

[7] Cf. 'Much of the history of innovation in the English primary iron industry in the eighteenth century can be explained by reference to fuel on the same lines as American developments in the nineteenth century have been explained by reference to labour'. Habakkuk, *op. cit.*, p. 157.

[8] Foreign competition, despite tariff protection, set an upper limit to which higher costs could be passed on to the manufacturer.

[9] In 1709, the Coalbrookdale furnace was yielding only 1·4 tons of pig iron a

appeared to be relatively cheap and inexhaustible, the incentive to fuel economy—from the hot blast of J. B. Neilson to the Bessemer process—continued to be a most powerful active force behind technological innovation.

The plant in an iron and steel works was certainly capital-intensive and here again is a problem of substitution of machinery for skilled and unskilled labour. Nevertheless, the degree of possibility of substituting the factors of production in the iron and steel industry in the nineteenth century was far more restricted than it is today. Moreover, the cheapness of pauper labour was said to be an explanation of the backwardness of the iron industry in Britain.[10] Nevertheless, there were abortive attempts to introduce some mechanisation in the wrought iron branch of the industry, e.g. mechanical puddling from the 1830s onwards. Here, humanitarian as well as economic reasons were apparent.[11] When, in the later 1860s, the technical problems were overcome, then the strength of union resistance to mechanisation was able to cancel out its advantages. The outcome of this situation *vis-à-vis* capital and labour was that during the period great improvements were made in the blast furnace to give greater production, higher efficiency, greater regularity and longer periods of productivity. Here capital expenditure could yield returns free from the traditional impost of labour.

The historian, and particularly the business historian, asks who were the entrepreneurs on whose judgment depended the critical decision as to the timing and the scale of any technological innovations. Of course, for any single industry numbering a large number of firms over a longish period of time this is very difficult or next to impossible to determine. 'It would demand' (I quote from Miss Erickson, who has faced this problem for her study of British business leaders) 'intimate details of how decisions were actually made within the individual firm. . . .'[12] The adoption of the Bessemer and open-hearth processes which created the heavy steel industry in

---

week or a total of 81 tons in the year. Even in 1719, when the make had risen to about 200 tons, this still compared unfavourably with the 700 tons output—and even exceeding 1000 tons output—of some of the Forest of Dean furnaces. In Shropshire the average capacity in 1720 was 380 tons. It does not, therefore, seem very realistic to suggest that, at Coalbrookdale, this situation was met by building another coke-fired furnace.

[10] This was denied by Sir Lowthian Bell (Habakkuk, *op. cit.*, p. 201).

[11] If labour was cheap, it was also alleged that it was expensive in its lack of discipline and untrustworthiness.

[12] C. Erickson, *British Industrialists: Steel and Hosiery 1850–1950* (C.U.P., 1959).

the years 1856–79 does, however, constitute a major innovation involving the entire structure of the industry. It provides, too, a testing ground to ask the following questions:[13]

> Was the innovator 'almost as often' an inventor as he was a capitalist?
>
> Where did he obtain capital for making changes?
>
> Did the innovators as a class come from virtually all ranks of society?
>
> Whether or not firms which first adopted the new methods of making steel amassed the 'surplus gains' which Schumpeter regarded as the 'distinctive' returns on the exercise of the function of entrepreneurship?

As Miss Erickson suggests, 'it seems likely that during the critical phase of the modern steel industry, when new firms, new areas and new processes were going into production, a different class of men would have been attracted to it and would have found more opportunities in it than in later phases, when the industry was more conservative. . . .' This assumption that such differences would be found has been made by historians of the industry.[14] 'Is it true,' she asks, 'that there was an opportunity for bold and able men to overcome a want of the relatively large amount of capital for a small Bessemer converter plant?' Alfred Marshall, we are reminded, taught us to expect to find such cases of strikingly successful careers in the face of heavy odds during periods of revolutionary change in industry. The history of the steel industry offers an opportunity to test this hypothesis. Miss Erickson's conclusions may be set out briefly:[15]

(i) The relatively young firms appeared most frequently among the Bessemer innovators.
(ii) Most of these firms had developed markets for heavy steel products; connections with the railway and armaments industries were of the greatest importance.
(iii) All the Bessemer and open-hearth leaders had large internal capital resources before embarking on the innovation.
(iv) Conversely, reliance on the outside capital market led to 'the most conspicuous abortive attempts to enter the steel trade'.[16]

[13] *Op. cit.*, p. 141.
[14] Erickson, *op. cit.*, p. 9.
[15] *Op. cit.*, p. 165.
[16] Miss Erickson seems to suggest that these company promotions created absent and inactive boards of directors, i.e. there was no real entrepreneur represented.

(v) Innovation proved risky, leading to a high mortality rate. Of the group of 31 firms studied in this particular phase of the research, 14 had failed and sold out to new companies by 1900.

When we look at the results of Miss Erickson's studies for the light they throw on the connection of technical experts with the innovating steel firms, we find that although the founders of these companies were mainly commercial in origin (the founders had received their apprenticeship in a merchant's counting house followed by a venture with independent business), necessarily there was a technical expert in a leading position. As one would expect, the technical demands of perfecting and controlling the new processes, depending more and more on chemical and metallurgical knowledge, made this association a critical condition of success. The career patterns and technical experience of these men were clearly directed towards the emergence of the professional salaried expert.[17]

These conclusions possibly need some development—one would like to know more, for instance, about the reasons for the failure of this significantly large group of innovatory firms. Besides the risks of the technical innovation, the steel industry also experienced the impact of a financial reorganisation of its capital through the adoption of the limited liability company structure. The influence of investors was reflected, for instance, in the number of 'outside' directors with and without executive experience. But the problem of the retention of family control of firms through the generations has been present from the beginnings of the industrial dynasties in the eighteenth century. In this survey of recent literature, and particularly the business histories which will now be examined, this problem is exhibited in its most acute forms.

Michael Flinn has given us a remarkable example of how to write a business history without the records of the firm.[18] The author has assembled a detailed structure out of a widely scattered array of external sources, among them the rich Swedish travel diaries of Angerstein and his successors. It is both a biography of the Crowley family and a study of the commercial and industrial organization of a highly integrated business (but which excluded blast furnaces).[19] Flinn is able to assemble evidence to indicate the reasons behind the

[17] Erickson, *op. cit.*, p. 170.

[18] *Men of Iron: the Crowleys in the Early Iron Industry* (Edinburgh, 1962). The most voluminous and detailed records still extant are, of course, the famous Law Book regulating the domestic and industrial activities of the workers at Swalwell.

[19] Except for a short period lease of the Ashburnham furnace in Sussex between 1730 and 1740. He does not say why this was given up.

decision to locate the headquarters of the business in the North-east of England; the policies determining the role of the various branches of the firm in the North and the Midlands; the problems of obtaining raw material—timber, charcoal and iron—from England and overseas; and most important, the marketing of the products of the forges and slitting-mills from the factories and from the London warehouse; again for an inland and overseas trade. Above all, Flinn is able to assess the significance of Ambrose Crowley III's decision to establish the main centre of the firm in the North-east in the context of the national economy at the time. 'Ambrose III's supreme stroke of genius', we are told, 'had been to appreciate the advantages of the North-east under the transport and market conditions of the late seventeenth century.'[20] Again, the decline of the firm is linked with the rise of a native wrought-iron industry. 'The centres of consumption had shifted to the Midlands; the replacement of Swedish and Russian iron by a superior and cheaper native product after the technical advances of the eighteenth century destroyed the advantages of Newcastle's convenience for the Baltic trade; while improved inland transport lifted the last barrier to the economic advance of the Midlands.'[21] This is the setting for this valuable account of an outstanding enterprise.

Yet, as Flinn is the first to admit, the lack of certain kinds of records inevitably leaves unanswered many questions an economic historian wishes to ask and which, in the main, can only be answered from a firm's own records. Lacking regular accounts, of course, there can be no close examination of how the capital of the enterprise was deployed, nor of its profitability in response to the changing conditions of the market. The composition of the firm's assets is revealed when a Crowley dies or when a marriage settlement has to be arranged. But, fortunate as this may be for the historian, these natural terms of family interest spell also the decline of family control. In 1782, when Theodosia died, after a twenty-year period as head of the Crowley family, Isaiah Millington, 'the last of a succession of salaried managers', entered into a partnership with the Crowleys. Millingtons administered the business from 1782 until its final collapse in 1863.[22] And it is from this period local competition comes into the picture and here Flinn is forced, by his lack of records, to hazard a guess that perhaps the vigorous and enterprising leadership of earlier years was not there. 'The factories at Swalwell and Winlaton mill *seem* not to have been able to offer effective resistance to their energetic competitors.'[23] I do not need to

[20] Flinn, *op. cit.*, p. 95. [21] *Ibid.* [22] Flinn, *op. cit.*, p. 9.
[23] Flinn, *op. cit.*, p. 93.

say that for the business historian such problems of cost and price policy are of great moment.

These figure prominently, of course, in J. P. Addis's book *The Crawshay Dynasty*.[24] However, the very title of the book does suggest that family control was held together generation after generation from Richard Crawshay's association with Anthony Bacon's interests in Cyfarthfa from 1777 until 1890 when the family business became a limited liability company. Even then the three Crawshay brothers of that generation—the great-great-grandsons of the founder—were permanent directors. However, only one Crawshay had lived at Cyfarthfa Castle and he had moved to Caversham before 1890. Moreover, this reorganisation, to make possible the conversion of the conservative ironworks into a modern steelworks, like Ebbw Vale, merely heralded the take-over of Cyfarthfa by the newly-amalgamated Guest, Keen, Dowlais and Patent Bolt Company. As Addis says, 'this was virtually the end of Cyfarthfa'.[25] The reasons for this demise can be put down simply as the result of the Welsh ironmasters' resignation to powerful exterior forces and the recognition that the days of prosperity for the South Wales coal- and iron-field were running to an end. For during the early 1860s, a time of dislocation of trade through the American Civil War and of increasing Continental competition, William Crawshay had voiced the opinion that 'the star of the iron trade was fast setting'.[26] The Crawshays should get out of the iron industry, he advised. In 1875, after several unsuccessful negotiations to sell the works, Cyfarthfa was closed for four years as a result of labour troubles.[27] None of these circumstances were new; what was changed was the determination of the Crawshays to take up arms against this sea of troubles. Unfortunately, too, this period of crisis is not well documented and, therefore, the business historian is unable to inform us about the economics of adjustment.

The South Wales iron industry does epitomise the main phases of development of the British industry as a whole. First the establishment of the industry in the Elizabethan era to obtain supplies of timber, then the new point of departure, the establishment of the coke-fired furnace on the eve of the Industrial Revolution; then the real take-off through the perfecting and exploitation of Cort's puddling and rolling processes to make wrought iron, which took

[24] University of Wales Press, Cardiff, 1957.

[25] Addis, *op. cit.*, p. 150.

[26] Quoted in J. H. Morris and L. J. Williams, *The South Wales Coal Industry 1841–75* (University of Wales Press, Cardiff, 1953).

[27] Addis, *op. cit.*, p. 147.

the industry to its peak of prosperity. Decline sets in with the introduction of the Bessemer steel converter demanding phosphorus-free ore. This shift in the balance of the costs of production was paving the way for the move of the South Wales industry to the coast.

In 1890, when Lady Charlotte Schreiber paid her last visit to Dowlais, this move to Cardiff had already been decided upon. And at Dowlais, through the collection of Dowlais records, brought together and edited by Miss Elsas,[28] we can trace again the course of this regional expansion and decline. But for many of us the fascination of the Dowlais story lies in the brave efforts of Lady Charlotte to be an ironmistress. She did, as we know, pay the closest possible attention to the day-to-day working of Dowlais as well as to the welfare of the town of Merthyr Tydfil. Her diaries[29] reveal her inspecting the mines—not in state but travelling on loads of coal—supervising the payment of wages, negotiating with customers over contracts and with her neighbour ironmasters over the tricky decisions affecting labour. This came about largely as the result of the death of Sir John Josiah Guest in 1852. Then a board of trustees under the direction of William Menelaus, engineer, was appointed to manage the affairs of Dowlais. The technician had come into his own.

Carron, by contrast, has remained an independent concern. In one sense the classical *locus* of the combination of scientific empiricism, capital and business acumen brought together by the captains of industry of the Industrial Revolution—Garbett and Roebuck—it is yet another starting-point for the main stream of industrial history. Its furnaces, of course, did not dramatically trumpet the beginning of the Industrial Revolution by being fired on 1st January as some books have suggested. However, apart from that auspicious commencement, Carron's history[30] does not demonstrate very much of that spirit of innovation: it is true that the carronade, that most effective piece of light armament for shipping, which was put on the market in 1778, could perhaps be described as 'a great innovation' (Campbell claims this for it): certainly it was an important product the sales of which prevented Carron from being a spectacular failure. Yet the productive techniques, apparently, were extremely conservative. Carron, for instance, had not fully gone over to the hot-blast furnace as late as fifty years after that innovation.

Campbell, incidentally, is inclined to blame the general neglect in

[28] M. Elsas, *Iron in the Making, the Dowlais Iron Company Letters 1782–1860* (Cardiff, 1960).

[29] Edited by Earl of Bessborough (London, 1950 and 1952).

[30] R. H. Campbell, *Carron Company* (Oliver & Boyd, Edinburgh, 1961).

Scotland of technical progress and, more convincingly, the preoccupation of the company with litigation. That litigation[31] which arose from a systematic policy of 'managerial machination'—actually the manipulation and concentration of the company's wealth into the hands of the Stainton family pursuing a policy of nepotism among the Stainton partners—could well illustrate Charles Wilson's 'Economic Crime Club' brand of business history. In 1865, after twelve or more years' litigation for the secret and undivided profits to be disgorged, a total of some £320,000 was paid to the Carron Company.[32] Carron's historian commends this policy of accumulation and Stainton's wish to keep out the investing 'Edinburgh writer bankers' bent on speculation; but, of course, he does admit that the result of this was these liquid reserves were held at the cost of capital development. Moreover, so much effort was spent in uncovering these machinations that 'for fifteen years little else could be achieved'. Yet Carron survived this neglect and entered a new phase of development as a modern company employing professional managers. Now, as Campbell reveals, the major shareholders were losing interest in the daily affairs of Carron. 'They were increasingly attracted by a more leisured life, or often by one actively devoted to liberal pursuits.' Previously some members of the Stainton family had managed to combine ornithology or orchid-growing with business. Now the tasks of large-scale investment and construction posed increasingly difficult and complex problems demanding full-time attention. So Carron came under the rule of full-time professional employees as managers and senior officials.[33]

This survey of some recent books written both as business histories and industrial histories is intended to open up two main areas of discussion, innovation and management in the iron and steel

[31] Campbell, *op. cit.*, pp. 194–8.

[32] The Staintons also apparently managed to pull the wool over the eyes of the Inspector of Taxes. In 1844 that official complained he could not understand why the Carron profits were far below those of other iron works doing less than one-quarter of Carron's turnover. He was given the same abstract of the accounts which had deceived the partners.

[33] Campbell, *op. cit.*, p. 241. The dependence of a business concern upon a salaried manager is not, of course, a modern development. Inevitably where family partnership interests were spread over an area of enterprise, then, as in the case of the Sheffield ironmasters in the early part of the eighteenth century, the management devolved on 'The Clerk'. He himself was a partner (a member of the Fell family) and responsible for supplies of raw materials, management at the works and for sales. Fell had an entertainment allowance for this purpose. G. G. Hopkinson, 'The charcoal industry in the Sheffield region, 1588–1775', *Trans. Hunter Arch. Soc.*, vol. VIII, Part III, 1958.

industry.[34] In concentrating on these, several important aspects of the industry's development have been neglected, chiefly the role of labour. In this respect this interpretation has been partial. Nevertheless, this does come about as well from the nature of the sources available for writing the business histories we have mentioned. It is clear, I suggest, that the writing of business histories, even with the widest context of industrial development, can be no substitute for the writing of an industrial history. Its sum is surely more than the constituent parts of the histories of individual firms.

How might one envisage such a history of industry, which is far removed from the kind of technological history published of late? Here, I suggest, it is possible that Rostow's framework of economic growth, which has not commanded the widest support from economic historians, might seem more feasible on this limited scale, the study of one industry.

Rostow indicated some of the range of factors that might be relevant to an investigation bearing upon the 'propensity to apply innovations as between different industries'. These would include: 'the age of the industry and its rate of growth; the sort of management it possesses; the degree and character of the monopolistic elements in the organisation; its outlook on its future prospects; and so forth'. He goes on to say: 'A special kind of mixed economic social, political, *and perhaps even biographical analysis* would be required to explain the response in a particular industry to the potentialities presented by the flow (and pool) of possible innovations.'[35] One might add that for the iron and steel industry one should also take into account the geographical factors determining the relocation of the industry.[36]

[34] It would be interesting, too, to trace the development of entrepreneurship in the industry through the stages suggested by A. H. Cole: the empirical, the rational and the cognitive, as the growing complexity of business problems and the proliferation of business transactions per day made the expert necessary for observing and controlling the business environment. A. H. Cole, 'An approach to the study of entrepreneurship' in F. C. Lane and J. C. Riemersma, *Enterprise and Secular Change* (London, 1953).

[35] W. W. Rostow, *The Process of Economic Growth* (Oxford, 1953), p. 127.

[36] See H. G. Roepke, 'Movements of the British iron and steel industry, 1720–1951', *Illinois Studies in the Social Sciences*, Vol. 36 (Illinois, Urbana, 1956).

CHAPTER II

Iron in the Economy of England and Wales in the Eighteenth Century

Professor Asa Briggs has spoken of the modern steel industry as 'the core of a changing economy'. During the Industrial Revolution, when iron and steel was established as a large-scale, coal-burning industry basic to Britain's industrialisation, this truth was only then being worked out. The alliance of coal and iron, as we know, forged the superstructure of an industrial civilisation. Previously, iron smelting and forging had been a rural, somewhat scattered, industry, and its products, horse-shoes, nails, chains, hoes and so on, had been chiefly used by the yeoman farmer. Now, when that Industrial Revolution was under way, the furnaces and forges, rolling mills and foundries, would be concentrated together on the coalfield; the cast iron and wrought iron would be 'manufactured' into pistons, pinions, cog-wheels, rails and girders for the new factories and their power-driven equipment.

It is true that these changes were a break with the past; timber, the traditional raw material for construction and engineering, was being superseded by the lighter, stronger and less clumsy metal. In this sense, then, the product of Abraham Darby's blast furnace at Coalbrookdale was a new product, capable of being cast into precise intricate shapes. When one compares the innovatory iron bridge across the River Severn with the ancient stone structure up-stream at Atcham, there is the evidence of the new age before one's eyes.

Yet, for all that, the rise of the iron industry in the eighteenth century was not a new departure; nor for that matter, did it represent the triumph of native skill over well-established foreign competition, relieving this country of an abject dependence, perilous in times of war, a drain of the nation's wealth when commerce was untrammelled. It is true that eventually by the end of the century Cort's puddling furnaces and rolling mills were able to supply about one-half of Britain's needs of wrought iron, and that the blast furnaces in South Wales and in the Midlands were capable of an output of more than 150,000 tons of pig iron; however, the mercantilist goal of self-sufficiency was far from accomplished. This expansion, nevertheless, was itself taking place in a context of a rising population, and during the fitting out of England's capital equipment—canals, bridges, roads and even the enclosure of the open fields.[1] The Swedish historian of the iron industry has described most admirably the conditions for a booming expansion.

> 'England's population rose in the eighteenth century from five and a half to nine millions—i.e. by about 80%. A corresponding increase in the consumption of bar iron would only have meant that iron supplies remained relatively unchanged. But in fact iron consumption in England increased more rapidly than population. The tentative figures . . . suggest an increase of between one-third and one-half in the *per capita* consumption of bar iron in England.'[2]

The mounting figures of imports can be most readily seen from this table:

BAR IRON IMPORTS

Annual averages, units of 1000 tons[3]

1700–1709	16·0	1735–1744	24·2	1770–1779	44·5
1705–1714	16·3	1740–1749	22·5	1775–1784	43·0
1710–1719	17·3	1745–1754	26·6	1780–1789	44·1
1715–1724	19·0	1750–1759	29·3	1785–1794	—
1720–1729	19·7	1755–1764	33·0	1790–1799	49·9
1725–1734	21·5	1760–1769	39·7	1795–1804	43·0
1730–1739	25·5	1765–1774	44·9		

But, as Hildebrand says, the change is much more drastic than appears from these figures. There was a substitution of cast iron instead of wrought iron, as well as for purposes where formerly no iron was used at all. 'Only when viewed in this setting do the full

[1] Dr. A. H. John has reminded me that home demand was always far more important to the iron industry than the more spectacular call for arms during the several and not infrequent wars of the eighteenth century. It is perhaps rather pointless trying to assess actual quantities; but in the agriculture of the enclosures each field needed a gate; and a gate, with its heavy hinges and catch, might account for 30 lb. of iron. For the cultivation of the fields, the iron plough was coming into popularity, and so on.

[2] K. G. Hildebrand, 'Foreign markets for Swedish iron in the eighteenth century', *Scand. Econ. History Review*, Vol, 6, 1958.

[3] Source: extract from Table 15, Selected indications of eighteenth-century industrial growth, P. Deane and W. A. Cole, *British Economic Growth, 1688–1959* (Cambridge, 1962), p. 51. The annual figures may be obtained from E. B. Schumpeter, *English Overseas Trade Statistics, 1697–1808* (1960).

implications of the increased demand for malleable iron become apparent; it was part of a more general advance on the road to the industrial iron age of the nineteenth century.' We have spoken already of the changes in agriculture. Professor Hildebrand quotes an eighteenth-century ballad bemoaning the new-fangled demand for ploughs made of cast and wrought iron:[4]

> 'They mun all be of iron, and wood there's no trade for.
> Why, what do the fools think as Ash Trees were made for?'

Professor Court cites a striking example of this penetration of the countryside by iron. A single ironmonger in 'a deep country district like the Vale of Evesham held a stock of three tons of nails at a time for "the country sales" '.[5] In the towns, another class of iron-monger, the travelling middlemen between the merchants and the blacksmith, peddled a large array of domestic ironmongery. Angerstein, the Swedish travelling expert, records the zeal of these dealers anxious to show him samples of locks, hinges, buckles, corkscrews, tailor's irons, curling irons, surface axes and candle snuffers.

Transport between town and country—whether it was by the lumbering farm horse pulling the produce-laden cart to market; or the sure-footed pack horse heading his way across the moorland tracks; or again the hard-driven team of coach horses expending both horse-flesh and horseshoes at a rapid rate—provided a market for the dealer serving the blacksmith with horseshoe iron and tyre iron for the coach and wagon wheels.

However the great works of construction, the famous bridges of the period were perhaps, at this date, the most conspicuous examples of the demand for iron, and in the 1790s it was estimated that the bridge over the River Wear accounted for 55 tons of bar iron and 206 tons of cast iron—yet, as we shall see in a later chapter, by the 1790s architects were using cast iron beams and pillars in the structures of mills and factories. In fact, as Hildebrand points out, the demand for iron was determined by the general level of investment in housing, shipping, mining and transport as well as by a steadily

[4] Hildebrand, *loc. cit.*, citing G. E. Fussell, *The Farmer's Tools, 1500–1900* (1952). See also P. Mantoux, *The Industrial Revolution* (1928), pp. 321 ff. for a vivid picture of the increasing use of iron in other fields.

[5] Cited by A. H. John, 'Aspects of economic growth in the eighteenth century', in *Essays in Economic History*, Vol. II (ed. Carus-Wilson) London, 1962, p. 368. John stresses the close connection between agricultural improvement and the development of the Birmingham and Sheffield areas.

increasing export trade. This export trade of such diverse objects as palisades for the American frontier, or Sheffield knives and scissors for Europe, probably represented about one-quarter of the supplies of Britain's imported bar iron.[6]

There is also the case of the significant rise of the tin-plate industry in South Wales, Yorkshire, the Midlands and elsewhere, which brought into being forges and rolling mills to produce the wrought iron plate for tinning. In 1710 over two million sheets were shipped to this country from Holland and Germany: in 1750 this import trade had slumped down to a mere 1611 sheets.[7] By the beginning of the next century, in 1805, when an export business had been built up, the annual production had risen up to 80,000 boxes (= 18 million sheets), equivalent to about 4000 tons.[8] As A. H. John remarks, the domestic products were the export products too, even though as yet the quantities were not great.

Finally, in this survey of the demand for iron, the importance of the rapidly expanding British merchant and naval fleets should not be overlooked. We do know that both the number and tonnage of shipping registered in the United Kingdom just about doubled during the years 1788–1815. (The number of ships increased from 12,464 to 22,026 in 1816.) It will be readily appreciated that the requirements of iron for naval and marine stores—for anchors, chains, shackles, bolts, ballast, etc.—would be considerable. It is, of course, not practicable to estimate even an average amount of iron to be found on board one ship.[9]

To give an overall impression of production, trade and consumption of iron in Britain, then, during the eighteenth century, we may be guided by the trends shown in Professor Usher's table compiled to show this base of industrialization.[10]

These figures, which are no more than an index to the demand for iron, show that home consumption in the last twenty years of the eighteenth century—the take-off—about doubled. Although this demand was being met in some degree by the revolutionary capacities of the new blast furnaces, driven by their Wilkinson blast

[6] *Ibid.* In the 1750s an annual average of 1600 tons of nails were shipped abroad; the export of wrought iron mounted up to 17,440 tons by the 1790s (Hildebrand, *loc. cit.*, p. 19).

[7] W. E. Minchinton, *The British Tinplate Industry: A History* (Oxford, 1957), p. 114.

[8] *Op. cit.*, p. 19.

[9] Figures of shipping taken from B. R. Mitchell, *Historical Statistics* (Cambridge, 1962), p. 217.

[10] A. P. Usher, 'The industrialisation of modern Britain', *Technology and Culture* (Spring, 1960).

	Domestic production of pig iron (1000 tons)	Total supply of pig iron (1000 tons)	Home consumption per capita (lb)	Total supply per capita (lb)
1717	18·1	35·1		
1720	17·3	35·8		9·85
1736	17·5	47·7		13·1
1850	26·0	73·0		18·8
1770	32·0	103·6		24·3
1788	68·3	116·0	20·6	24·1
1800	156·0	213·0	30·8	40·5
1806	258·0	297·4	42·5	52·6

engines, such as were to be found at Dowlais or Bradley, it was still necessary to import even larger amounts of bar iron from Sweden and Russia. Hildebrand estimates that, in the 1780s, something like 66 per cent of the total supply of bar iron was foreign; it is some measure of the growing strength of the domestic industry that by the end of the century this proportion was falling back to 50 per cent.[11] This dependence on Swedish iron for its special qualities for some purposes, as conversion into steel or the manufacture of anchors, of course, continued and increased until 1913, when these imports began to decline. At that time, 200,000 tons of bar iron were shipped to Britain, the re-export trade being negligible. By then, the import and export trade in iron and steel was enormously complicated and the categories in the official statistics do not enable us to make a precise comparison with the import situation at the beginning of the nineteenth century, for example. However, the main trend of the industry's development in the eighteenth century was carried on and it was necessary to supplement the increasing production at home for the conversion of semi-finished products to higher value products (e.g. cannon in the eighteenth century; railway material in the nineteenth; special alloy steels in this) by a greater supply of imported bar and wrought iron and steel.

The sources of that bar iron in the eighteenth century were mainly from two countries, Sweden and Russia. Professor Hildebrand's table readily sets out the changing supply factors operating at that time.

[11] Hildebrand, *loc. cit.*, p. 10. In 1878 the imports of bar iron exceeded 100,000 tons: however, 60,000 were re-exported and the balance accounting for $\frac{40{,}000}{1} \times \frac{10}{7}$ (Usher's factor of conversion) equals *c.* 57,000 tons of pig iron. This might, in turn, represent less than 0·1 per cent of the total British pig iron output. Figures taken from B. R. Mitchell, *Historical Statistics* (Cambridge, 1962), p. 141.

SOURCES OF ENGLISH IMPORTS OF BAR IRON 1699–1799
(Annual averages over 10-year periods)[12]

	Sweden	%	Russia		Others		Total
1700–09	14·100	88	20	—	1950	12	16·050
1710–19	10·500	65	20	—	5440	35	16·250
1720–29	15·000	76	400	2	4250	22	19·650
1730–39	19·000	74	3·300	13	3250	13	25·550
1740–49	16·800	75	3·600	16	2100	9	22·500
1750–59	18·700	64	8·100	27·5	2550	8·5	29·300
1760–69	19·600	49	17·400	44	2700	7	39·700
1720–79	16·700	38	25·300	57	2100	5	44·100
1780–89	15·300	34·5	27·800	63	1100	25	44·200
1790–99	18·200	40	26·300	58	1000	2	45·561

The ebb and flow in this spring tide of mounting imports of a vital raw material of industry was influenced in part by the external conditions of supply in which political considerations were often a major force. Bounties on exports and increasing tariffs imposed to build up a shore defence to protect home industry also helped determine the rate at which the tide of orders might rise. When supplies of home-made iron were forthcoming to compete with Swedish bar iron and the Russian brands at the wholesale merchant ironmongers' yards, the known quality ore reputation of the iron was pre-eminently the decisive factor whether a trial purchase should be made. When the iron was being used for traditional and long-standing purposes then quality alone told: at least, considerations of cost would not count for very much. Later, when it was found, for instance, that one kind of iron gave economies on account of greater strength according to weight, as did wrought iron rails,[13] then the relative prices of the different types of iron and as between foreign and British iron (when the qualities were judged by experts to be equal for certain uses) became more important as a factor encouraging or discouraging home production.[14] Unfortunately, the working

[12] These figures are computed from the Customs Records, 3 and 17 (P.R.O.), by K. G. Hildebrand, 'Foreign markets for Swedish iron in the eighteenth century', *Scand. Econ. History Review*, Vol. 6, Nos. 1 and 2, 1958, p. 10. The new importance of bar iron in Scandinavian trade is a central thesis of S. E. Astrom, *From Cloth to Iron* (Helsingfors, 1963).

[13] When the Stockton and Darlington Management Committee in 1821, on Stephenson's recommendation, decided to introduce wrought iron rails instead of cast iron (L. T. C. Rolt, *George and Robert Stephenson, the Railway Revolution* (London, 1960), p. 74).

[14] As Michael Flinn has shown in his paper 'William Wood and the coke-smelting process' (*Trans. Newcomen Society*, Vol. XXXIV, 1961–2 (1963)), when Wood's claims were being investigated in the 1730s, they were put to a rigorous examination of both capital and running costs. The traditional methods, the blast furnace and forge, were able to produce much cheaper iron.

out of this cost factor cannot be precisely demonstrated for this period; however, an index of the price of imports of bar iron for the eighteenth century does show a very clear rise as the pressure of demand bore on the limited Swedish supplies.

IMPORTS OF BAR IRON 1750–99:
PRICE INDEX
1753 = 100,[15] 1799 = 167 (3 years)

1750	86	1775	95
1755	96	1780	107
1760	103	1785	116
1765	93	1790	115
1770	98	1795	132

Nevertheless, despite the increased demand, the outcome of the 'new element' as Hildebrand terms it—the success of puddling and the heavy tariff protection of English iron—created, in this author's words, 'a price gap of unparalleled dimensions between the English product, now being made more cheaply, and the imported charcoal iron'.[16]

During the Napoleonic wars this was never of dramatic proportions, for the pressure of demand and the favourable rates o exchange on the Hamburg Banco kept the margin between the traditional product and its 'new' competitor down to a reasonable limit. If there is a turning-point when the native product competed on equal terms with the imported bar iron (still, of course, to be handicapped further with a tariff levy to bear), then from a set of figures computed by Svedenstierna, the Swedish metallurgist, written in 1816,[17] it appears that only after the year 1803 did the new technology begin to assert itself in price competition. Just then a ton of English bar iron cost, according to Svedenstierna, £19 10s., the same as the cost (minus duty) of the Swedish iron: thereafter, with the exception of a high English price in the year 1806, the puddled bar was always cheaper—by more than £2 per ton.

In fact, British puddled iron, containing more sulphur than the

[15] Deane and Cole, *op. cit.*, p. 84. The authors do not say if this index includes the import duty; however, the critical increases in duty, amounting to 19s. 3d. and bringing the total import duty to £3 15s. 5d. per ton, were imposed in the years 1796–8.

[16] Hildebrand, *loc. cit.*, p. 37.

[17] E. T. Svedenstierna, 'Om Vexelkursen och jern-prisen in England och Sverige under Sist förflutna åren', in *Jern Kontorets Annaler*, 1817, (2). Table facing p. 160. The price of Russian iron, always more in line with English bar iron, did not lose the advantage of cheapness until 1807. Then one ton of Russian bar cost £16 3s. against £15 10s. for the English.

	English bar iron prices £ sterling	Russian bar iron prices £ sterling	Swedish bar iron prices £ sterling
1799	18/5	17/15	21/2
1800	18/10	18/2	21/7
1801	20	20/10	20/17
1802	18	15/2	18/10
1803	19/10	15/2	19/10
1804	17/–	16/13	19/17
1805	17/10	16/11	19/13
1806	18/10	17/12	19/5
1807	15/10	16/3	17/15
1808	15/10	16/8	17/15
1809	15/10	16/6	17/7
1810	15/10	16/6	17/7

Source: Svedenstierna, *loc. cit.*

Swedish iron, was better and cheaper and more suitable for constructional purposes. As Arpi, writing about the Swedish Ironmasters' Association, points out: 'the new process confronted the Swedish iron industry with competition of a quite unprecedented nature'.[18]

The outcome of this situation, as we know, was the Swedish specialisation in the production of those sorts of iron needed for steel-making; the Russian contribution dwindled away to a mere 1710 tons in 1817. According to Wahlström, Russian export ceased altogether early in the nineteenth century. This was mainly on account of the iron's unsuitable quality for the steel makers.

[18] G. Arpi, 'The Swedish Ironmasters' Association', *Scand. Econ. History Review*, Vol. 8, 1960.

CHAPTER III

The Rise of the Coke-iron Industry

The problems of describing and explaining the course of industrialisation in England in the eighteenth century are still being examined by economic historians. It would be a considerable exercise to assess the significance of recent discussions relevant to this enquiry. Through the influence of Rostow a great deal of attention has been focused on the 'take-off' for industrialisation—a pattern of analysis into which the iron industry, with its spectacular growth from the 1780s, fits quite neatly. Indeed, much of this book will be devoted to examining the working out of technical progress and innovation springing from the metallurgical advances of that period. However, both for the economic and the industrial historian, there is as much significance to be attached to the period of gestation when the early stages of industrial growth were under way. The history of the iron industry illustrates to a striking degree the most important features of these economic changes.

As one recent commentator has observed, growth comes forth at this time from a number of 'growth points'; in other words, until a national economy has been linked together by transport and communications, 'we should see the country as a combination of largely regional economies with certain links and certain well-marked differences'.[1] Now, the iron industry in the eighteenth century was an industry on the move—T. S. Ashton describes the pattern of development in terms of a centrifugal migration from the sixteenth century onwards in order to escape 'the tyranny of wood and water'. It is true, as Ashton points out, that first of all the industry gravitated to the north-west of England and South Wales, where more plentiful supplies of charcoal were at hand (and then, late on, in one or two instances, to Scotland[2]). This hypothesis has been developed more

[1] D. C. Eversley puts his viewpoint in an unpublished essay: 'In Pudding-time: the early stages of industrialisation in England 1730–1780.'

[2] H. M. Cadell, *The Story of the Forth* (Glasgow, 1913), and Ashton, *op. cit.*, p. 17, citing A. Fell, *The Early Iron Industry in Furness* (1905), which gives details of the Bonawe, Goatfield and Invergary furnaces. See also *The Diaries of James Losh* (edited by E. Hughes), Surtees Society, Vol. 171, p. xiv.

fully by H. G. Roepke, who describes the charcoal iron industry as subject to migratory and intermittent operations. 'It follows', he argues, 'that the details of its distribution could change rather quickly.'[3] The factors influencing the location of the industry, according to this argument, were primarily the shortage of timber fuel, secondly the competition of foreign iron in a market which was as yet relatively inelastic.

The conclusion of this movement, according to Ashton in particular but supported by many other historians, was the concentration of the ironworks and the coalfields as, first, coke was used for smelting and, more decisively, as Cort's puddling processes necessitated the re-integration of the smelting and refining processes. In Ashton's words, 'in Staffordshire, Yorkshire, and above all in South Wales, all processes, from the mining of coal and ore to the slitting of rods and the production of finished wares, were henceforth carried on in the same locality, by the same firm, and most of them in single establishments'.[4] Here, then, is one manifestation of an industry and an economy coming into being.

However, the history of the British iron industry in the eighteenth century, at least up till the 1780s, is, indeed, the subject of much conjecture. The picture of an industry declining until it is rescued by the discovery of coke-smelting and the harnessing of steam power to the blast of the furnace is being modified. Only the briefest account of what appear to be the main findings of recent work on this subject will be attempted here: however, one is all too well aware that the definitive histories of the charcoal industry are urgently needed in order to give us the proper perspectives for an interpretation of the 'take-off'.[5]

If there was a wholesale shortage of charcoal, so that furnaces had to be abandoned or used at longer intervals until stocks of fuel were again accumulated, then we would expect to see an industry in decline with smaller and smaller outputs. And this is the traditional picture of the industry, stagnating if not disappearing entirely. To quote Ashton again, 'the industry of smelting and refining was literally fleeing to the wilderness to escape destruction'.[6]

Yet there is another school of opinion, based in its assertions on the business records of the forgemasters and on the evident fortunes

[3] H. G. Roepke, 'Movements of the British iron and steel industry, 1720–1951', *Illinois Studies in the Social Sciences*, Vol. 36 (Illinois, Urbana, 1956), Ch. II.

[4] T. S. Ashton, *An Economic History of England. The Eighteenth Century* (London, 1961), p. 117.

[5] It is to be hoped that the researches of Messrs. Downes and Hammersley and G. G. Hopkinson will be published before long.

[6] Ashton, *op. cit.*, p. 15.

of the ironmasters, such as the Foleys, Lloyds, Darbys, Knights, etc., which presents the industry in a very different light. The central point of these arguments, expressed by B. L. C. Johnson, stresses the continuity of the industry and the importance of coal-fields to the ironmaster as a source of iron ore and fuel for the chafing process as far back as 1585.[7] Again, this interpretation of movement in the iron industry demonstrates the working of an economy: the transportation of the pig iron from the Forest of Dean for conversion to the forges of the Stour valley—from an area where charcoal was available for smelting to the midlands, where coal could be used for some of the intermediate and finishing processes. This centripetal field, the Midlands secondary iron industries, drew its power from coal. Moreover, this market for iron exerted its influence as far afield as Cheshire and North Staffordshire.

This articulation of geographical specialisation by trade between furnace and forge and forge and slitting mill is certainly a more convincing picture than an interpretation taking into account merely the fuel problem.[8] Fuel (viz., charcoal) which could be bought at a cheaper price in the Stour valley than in Shropshire, where Abraham Darby was stimulated to perfect the process of substitution of coal for charcoal) was of course a limiting factor. It limited the expansion of the iron industry in the Birmingham region; elsewhere the ironmasters came to an arrangement and limitation of output,[9] but the supply position did not, at this time, bring the industry to a standstill.

At this point it would be highly desirable to have a set of accurate statistics showing the number of furnaces, forges, and the output of pig and bar iron during the eighteenth century. As is well known, however, these are very few and, as Michael Flinn has shown, are produced by partisan sources; they are unlikely to be accurate and there are wide discrepancies between the various estimates for the period they cover, forty years only, to 1757.[10] It is possible that the

[7] B. L. C. Johnson, 'L'influence des bassins houillers sur l'emplacement des usines à fer en Angleterre avant circa 1717', *Actes du colloque internationale, le Fer à travers les Ages* (Nancy 1956).

[8] See also Johnson's 'The Midland iron industry in the early eighteenth century', *Business History*, Vol. II, No. 2, June 1960. Here his position is made quite clear. 'Significant as they were to the detailed location of furnaces and forges, the occurrence of iron ores, of coppice woodland for charcoal making, of streams of sufficiently reliable flow to drive bellows wheels or forge hammers cannot account for the regional pattern of the industry.' The two facts of outstanding importance to the understanding of the iron trade of the period are: water-communications and markets.

[9] M. W. Flinn, 'The Growth of the English iron industry 1660–1760', *Econ. Hist. Rev.*, 2nd Ser, Vol. XI, No. 1, 1958.

[10] Flinn, *ibid.*

industry had been expanding slowly, with an output of 17,350 tons of pig iron in 1720, to something more than 30,000 tons in 1760 when the coke furnaces were coming into their own.[11] However, as Flinn rightly points out, 'estimates of total production cannot be more than wild guesses'.[12] We would agree with his modest conclusion: 'It is hard to see that the changes in the industry at this period can be interpreted as stagnation or decline. . . . It seems more probable that there was a significant increase in capacity in all branches of the industry. That new investment on such a scale took place at all suggests a belief in a marginal productivity of capital high enough to be inconsistent with the picture of an industry slowly being crushed between the millstones of growing foreign competition and rising costs.'[13]

The question of rising costs is, of course, centred on the problem of the shortage of timber for charcoal, and again Flinn has convincingly rebutted the argument that this was responsible for the supposed decline of the industry before 1760.[14] Admittedly, one can cite several instances of shortage of charcoal fuel preventing furnaces from being blown all the year round; on occasion even where there had been an accumulation over a period of time, it was found that charcoal had perished in store and was incapable of supporting the charge in the furnace. There is evidence, too, that the ironmasters had to range far afield for fuel. Marshall mentions that the Back-

[11] This figure is taken from P. Deane and W. A. Cole, *British Economic Growth, 1688–1959*; however, Flinn is less definite and he estimates, from a list of 43 new furnaces (and 20 which ceased to operate) a net increase of upwards of 10,000 tons of pig iron in the period 1660–1760.

[12] Dr. R. Mott, 'The Coalbrookdale Group Horsehay Works' Pt. II, *T.N.S.*, Vol. XXXII, 1959–60, makes the following estimate of the production of pig iron, bar iron made by charcoal and coke:

PRODUCTION IN GREAT BRITAIN

Year	Pig iron		Bar iron	
	Charcoal	Coke	Charcoal	Coke
1720	20,500	400	14,900	nil
1750	24,500*	2,560	18,800	100
1760	—	10,000	—	3,000
1788	14,000	54,000	11,000†	22,000
1800	—	—	—	125,000

* This is computed from the amount of bar iron produced using a conversion factor of 1·3 tons of pig iron equal 1 ton of bar iron.

† Elsewhere (*loc. cit.*, p. 43) Mott says 16,400 tons of charcoal bar iron. The source of these figures is evidently Hulme's list and the Coalbrookdale records.

[13] Flinn, *loc. cit.*, p. 148.

[14] *Ibid.*, pp. 148 ff., and 'Timber and the advance of technology: A reconsideration'. *Annals of Science*, Vol. 15, No. 2, June 1959.

barrow partners were drawing their supplies from an area of perhaps 150 square miles.[15] The Backbarrow partners were successful for a time in arriving at an agreement with neighbouring ironmasters to restrain themselves from offering high prices to the owners of the woods. In Yorkshire, too, among the Spencer group there was a move, apparently put into effect in 1728, to allocate woodlands to the nearest forge or furnace and to allot quotas of outputs to the five furnaces in the group.[16] Here one complaint from the landlord's agent, responsible for the woods, was that the digging of coal pits was destroying all 'the spring wood'.[17]

One proviso of the Spencer agreement took care of the contingency that coke (mentioned as being used for malting grain) might be used for smelting the iron:

> 'When any Pigs or other cast iron is blown at any furnace with Pitt Coles, it is to be reckoned as Part of Proportion only so far as the Quantity of Charcoal spent along with the Pitt Coles may amount to.'[18]

It would be unwise to argue from the fact that even in this one instance the circumstances of scarce supplies and rising costs were not sufficiently marked to have made this more than a possibility for the future; however, it does give point to Flinn's thesis that 'it is . . . only true in a special sense that rising charcoal costs hindered the growth of the English iron industry'.[19] It was possible for a group of ironworks to maintain a furnace in blast, drawing its supplies of fuel from its economic area, for a long time, e.g. the Halesowen furnace from 1616 to 1772. Moreover, output could be maintained at a level at a price which could just compete with Swedish imports.[20]

[15] J. D. Marshall, *Furness and the Industrial Revolution* (Barrow-in-Furness, 1958), p. 23. In America at the same time, where of course there was an abundance of timbered land, the Hopwell plantation and ironworks in 1776 covered 10,000 acres. It was consuming about 600 acres of woodland a year, that is an acre of timber to make 2 tons of iron a day. (J. B. Austin, 'Three centuries of progress—American steel industry,' *J.I.S.I.*, Vol. 200, Pt. 3, March 1962.)

[16] R. Butler, *The History of Kirkstall Forge* (York, 1954), p. 9.

[17] *Op. cit.*, p. 10, letter to Lord Cardigan from his agent, William Elmsall, December 1731.

[18] *Op. cit.*, p. 9.

[19] Flinn, *loc. cit.*, p. 150.

[20] Butler, *op. cit.*, Appendix B, p. 239, quoting William Spencer's Memorandum of 1737 on the subject of Swedish imports, mentions that 'every sort of tough iron in England is preferable to Swedish—none sold under £15 10s. and some £16 p.tun. No Swedish iron is sold . . . under £14 and as it will not work into nails as easy as our own, this branch of our Trade is not affected by the Swedish.'

But it was not possible to increase production without raising costs above this competitive level.

This does not mean to say, of course, that the potentialities of coal as a fuel for the smelting and fining processes were not realised. Indeed, the triumph of Abraham Darby in 1709, a technological achievement which needed many more years for it to be translated into an economic innovation, represents the culmination of a century of theorising, experimentation and patent-seeking;[21] moreover, the full potentialities of coal in the forge were not to be realised until Henry Cort's puddling process was adopted in the 1780s and 1790s. The challenge to the inventors and projectors of companies and patents still remained; and the history of this formative period in the new metallurgy remains to be written.[22] It is in this context of ignorance and, as Flinn remarks, the puzzling case of the eighteenth-century ironmasters being sceptical of or blind to the Coalbrookdale experiments, that there has developed a major controversy over the nature of Abraham Darby's invention. For the purposes of this chapter, fortunately, the fine points of the discussion over the exact dating of Darby's success do not concern us. One is inclined to agree with Dr. Schubert when, after a careful examination of evidence, he says 'it is impossible to fix a precise date' for the first successful application[23] of coke as a fuel for iron smelting. Nevertheless,

[21] H. R. Schubert, *History of the British Iron and Steel Industry c. 450 B.C. to 1775 A.D.* (London, 1957), pp. 226–9, pp. 331–5.

[22] Unfortunately, Dr. Schubert has not dealt with this phase of the changeover to coke. Recently, through the kindness of Mr. Robin Craig of the University College, London, I have become aware how alive this problem was, as reflected in the petitions for patents. Craig's files begin with the charter granted in 1693 to the company for making iron with pit coal. This charter came into the possession of Sir Thomas Mackworth in 1728 (when among the petitioners was one John Darby). The connection between the experiments to smelt copper with coal at Bristol in the 1680s and Darby's own associations with Bristol, and the eventual Coalbrookdale experiment, has been sketched in by J. R. Harris, 'The introduction of coal into iron smelting', *Edgar Allen* News, Vol. 37, September 1958 p. 435. Other petitions for patents came from: Robert Lydall, Dec. 1716; Thomas Forrest, Mar 1720; Jonathan Brown, Apr. 1720; and William Wright (who claimed to be 'concerned in greater numbers of iron works than any other partnership in England'); William Fallowfield, Jan. 1726/7; William Wood, Aug. 1728; John Payne, 1728.

(The involved history of Wood's patent and the counter-claim of his rivals is dealt with in M. W. Flinn, 'William Wood and the coke smelting process" *Trans. Newcomen Society*, Vol. XXXIV, 1961–2). Later, in the 1750s, we have the petitions of: Isaac Wilkinson, 1757–8; John Wood, of Wednesbury, 1760; John Roebuck, 1762; the Cranage Brothers, 1766; John Chadwick and Edward Thomas (for part); John Cockshutt, 1771; Richard Jesson and John Wright, 1772.

[23] See R. A. Mott, 'Abraham Darby I and II and the coal industry', *T.N.S.*,

whatever weight may be put on the evidence as regards the technical success of the process and the economy achieved, the case that Coalbrookdale made the charcoal furnaces of the Forest of Dean and elsewhere obsolete is impossible to maintain. Indeed, it does appear that for a period up to twenty years the output of the Coalbrookdale furnace was smaller than the most efficient charcoal furnaces, even in Shropshire.[24] As a fuel, where it was available, charcoal was apparently not much dearer than coked coal.[25] It is certain that Swedish observers were unimpressed by these developments of Coalbrookdale; in a review of these experiments, it was reported that 'twice as much stone coal has to be used. . . . Moreover, the iron which results cannot be wrought except when added in small quantities to better iron, although it is said to be well suited for casting.'[26] But if the Swedes were perhaps too complacent in thinking that such attempts would 'cost too much to be ever worthwhile',[27] there can be no doubt that, as Flinn says, the technical limitations of the coke-smelting process of Abraham Darby I were severe, and

Vol. XXXI, pp. 49–86; M. W. Flinn, 'Abraham Darby and the coke smelting process', *Economica*, 1959, pp. 54–9; H. R. Schubert, 'Abraham Darby and the beginnings of the coke-iron industry', *Journal of the Iron and Steel Institute*, Vol. 139, 1959.

[24] Cf. Mott, *loc. cit.*, p. 73, 'Thus it is clear that, by the last quarter of 1709, the process was a financial success. By 1715 the Coalbrookdale Co., with its two furnaces, could compare in production with most (single) charcoal furnace plants; by 1737 each coke blast furnace could be compared favourably with the average-size charcoal blast furnace . . . the fuel cost was half that of a contemporary charcoal furnace and the supplies of fuel available were such that plants of two, three or four furnaces could be built on favourable sites, though they were limited by water supply.' It is strange that the economic weakness of this argument has not been stressed by those scholars sceptical of Mott's claim.

[25] See B. L. C. Johnson, 'The Midland iron industry in the early eighteenth century', *Business History*, Vol. II, No. 2, 1960. Johnson gives some tentative figures showing that, taking in Mott's figures of costs of charcoal in Coalbrookdale (which appear to be favourable to Mott's calculations), the saving per ton of pig iron using coke was £1 4s. 6d. approx. Dr. G. F. Hammersley in a private communication intimates to me that 'I have nowhere else come across charcoal costs that high (i.e. £6 13s. 3d. per ton of pig) nor as great a consumption of charcoal in smelting since the mid-seventeenth century. . . . I have seen no accounts to make me think that charcoal at the furnace normally costs more than £4 per ton of pig; mostly it would have come no dearer than coked coal and may have been cheaper'. His conclusion is most pertinent: 'If this is anywhere right then it would have been silly for anyone to convert to coke (smelting) to produce worse bar iron.'

[26] E. F. Heckscher, *An Economic History of Sweden* (trans. B. Ohlin), citing *Report to Mining Board* 1774.

[27] Heckscher, *op. cit.*, p. 180, quoting *Report* by Carleson, Clerk of the Board of Ordnance, 1729.

widely known amongst practising ironmasters. 'The process would, apparently, make little iron beyond what was suited to the specialised needs of the pot founder.' Moreover, 'its limitations do explain the continuance, unabated, of the quest for an alternative to charcoal-smelting for the best part of half a century after Darby's momentous discovery'.[28] Therefore the introduction of the coke-smelting process in the first decade of the eighteenth century had little effect on the iron trade as a whole. Even thirty to forty years after coke had come into regular use in the Darby's blast furnaces, there were but seventeen coke furnaces working in Britain, and the greater part of the century's requirements for pig iron (for conversion into bar, hoop and strip iron) must still have come from charcoal furnaces.[29] The innovation of real significance—the manufacture of a coke pig iron suitable for use in the forge for refining into bar iron of good quality—was another Coalbrookdale improvement. From 1729, according to Dr. Mott, the Coalbrookdale forge had made bar iron of inferior but marketable quality from coke pig iron. However, the final phase in the improvement of the product occurred during Abraham Darby II's management from 1748 onwards; success depended apparently upon the most careful selection of the clod coal and certain kinds of ironstone (Black, Ball and Penny ironstone) for the furnace charges. Thus was the phosphorus content of the pig iron kept to a minimum and the resulting bar iron was only 'a little cold-short'.

As Mott admits, 'the success of Abraham Darby II in making coke-iron for refining to bar iron of good quality had a more profound effect than did the success of Abraham Darby I'.[30] The opening up of the iron industries on the coalfields in the Black Country, South Wales and Scotland (Carron) was the culmination of this struggle for technical success.

Charcoal remained the fuel for the finery for most of the eighteenth century, and since this process consumed more charcoal than the actual smelting—according to Mushet, 50 per cent more[31]—it is not surprising that a long series of partly successful efforts to use coal or coke were directed towards making economies there. The difficulty

[28] Flinn, *loc. cit.*

[29] B. L. C. Johnson, *loc. cit.*, p. 67. If, as other economic historians suggest, the domestic market for cast-iron products only began to reflect the growth of population in the second half of the century, then this would be the reason why coke-smelting was so slowly developed. It is significant that a substantial part of Coalbrookdale's expansion depended on the export trade through Nehemiah Champion in Bristol. (Raistrick, *Dynasty of Ironfounders* (London, 1953), p. 55.)

[30] Mott, *loc. cit.*, p. 84.

[31] D. Mushet, *Papers on Iron and Steel* (London, 1840), p. 407.

lay in the nature of the coke-iron, as Dr. Richard Watson in his *Chemical Essays* makes clear: 'Iron thus produced is very fragile and coarse and cannot, without great difficulty, be rendered as malleable as that which is fluxed with charcoal, it may answer nearly as well for casting, but it cannot be made into good bar iron by any art hitherto known.'[32]

It is known that there were a number of experiments. In 1762 and 1766 a premium for a successful method was held out as some inducement. In 1762 Dr. Roebuck of the Carron ironworks anticipated the essentials of the method of puddling.[33] His patent specifications speak of reducing the iron to a loop, and Campbell in his history of Carron Company, confirms that the product was 'as best could be wished for nails, hoops and every trial we have given it'. However, for the manufacture of steel, when the Carron partners estimated the profits might amount to 20 per cent if successful, after a promising trial at Benjamin Roebuck's steel furnace in Sheffield, the innovation fizzled out. Mantoux suggests that the metal was not pure enough to compete with the Swedish iron. Campbell says: 'Subsequent events confirm that he had failed to make the advance originally thought. No more reference is made to the manufacture of steel from the coke-forged bar iron, and as the Company's late history shows, success with coke in the forge had to wait on the work of Henry Cort.'[34]

Cadell, a Carron partner, priding himself on the readiness of the company to experiment, had once asserted that 'ironmasters in general are conceited and not ready to improve'. However, it is apparent that a number of independent workers were approaching the goal by different techniques.[35] John Wilkinson, in a letter of 1762 claimed success: 'There has been for the six years past great quantities of bar iron made from coaks. We call it ordinary or mill iron; (it

[32] *Op. cit.*, Vol II. p. 340 (3rd edition).

[33] Patent No. 780, 25 October 1762. *Abridgement of Specifications relating to the Manufacture of Iron and Steel*, p. 9. P.R.O. 44/264 Petition of John Roebuck of Birmingham.

'Sheweth, that your Petitioner, after much study, labour and expence, tha invented a way of making malleable iron from pig or sow metal commonly called cast iron without the use of charcoal or of an air furnace, or day pots moulds or any other close vessell for the purpose of refining such iron, o without any complex machinery whatsoever, but only with raw pit coal, in hearths and with a machinery of the very same kind as is commonly used in making such malleable iron with charcoal. . . . '

[34] R. H. Campbell, *Carron Company* (Edinburgh, 1961), pp. 55–6.

[35] E.g. by John Cockshutt (Patent No. 988, 1771), the brother of James, who played a prominent part in the development of Cort's process at Cyfarthfa in South Wales. Cockshutt, however, used charcoal in a new and improved refinery

is slit into rods and used in the nail manufactory. . . .'[36]

Then there were the processes deriving from the patents of the Wood brothers of 1761 and 1763, to which Roebuck was doubtless referring in his petition, for a reverberatory furnace was used and the iron for finery put into crucibles to prevent contamination by the coke-fire[37] in the Midlands and South Wales; the 'stamping and potting' process of Richard Jesson and John Wright (Patent No. 1054, October 1774) established itself. The essentials of the stamping process lay in the use of a coke-fired refinery; a stamping, however, was necessary to break the refined pig iron into small 'cakes' for the final process of refining in the pots or crucibles.[38] In 1784, when Cort's process was being discussed, Watt stated that the iron wrought in the Black Country was mostly made by Wright and Jesson's process.[39] In 1795, when, in Shropshire, Cort's puddling processes were in operation alongside the potting and stamping method, at Ketley and Coalbrookdale respectively, they were then described by Josiah Gilpin as the 'New' and the 'Old Way'.[40]

Yet it was at Coalbrookdale that the most successful attempts to anticipate the puddling process were made.[41] In the 1760s, the Cranage brothers evolved a process which was put in use at Coalbrookdale. They used the reverberatory furnace (which had been used long before that) and their stirring of the iron in the furnace

[36] Royal Society of Arts Guard Book VI, letter No. 129. See also *Journal of the Royal Society of Arts*, Vol. 88, 1938–40, article by M. G. Jones, 'The charcoal iron industry', pp. 41–50, 50–7.

[37] Robert Erskine (*Erskine MSS*, Library New Jersey History Society) in a letter dated October 1770 describes (at second hand) the process as worked by Anthony Bacon in South Wales: 'They let the metal run from the furnace into water to granulate it, and what is not small enough is pounded under Stampers . . . it is granulated to be made malleable with raw coal. For this purpose a number of crucibles or pots about 6 inches deep and a foot wide are prepared, these vessels are filled with small iron . . . the pots are put in an oven or reverberatory furnace, and when the metal begins to run break. The breaking of the pots is the signal to take it out . . . it is now cooled and again granulated, and the operation repeated a second time when it is taken to the hammer.'

[38] See Marchant de la Houlière's *Report to the French Government on British Methods of Smelting Iron Ore with Coke and Casting Naval Cannon* (1775) ranslated by W. H. Chaloner.

[39] Quoted by Chaloner, *op. cit.*, James Watt—Sir James Dalrymple, 25 August 784.

[40] Harold B. Hancock and N. B. Wilkinson, 'Joshua Gilpin: An American anufacturer in England and Wales, 1795–1801', Pt. I, *T.N.S.*, Vol. XXXII, 59–60.

[41] *Op. cit.*, pp. 86–7. Raistrick says this discovery encouraged the Dale Comny to extend their forges and to continue working forges at Bridgnorth and verpool.

brought the iron to 'nature'; what is more uncertain is the degree of control they had over the finished product. Ashton suggests that their iron was only suitable for nail iron.[42] Certainly a letter quoted by Samuel Smiles indicates that a certain quantity of the iron was sold—£247 worth.[43] Yet a writer in the *Weale MSS* says that he 'learnt from Thomas Cranage lately . . . (that) he and his brother (who is dead) lost money by it by neglecting their proper work, for which they would have been better paid—he said they lost over £20 but that he should not have minded if the thing had answered, it failed, however, and was totally given up'.[44] The reason for the abandonment of the process is suggested by Joseph Reynolds: 'on further trials many difficulties arose. The bottoms of the furnaces were destroyed by the heat, and the quality of the iron varied.'[45]

It is not surprising, in view of the first reason given here, that the next experiments at Coalbrookdale took place in a furnace with a flow of water underneath to cool the bottom where the puddling took place. The *Weale MSS* again are most informative about these experiments of Peter Onions, who took out a patent on 7 May 1783 (No. 1370). He, unlike Cort, prescribed a blast 'for exciting the fire . . . to introduce the effect of the common finery',[46] but, once more, the goal of his efforts was to bring the iron 'to nature', as in puddling.

The Coalbrookdale Company encouraged Onions in his efforts. Two letters in the Weale collection from William Reynolds to Henry Cort in January and February 1784 confirm this. In the first, Reynolds admits, 'I am much chagrined that Peter Onions has not yet been able to succeed in his furnaces, he is going to try in one built some time since by the Dale Company, which I hope will prove more propitious to his wishes'.[47] In the second, he reports that 'Onions has not succeeded, and has with us, entirely given up the point, however, he seems confident of succeeding elsewhere, but this I doubt. . . .'[48] Onions's further attempts were made at Dowlais, where quite possibly the original method was tried out; however, they were none too successful in comparison with the improved Cort's puddling processes. Moreover, Onions made no mention of rolling. However,

[42] Ashton, *op. cit.*, p. 90, quoting R. Reynolds' letter of 1784: 'The nail trade would have been lost to this country had it not been found practicable to make nails with iron made with pit coal.'

[43] Smiles, *Industrial Biography, The Ironworkers* (London, 1908), 9.88.

[44] *Weale MSS* (*c.* 1810). According to Dr. Mott, James Weale collected this material for a book never completed.

[45] Smiles, *op. cit.*, p. 88.

[46] *Weale MSS.*

[47] *Ibid.*, William Reynolds—Henry Cort, 17 January 1784.

[48] *Ibid.*, William Reynolds—Henry Cort, 17 February 1784.

it appears that Coalbrookdale was much involved in the new processes, for Richard Reynolds stated in a petition to Earl Gower, President of the Council, in 1784: '. . . We have made, or rather are making the Alteration at Donnington Wood, Ketley etc., which we expect to compleat in the Present Year, but not at a less expence than 20,000 Pounds.' In setting out the conditions which prompted this enterprise it is significant that Reynolds mentioned that 'the only chance we have of making Iron as cheap as it can be imported, is the low price of our Fuel. . . .' Hence the experiments 'to make Bar Iron with Pit Coal'.[49]

HENRY CORT AND THE PUDDLING PROCESS

The contribution of Henry Cort (1742–1800) to the metallurgical arts has, until recently, been discussed in terms of the personalities concerned. The case of Henry Cort, the so-called inventor of puddling, is certainly a fascinating one and every investigator of this subject is tempted to probe a little deeper into the body of facts assembled by Cort's son, Richard, and by Richard Webster.[50] So skilful was their pleading for belated justice to the descendants of Cort that their statements of the circumstances have very largely moulded the accounts of John Percy and Samuel Smiles upon this subject.

Professor Ashton, of course, has given the main outline of the career and achievements of Henry Cort in his book *Iron and Steel in the Industrial Revolution*. In this he escaped from the fatal fascination of the alleged large-scale defalcations of Lord Melville and his underlings, and—of much more interest to the industrial historian—drew attention to Cort's relationship with Boulton and Watt.[51]

There is little need to repeat here the biographical details of Cort's family and early life. Cort, according to Hulme, was not born in 1740, as his gravestone has it, but in 1742, being most probably an illegitimate son of a Mayor of Kendal, born at Ellel, near Lancaster.

[49] *Rathbone MSS* (Liverpool University), Draft Petition to Earl Gower, President of the Council, dated 5 of 7th mo. 1784.

[50] Richard Cort, 'Review of the Report of the Committee of the House of Commons on the Petition of Mr. Cort for services rendered to the iron trade, etc.', *Journal of the Royal Society of Arts*, 20, 27 July and 3 August, 1855. Richard Webster, 'Memoir of Henry Cort', *Mechanics Magazine*, New Ser., Vols. I and II.

[51] Ashton, *op. cit.*, pp. 90–7. Dr. Mott has made use of the *Weale MSS.* and particularly those papers dealing with Cort in his exhaustive paper 'The Coalbrookdale Group Horsehay Works', Part II, *T.N.S.*, Vol. XXXII, 1959–60. He, too, discusses the transition from the traditional fining process using charcoal to Cort's puddling process.

Hulme points out that the affluence of the inventor in his early life resulted from this circumstance, since he succeeded to his father's building business.

It was by a second marriage—to Elizabeth Heysham, a niece of a contractor to the Portsmouth Dockyard—that Cort obtained his introduction to the iron industry. A few years previously, in 1765, according to the directories and guides of London, Cort was established as a Naval Agent in Crutched Friars. It was as a result of foreclosing upon a loan that Cort 'found it expedient to take upon himself the management of the iron works and the supply of the contract had been assigned to him by Mr. Morgan'.[52]

This was the Fontley Forge, two miles from Fareham, which had been a charcoal forge since at least the early eighteenth century, with an output of 200 tons per annum. He also possessed, some time before 1779, a warehouse and wharf at Gosport for receiving coal and iron by ship. It was at Fontley, where his rolling mill replaced forge-hammers, that Cort's experiments were carried out.

It was in 1780, according to Weale,[53] that Cort entered into a contract to make ironwork for the Victualling Board, and in consequence built his mill 'at very considerable expense'. Weale reports, as a result of this contract and the unfair competition of several London contractors, that 'the ultimate result of which proceeding was that he sustained a loss of nearly £10,000 by the contract'.[54]

The next development was that Cort himself was indebted to Adam Jellicoe, described by Weale as 'the agent of Mr. Cort in receiving money there (at Portsmouth) due to officers for whom he was concerned'. In 1781 Cort entered into an equal partnership with Samuel Jellicoe, son of his creditor.[55]

Weale's account continues to explain that in the years 1781–3 the forge continued to make losses because large stocks of old iron hoops, which Cort had contracted to receive, were accumulating. Thus, 'under these circumstances, Mr. Cort entered upon a variety of experiments, with a view to the profitable conversion of their large stocks of old hoops, and, at length, discovered and perfected the process for which he obtained his first patent'. This was the patent for rolling with grooved rollers.[56]

[52] Quoted in Dickinson, *op. cit.*, p. 32.

[53] James Weale, MS. *Account of the Iron and Steel Trade, 1779–1811.*

[54] Dickinson, *op. cit.*, p. 32.

[55] Webster, *op. cit.*, Vol. II, p. 53.

[56] Patent No. 1351 dated 17 January 1783. Although Cort could take the credit for making the rolling-mill an integral part of his puddling process, it is by no means clear how this machine was evolved and exploited during this period. As Wolf points out in his *History of Science, Technology and Philosophy in the XVIth*

It is interesting to note that in 1782 (and indeed, three years earlier, in 1779) Cort approached Boulton and Watt about employing a steam engine.[57] But the proposal never came to anything and according to Dr. Dickinson it is certain that Cort never installed an engine. Thus Cort never took the revolutionary step of applying steam power to the rolls.[58] In view of this, Jellicoe's and Cort's estimates that his experiments cost at least £10,000 must have been somewhat inflated. Thus, as Dr. Dickinson points out, at this stage Cort was using the old-established method of faggoting and welding, but using rolls instead of hammers. Moreover, as his patent specifications tell us, he was at this time using the existing refinery process.

The next step forward was the one which many others had attempted before Cort—the attempts which have been described earlier. Cort's contract with the Navy Board was confirmed in 1782, and it seems that Cort was encouraged to try converting ships' cast iron ballast, or kentledge, into wrought iron. In this way Cort was prompted to carry the application of coal in the production of iron one stage further and to take out his patent, No. 1420, in February 1784. This gives a very clear account of the actual processes involved in Cort's puddling process.

The process was a success; the Navy Board at Portsmouth approved of a wrought iron bar made from ballast iron on 16 July 1784,[59] and contracts followed for 'converting their unsizeable Sweeds (sic) iron' into suitable sizes by his improved method after trials at all the Naval Dockyards. In 1787 and 1788, contracts were received for 300 tons in all, and by 1789 the Board was stipulating that iron for their contracts be made 'agreeable to Cort and Jellicoe's

---

and XVIIth Centuries, three reasons retarded the adoption of the power-driven rolling mill. In the first place, it was the high temperature of the iron. Secondly, there was a need for large water-wheels, until steam power filled the deficiency. Thirdly, the life of the rolls was short; hence, until improvements in case-hardening were made, rolling was an expensive process.

Foreign observers commonly referred to Cort and Purnell's process. William Purnell, ironmaster of Dursley, Gloucestershire, did not patent his process, which included rolling, until June 1787. However, this was in respect of his father's invention, made some time earlier. According to Dickinson, a patent taken out by Purnell was dated 1766, which showed grooved rolls, nut pinions and coupling boxes. However, William Purnell's specification to be found in the *Repertory of Arts*, Vol. IV, has no mention of grooves in his description of his 16-ft.-long rolls. There is no evidence to suggest that Purnell was associated with Cort in his experiments.

[57] The 1782 letter is quoted in Ashton, *op. cit.*, p. 91.

[58] In 1782, again, steam power had been applied to the forge hammer. (Ashton, *op. cit.*, p. 91.)

[59] Minute from the Navy Board, 16 July 1784.

patent'. It seems likely that Cort was, just then, a man with influence and obviously thought that the official adoption of his iron would boost the sale of licences for the patented processes. Also he was fully alive to the value of advertisement and *A brief state of the facts relative to the New Method of making Bar Iron with Raw Pit Coal and Grooved Rollers, discovered and Brought to Perfection by Mr. Henry Cort of Gosport, March 1787* was published by him.[60] However, even if Cort was successful in establishing personal contacts—Sir John Dalrymple describing him as 'a most ingenious and ingenuous sort of man'—it was the practical possibilities of the patents which attracted the attention of the ironmasters, and which made the puddling process a matter of great importance to the subsequent development of the industry.

In May 1783 Cort wrote to Boulton and Watt, referring to a previous visit to Soho, saying: 'We had some conversation on the subject of Iron—I intimated I had solicited a Patent for my invention of manufacturing iron on an improved method. . . . And you were kind enough to say you would mention me to Mr. Wilkinson.' So he sent on a letter to be forwarded to Wilkinson, putting forward his claims.[61] In 1784 a demonstration was given to the Midlands ironmasters at a forge at West Bromwich, and Dickinson quotes a report of William Playfair, a former employee of Watt: '(He) was present in 1784 at the first great Exhibition of the Rolling Mill with a welding heat on Mr. Cort's plan, at a Mill near Stourbridge belonging to Mr. Humphrey. All the ironmasters for 20 miles were invited.'[62]

Indeed, it seems likely that several trials were made in the district later on. Among the *Weale MSS* are transcripts of letters from William Reynolds to Henry Cort and they refer to trials near 'the Dale' and at Wednesbury and Shutt End.

The unsuccessful experiments of Peter Onions at Coalbrookdale have already been mentioned; whereupon Reynolds, after seeing the demonstrations,[63] was anxious to try Cort's methods. Among the *Weale MSS* there is a copy of an affidavit in Chancery sworn at

[60] A manuscript copy of this found its way in to the Earl of Liverpool's papers in the British Museum.

[61] Dickinson, *op. cit.*, p. 28. Dickinson, however, is wrong in saying that Wilkinson never adopted Cort's processes.

[62] *Monthly Magazine*, Vol. XLVIII (1819), p. 234 n.

[63] Certificate: 'Mr. Reynolds saw H. Cort make tough iron and cold short iron from the same materials at one heat, by a variation in the process . . . ' (Ketley, 18 December 1784). Reynolds also wrote in another letter (26 January 1785): 'If he had been in the Dale Company at the time of the Cranages' work, he should have pursued experiments . . . that he did not doubt he should himself have attained to the discovery made by Mr. Cort' (*Weale MSS*).

Gosport, 16 February 1785, by one Henry Foxall of Fontley Mills, a furnaceman, which gives, in considerable detail, the circumstances at Ketley. Reynolds, it was alleged, had told Foxall that they had stopped building eight fineries pending trials at Shutt End. Six workmen of Reynolds were instructed in December 1784 by Foxall and a finery man, named Bartholomew, in the new process which was to supersede the making of iron in a 'Killing Furnace'. Finally it was claimed, upon Reynolds' own word, that Cort's process gave a greater yield than the earlier method of refining. It was stated that only 30 cwt. as against 32 cwt. of iron were required to make a ton of bars.

The puddling process was adopted at Ketley, and an agreement was made to pay Cort ten shillings per ton of wrought iron produced. However, later on Reynolds was inclined to support objections against the originality of Cort's patents. This is not surprising in view of the company's experiments with Onions's patent. By 1812, when the question had been reopened by the Select Committee on Coningsby Cort's petition for a pension, Joseph Reynolds wrote in a letter to Phillip Price of Birmingham: '. . . an air furnace and pit coal are used in both modes (i.e. Cort's process and the present method of making wrought iron) . . . and were used in the making of pig iron into malleable iron long prior to the date of this Patent. . . .' Nevertheless, he conceded the fact that rollers were 'the principal cause of the great increase of the British Iron Trade'.

However, Cort's demonstrations were not uniformly successful. Dickinson, again, quotes a letter of James Watt dated 6 June 1784. He criticised the iron on account of its cold-shortness, and Dickinson points out that Watt always specified charcoal-wrought iron for his engines, Nevertheless, Watt showed a certain sympathy for Cort: 'Mr. Cort has, you observe, been most illiberally treated by the trade; they are ignorant brutes; but he exposed himself to it by exposing his process to them before it was perfect and they saw his ignorance of the common operations of making iron, laughed at and despised him; yet they will contrive by some dirty evasion to use his process. . . .'[64]

Yet Cort appears to have made a great impression in Scotland. Admittedly, it was not the ironmasters who showed the enthusiastic interest of scientists like Dr. James Hutton, who wrote: '. . . By his perfect finery, the worst pig iron is made bar of the most valuable kind. . . . You will understand the value of this discovery, the land of coal and iron stone will be the land of bar iron.' With an imaginative flourish, Hutton indicated how the slit mill gave origin to the

[64] Dickinson, *op. cit.*, p. 39.

cotton mill, 'now the cotton mill may give a lesson to its parent, and learn Britain to spin iron for all the world'.[65]

Another enthusiast for the potentialities of Cort's processes was Sir John Dalrymple. In a letter to Lord Sheffield, he mentions that demonstrations took place at Edinburgh, Glasgow and elsewhere. So interested was he that, he says, 'I almost roasted my eyes out watching them'.[66] Dr. Black, whom Cort had also interested in the trials, accompanied Dalrymple and, with some reserve of judgment, said they 'appear compleat and easy and cheap in execution'. However, Dalrymple thought that Cort's patents, coming at the same time as Dundonald's tar extraction process and the Boulton and Watt engine which 'give the command of the iron trade of the world to Britain', would provide a fine opportunity to exploit the mineral resources of his estate near Edinburgh. Hence he published a pamphlet[67] advertising the assets of his property, stating at the same time the great advantages of Cort's puddling process. From this we can learn several important details of the economic advantages of the invention. The process had been 'tried out and approved at Edinburgh, Newcastle, West Bromwich, Cardigan and other places' and had given a uniform quality of iron. Different kinds of 'coak-pig have been used, among others the Tipton pig from Lord Dudley's ironstone; the Bradley pig, from the works of John Wilkinson Esq., the Ketley and Horsehay pigs from Messrs. Reynolds and Rathbone's, which are all smelted with coaked coal, and from all of them the best tough iron is uniformly produced, by the Cort's process, equal to any made from charcoal pigs'.[68] Production was speeded up it was claimed; two men could work 6 cwt. or loops in three hours, and 'one pair [of rolls] in a powerful mill will roll shingled slabs into bars, at the rate of a ton in 30 minutes'.

But although Cort had interested the influential Dr. Black (through his agent in Scotland, John Wauchope, Writer to the Signet), his invention was not adopted by the Carron Company. This is very surprising, because in the first place the company had been anxious to perfect a method of producing wrought iron and both Boulton's and Dr. Black's help had been sought. The advantages of Cort's process appeared attractive, for, according to Wauchope, the Carron Company could not 'make a Ton of Bars

[65] *Ibid.*, p. 38, quoting a letter, Dr. Hutton—Matthew Boulton, 24 May 1784.

[66] *Weale MSS*, Sir J. Dalrymple—Lord Sheffield, n.d. (?1785).

[67] *Sir John Dalrymple's proposals* (Edinburgh, 1784).

[68] Dalrymple's *Proposals*, p. 8: 'The Tipton and Bradley pigs were proved at Mr. Wright's forge at West Bromwich, the 9th and 10th of June; as the Horsehay and Ketley have been repeatedly at Mr. Cort's works.'

under £18. . . .'[69] But whilst Carron, after much persistence by Cort, was willing to grant the inventor a lease of the forge, it did not adopt the puddling process. Campbell[70] suggests that the reason for this lay in part with Gascoigne, the manager, who was convinced that he already had a satisfactory solution to the problem of manufacturing bar iron with coke. At all events, the Cort experiment at Carron was written off: 'We tried Cort's Plan and made iron far from being bad, but the extraordinary work rendered (it) a very losing process.' As Campbell says, 'Whatever happened Cort went his way, Carron its way: an opportunity was missed.'[71]

The puddling process was taken up and exploited in South Wales. Not only that, it was here that the imperfections of the invention were removed and that the process was brought to the stage of large-scale industrial production. In this Richard Crawshay was a leading figure, and the *Weale MSS* quote a considerable amount of correspondence revealing the intimate relations between Cort and the Welsh ironmaster. They indicate, too, that Crawshay was responsible for interesting other ironmasters in the invention. Amongst these was John Wilkinson of Bradley, whose pig iron had been tried by Cort in his demonstrations. However, Wilkinson, it seems, was critical of the process, and it has been suggested that he was conservative in this one respect and never adopted the puddling process. Yet the following extracts from letters between the two Jellicoes (father and son) show that Crawshay had interested Wilkinson in the process in 1787: '. . . He (R. Crawshay) even said that Mr. Wilkinson saw it in a better light than he had done and that he told Mr. C—y when he had got thoroughly into the business he should make a report to him. . . .' The second extract speaks of Wilkinson being won over: '(Crawshay) says he (Wilkinson) is now a convert and I really believe it may be owing in some measure to what Mr. C—y said to him when he met him lately at Birmingham'.[72]

At Cyfarthfa, Crawshay and his partner James Cockshutt provided the most powerful argument for its adoption—practical success. Towards this end, there was a long way to go between Cort's demonstrations and the overcoming of local difficulties.

In 1787, Crawshay entered into an agreement with Cort to pay

[69] *Weale MSS*, J. Wauchope—H. Cort, 30 January 1785.

[70] Campbell, *op. cit.*, p. 59, quoting a contemporary document: 'Mr. Cort should then have the entire direction of the Forge both as to the mode of his operations and as to the workmen that should perform them. . . . In this situation Mr. Cort would be perfectly at ease, undisturbed as to his pursuits and he alone would be answerable for their success.'

[71] *Op. cit.*, p. 60.

[72] *Weale MSS.*

ten shillings per ton for iron made in the puddling furnace. The reasons leading Crawshay to do this were a 'bottleneck' in the production of wrought iron; outside firms such as Reynolds (of Ketley) and the Homfrays (of Penydarren) were under contract to Crawshay to make bar iron for him. But stocks of pig iron were accumulating. In June of that year a visit was paid to the 'little mill of Fontley' and shortly afterwards Cort and his son William went to South Wales to supervise the construction of the forge.

Crawshay was very optimistic about the outcome of Cort's invention: 'My dear Cort, this will do—we shall now make Swedish Iron for Fire in England which heretofore we were obliged to have from Sweden—I would not but have seen the sight for £500. . . .'

In September Crawshay wrote to Cort, 'Cockshutt . . . purposes building 8 new furnaces on the Boring Mill stream.' In the same year an agreement was drawn up with one Thomas Edmunds to train him as a refiner.[73] He was to be paid 8s. (a week) until the furnace was built; 10s. 6d. 'until he shall be able to undertake the business by the ton when he shall be paid for himself and partners the sum of 14s. 6d. for every ton of good perfect blooms produced'. This was to be increased to 18s. 6d. if he mastered the art of shingling half blooms. He was to be allowed 30 cwt. of 'metal' to make 22 cwt. of perfect half-blooms. Finally, he engaged to teach other workmen and he was to receive a guinea for every man taught. In November Cort sent over to Cyfarthfa three men to teach the Welsh the new method.

By July of the next year Crawshay reported to Cort and Jellicoe, 'We now make about 2 tons weekly from six furnaces better iron than ever.' Yet one difficulty remained—the considerable consumption of fuel required to bring the iron 'to nature'. And although Crawshay was advising Cort to take 5s. a ton to get his patent into general use, and was interesting Gibbons, the Midlands ironmaster, in puddling as late as April 1789, the process had not been finally worked out in all its details.

In the previous December John Cooke of Kilnhurst Forge, near Doncaster, had written to Cort: '. . . I am informed that at Cyfarthfa one Forge is employed in your method, the other in the Shropshire mode of Fineries with coak, and stamping and balling as it is there called, and they cannot decide on the advantage of one over the other.' In 1789 Cockshutt was writing from Cyfarthfa that experiments were being conducted to try the effects of iron and sand

[73] *Weale MSS*, 1787. However, the copy of this document gives no definite date for its execution.

bottoms to the furnaces, but 'the iron bottoms occasioned such labour to the workman that we could not continue them'. The puddling process of Onions had not been abandoned, and that was being tried with and without blast. In this connection, the results of Onions's method, which had been adopted at nearby Dowlais, were compared with those at Cyfarthfa. However, the verdict was in Cort's favour: 'They do not near so well as we do.'

It was Crawshay's neighbour, Homfray, at Penydarren, who resorted to the fairly obvious solution to the difficulty. This was to continue to use a finery to refine the metal before putting it into the puddling furnace.[74] It is significant that the Select Committee reporting upon the merits of Cort's invention in 1812 called attention to this fact: 'No good malleable iron can at this time with certainty be made by the method claimed as the invention of Mr. Cort, unless the cast Iron has previously been converted into what is called finery or finer's metal, by a process since found out.'[75]

These, then, were the circumstances surrounding the development of the puddling process. Indeed, so intimately was this bound up with the improvements evolved at Cyfarthfa and Penydarren that the puddling process became known as the 'Welsh Method'. Here it was that the consequences of this innovation were most marked. The Homfrays admitted to the 1812 committee that they had produced 20,000 tons of puddled and rolled iron between 1788 and 1798.[76] Cort, indeed, in his review of the evidence, points out that the returns of the Glamorganshire Canal show that the Penydarren Company shipped 265,496 tons of wrought iron in the twenty-four year period up to 1812. A few years earlier the Bishop of Llandaff reported that in 1806 in his diocese 44 furnaces and forges were producing 73,580 tons of pig iron and 35,800 tons of bar iron.[77] In 1808, Crawshay says he was 'now making 10,000 tons a year, which takes 15,000 tons of pig iron of my own make'.[78] As early as 1794 Cort was informed that Crawshay was making and rolling an average of 80 tons per week.[79] As for the Dowlais Company, from their surviving records it does not apear that Cort's process was

[74] See Mantoux, *op. cit.*, p. 297, for the details of Robert Gardner's patent No. 1642 which modified the original process as conceived by Cort.

[75] *Report of Select Committee on Petition of C. Cort*, 1812.

[76] R. Cort, 'British iron manufacture', *Journal of the Royal Society of Arts*, Vol. 3 1855, p. 608.

[77] *Weale MSS*, letter of the Bishop of Llandaff—James Weale, 20 January 1808.

[78] *Weale MSS*, letter of R. Crawshay—James Weale, 14 December 1808.

[79] The *Weale MSS* point out that it was not until September 1789 that Crawshay began to work his new mill on Henry Cort's plan.

adopted until 1801—perhaps they still adhered to Onions's puddling process.

Thus there is abundant evidence that the puddling process was the keystone of the industrial structure of the Glamorgan iron smelting district. Here, too, as elsewhere it sometimes happened that puddling was sandwiched between the finery and the old type of hammer forge.[80] Nor was the question of the retention of the old method just a matter of capital invested in the old-type forge. In 1813 the writer of the article on iron in *Pantologia: A New Encyclopaedia* had this to say of the then established process: 'The advantage . . . is in the dispatch with which the iron is converted and comparatively small expense of fuel; at present the puddled iron is not so good in quality as the finery, though this circumstance should not deter manufacturers from using it, as it is scarcely possible that a newly-invested process should be at first brought to equal perfection which has exercised the ingenuity of manufacturers for ages past . . . even in the present state, the puddled iron made with rollers is by no means to be despised when its price and quality are compared.'[81] In this connection, among the 'Memoranda of Henry Cort' (in the *Weale MSS*) there is a comparison of making iron in the stamped method and Henry Cort's process dated 1787. If we can accept Cort's reckoning (one suspects it favours his own patent), then the relative cost of the manufacture of mill bars worth £16 a ton was £10 18s. 6d. for stamping and £9 11s. 6d. for Cort's puddled iron. Incidentally, the productive capacity of the process was not as great as sometimes quoted. Cort claims that three men in one furnace could work off upwards of 4 tons of pigs weekly.[82] It required four months for men to learn the methods of working the furnace, etc.

Undoubtedly, it was the conjunction of the rolling mill to the puddling processes which, as Joseph Reynolds in 1812 observed, was 'the principal cause of the great increase of the British Iron Trade'.[83] Since the 'ancony' stage could be dispensed with, production was

[80] A. H. John, *The Industrial Development of South Wales* (Cardiff, 1950), p. 157, citing Clydach in 1805.

[81] *Op. cit.*, (London, 1813), article, 'Iron'.

[82] A. Raistrick, *Dynasty of Ironfounders* (London, 1953), p. 227. 'Prior to the introduction of Cort's method, one ton of iron bars could be produced by 12 hours' work, but using Cort's process 12 hours was sufficient to make 5 tons of small or 15 tons of large iron bars.' This cannot refer to the process of puddling in one furnace. In the mid-nineteenth century the capacity for an eight-hour shift was five charges of 40 cwt. each.

[83] *Weale MSS*, letter of Jos. Reynolds—P. Price (Birmingham) 11 February 1812.

very much speeded up—partly through the use of the steam-engine. It was Boulton and Watt's rotative engine which enabled the reorganisation of the Horsehay and Ketley and Coalbrookdale forges described in detail by Arthur Raistrick.[84] Ashton records Cockshutt's negotiations with the Birmingham firm for supplying a steam engine to drive the rolling mill at Cyfarthfa in 1788.[85] Other ironmasters followed suit.

Thus the way was open for a great expansion of the wrought iron industry. Puddling furnaces could be multiplied as operators for them were trained. The power-driven rolling mill, indeed, with its voracious capacity had created a disequilibrium in the industry, so that the fineries and puddling furnaces had to be increased in number to keep in step with the progress made in the forge.

Most characteristically, John Wilkinson installed a massive rolling mill for his works at Bradley. The rolls were 4 feet in diameter and had an oscillating motion, driven by a steam engine with a 24-foot flywheel. The purpose of the oscillating motion was explained thus: 'This has the advantage that a single man suffices to roll the loop, because the iron comes back to him when the roll changes direction. . . .'[86] This 'Fine machine' served the whole forge at Bradley where there were twenty reverberatory furnaces (fourteen of them puddling furnaces). Over 420 tons of bar iron were made each month at Wilkinson's works. Indeed, such was the perfection of the Bradley works that Bonnard calculates that the capacity of the forge and rolling mills amounted in all to 18,000 tons a year—'a result approached by no other ironworks'.

[84] Raistrick, *op. cit.*, p. 266. Mott, *loc. cit.*, says it was in 1798 that the puddling process was adopted at Horsehay. Previously the forge had a capacity of 6 tons per week per furnace. By 1806–7, using the puddling process, output had been increased to over eight tons per furnace. The total production of the forge was increased fivefold to over 3250 tons of wrought iron a year.

[85] Ashton, *op. cit.*, p. 97.

[86] A. H. de Bonnard and R. O'Reilly, 'An account of the smelting of iron with coal in Great Britain, 1802–3', in *Annales des Arts et Manufactures*, Vol. XXIII, pp. 113–51; Vol. XXIV, pp. 44–62.

CHAPTER IV

The British Iron Industry during the Napoleonic Wars

'Waterloo may have been won on the playing fields of Eton, but the British supremacy in the economic war . . . was certainly engineered at Fontley Mill in Hampshire where Henry Cort revolutionised the art of making wrought iron.' So writes Dr. Mott.[1]

In 1788 a total of 105 charcoal forges in Britain were producing on average 3 tons of bar iron weekly. In the *Weale MSS* there is a list which gives the output by county: it is the last trace of an obsolescent industry:

LIST OF FORGES WHERE THE METHOD OF MAKING BARS UPON THE OLD PLAN STILL EXISTS, WITH THE NUMBER OF FINERIES AND QUANTITY OF IRON WHICH THEY ANNUALLY MADE, TAKEN AS AVERAGE, 1788

	Forges	**Fineries**	**Total output (tons)**
Cumberland, Westmorland and Durham	5	9	450
Yorkshire	11	25	1,750
Lancashire and Cheshire	8	11	770
Nottingham, Derbyshire, Warwick and Hereford	9	20	1,500
Shropshire	14	28	2,520
Staffordshire	10	13	1,170
Worcestershire	8	20	2,600
Gloucestershire	4	8	600
Sussex	4	4	120
Monmouthshire	12	26	2,340
South Wales	12	29	1,740
North Wales	7	12	600
Scotland	1	3	240
Total	105	208	16,400
There are now about 60 melting fineries making 5 tons of bars a week			15,600
Total quantity of bar iron made			32,000

[1] Mott, *loc. cit.*, p. 51.

From this it will be seen that the largest fineries—those in Worcestershire—were only producing 130 tons of bars a year. The average capacity was about 80 tons.[2]

The period which followed, commencing with the Napoleonic wars, was probably the period of most intensive development that the British trade ever experienced, stimulated as it was by the pressure of war and the probability that imports of bar iron from the Baltic would be stopped[3]. Mott[4] estimates that in 1800 the production of Britain's blast furnaces was sufficient to make 125,000 tons of bar iron and 20,000 tons of pig iron for direct casting.[5] This estimate is in agreement with the figures of output given in the *Weale MSS* which indicate that in 1810 there were 94 puddling furnaces producing 130,000 tons of iron made by the Cort processes. The list gives the following distribution:

27 furnaces in South Wales
25 „ „ Shropshire
6 „ „ Yorkshire
5 „ „ Carron[6]

However, the significance of Cort's innovation should be set into the context of the rapidly expanding coke-iron industry.

In 1788, according to David Mushet, the output of Britain's 85 furnaces was only 68,000 tons.[7] It is certain that this rather small

[2] This list has a special note about the once predominant Sussex forges and furnaces. Only Ashburnham and Heathfield produced pig iron—200 and 100 tons yearly. There were three works 'yet standing and possibly may work again, in case of war'. Four forges—Robertsbridge (50 tons); Biblehaur (?) (30 tons); Maresfield (30 tons); Burnash (30 tons)—'depend upon old waste iron collected in London and elsewhere'. One interesting reason is given for the decline of the charcoal iron industry there: it was more profitable to provide hop-poles than cordwood.

[3] Deane and Cole point out that between 1757 and 1788 the rate of increase can hardly have reached as much as 40 per cent per decade. Between 1788 and 1806 there was a rate of expansion of over 100 per cent. *Retained imports* of iron fell from 45,000 tons (average per annum for the period 1787–96) to 19,000 during 1806–10. *Exports* of pig iron equivalent (33 cwts to 1 ton of bar iron) rose from 45,000 tons 1796–1800 to 67,000 tons 1806–8. *Op. cit.*, p. 55.

[4] Mott, *loc. cit.*, p. 44.

[5] Mott, *loc. cit.*, p. 44.

[6] *Weale MSS.* The original list gives South Wales 23 and 4 in Plymouth. This is presumably an error for the Plymouth works near Merthyr Tydfil. It is unlikely, too, that the list showing four furnaces at Carron is correct. Campbell says (*Carron*, p. 205): 'Bar iron for use in the forge was imported until the beginning of the French Wars. . . . Carron bar irons, however, was used in the main only for nail making. . . . '

[7] D. Mushet, *Papers on Iron and Steel* (London, 1841), p. 44.

figure was supplemented by some of that 32,000 tons of bar iron converted from pig iron in the refinery. Unfortunately, the statistics do not make it clear to what extent the figures overlap, but it will have been noticed earlier that Cort refers to 60 smelting fineries making 5 tons of bars a week, that is, 15,600 tons of bar iron a year. These, it is possible, were the old-fashioned bloomeries; another writer speaks of their existence in Yorkshire as late as 1740. These, of course, did not make pig iron. There were, in 1788, 26 furnaces still operating with charcoal fuel and making 14,500 tons. Of these, the largest concentration was in the Glamorgan–Monmouth–Gloucester district (taking in the Forest of Dean) where there were ten furnaces with an output of 6500 tons. The only county, according to the return from which these figures are taken, not to have any charcoal furnaces at all was Staffordshire.

By 1796 the number of furnaces had increased to 125 and their production was something over 125,000 tons. That is to say, the average output of a blast furnace for the whole year was only 20 tons a week. Professor Ashton has, incidentally, indicated the distribution of the iron industry during the period 1788–1806,[8] and although it is true that the Midlands (Staffordshire and Shropshire) throughout these years was the predominant area, it does obscure the fact that South Wales had already overtaken Shropshire as the largest producer.[9] This was almost certainly the result of the successful exploitation of the puddling process which, putting the emphasis upon quality in the later processes, made the ironmasters anxious to expand the output of their blast furnaces.

Ten years later, when there was another scare among the ironmasters, who were threatened with a tax on coal, the following output statistics for the year 1805 were compiled:

Output and Furnaces 1805[10]

	Tons	In	Out	Total
1 South Wales	73,061	36	11	47
2 Shropshire	54,967	30	12	42
3 Staffordshire	48,560	32	10	42
4 Yorkshire	26,596	24	4	28
5 Derbyshire	8,994	11	6	17
6 Forest of Dean	4,173	6	—	6
7 North Wales	1,925	2	1	3

[8] Ashton, *op. cit.*, p. 98.

[9] In 1796 the Welsh production was 34,101 tons; the output of Shropshire was 32,969 tons. In 1788 the respective figures of output were 8,400 and 24,900 tons. It is necessary to issue a warning at this stage about the reliability of these statistics. At this early date none of the available sets of figures agrees with one another.

[10] *Boulton and Watt MSS*, List of Furnaces in Great Britain, 1806.

8 Cumberland	1,291	4	—	4
9 Scotland	23,240	18	10	28
	242,707	163	54	217
Returns not received, estimated	8,800	3	5	8[11]
	250,507	166	59	225

Thus in ten years the output of pig from the British blast furnaces had been doubled. This great expansion clearly shows that the foundations of the industry—the exploitation of coal, steam power, puddling and the mechanical finishing processes, hammering and rolling—laid by the pioneers of that century, Darby, Newcomen, Wilkinson, Watt and Cort, possessed a great potentiality. The sequence of development as it affected the individual ironworks has been well brought out by Dr. Raistrick in his history of Coalbrookdale. Iron was applied to the steam-engine; this in turn was applied to all processes of ironmaking, from mining at the beginning of the chain to hammering at the end; again, the increased output was put to the manufacture of more powerful means of production, and again the iron industry's capacity increased.

There was, nevertheless, another powerful stimulus to expansion. The results of the greedy demands of the wars of the eighteenth century upon the iron industry have been commented upon by Professor Ashton, but the last war of the century—the long struggle against revolutionary and Napoleonic France—far out-reached these. To appraise the influence of these political and military disturbances, which lasted over twenty years, is difficult, especially when we remember that these two decades coincided with that phase of Britain's economic development termed the Industrial Revolution. So interrelated are its processes that it is virtually impossible to separate out the two strands—the developments of peace and of war. However, some attempt can be made to trace the development of the iron industry directly connected with the manufacture of armaments.[12]

[11] Furnaces showing no output: Ashby (Leics.) 1 out, Layton (Leighton) (Lancs.) 1 in; Lymington (Northumberland) 1 in, 1 out; Hay (Haigh) (Lancs.) 1 in, 3 out. There were still 11 charcoal furnaces in use.

[12] See also J. U. Nef, *War and Human Progress* (London, 1950). Nef's view seems to be that the direct influence of war upon economic change has been, on balance negative. Dr. John in his studies of the eighteenth century inclines to the view that domestic expansion was far more important than the demands of war, particularly in the period after 1776. (See his article 'War and the English economy, 1700–1763', *Econ. Hist. Rev.*, Vol. VII, No. 3, April 1955.)

The first requisite is output figures for the war period. We do know at least 67 new blast furnaces were put into production in Great Britain between 1796 and 1802, and 22 more were under construction. Unfortunately there are no statistics of output for the later war years; indeed there is a regrettable gap from 1806 until 1820.[13] Thus there is no precise information to measure either the extent of the wartime boom or of the extreme post-war depression. However, it is now possible with the help of several incomplete series of output figures for some districts in 1812 and 1817[14] and figures of furnaces for 1815–16 for Staffordshire and Shropshire[15] to piece together some account of this period.

First, the extent of wartime expansion. Output figures are available for South Wales and Staffordshire for the year 1812:

		Furnaces		
	Tons	Total	In	Out
South Wales	116,750	64	45	19
Staffordshire	104,350	66	50	16

No output figures are available for the year 1815, and one can only estimate the expansion of the industry from the number of furnaces in blast. These are available for Staffordshire and Shropshire only when Butler recorded in September 1815 the following details:[16]

	No.	No. in blast
Staffordshire	71	55
Shropshire	37	25

Thus, if 40 tons per furnace per week be taken as an average output,[17] the Staffordshire make for 1815 would be in the region of 110,000 tons and the Shropshire make about 50,000 tons. According to this estimate, incidentally, Shropshire's output had declined from 55,000 tons in 1806; the number of furnaces in blast did in fact drop from 30 to 25.

It would be unwise, of course, to assume a uniform rate of increase

[13] Apart from an estimate of 30,000 tons for 1818.

[14] *Dowlais MSS*, Letter Book dated 1817, G. Gilpin—J. J. Guest.

[15] *Kirkstall Forge MSS*, Journal of a Tour among the Ironworks in Staffordshire, Shropshire and Wales by Thomas Butler, accompanied by Edmund Buckley of Manchester and Josh. Shaw of Leeds, in August 1815.

[16] *Ibid.*

[17] Notes which Butler added to the figures substantiate this estimate, for while Attwood's furnaces at Dudley Wood made 240 tons per week from 4 furnaces, Fereday and Co.'s Bradley furnaces produced 40 tons each, and others are mentioned as making 35 tons per week.

in furnace capacity in every district, for some must have had a much lower average than that of Staffordshire.

On the other hand, in South Wales the average weekly output per furnace was nearer 50 tons.[18] Indeed here, as in Staffordshire, the larger works like Dowlais possessed blast furnaces with a weekly capacity of 60 tons. Thus, if we assume that the number of these furnaces had increased by 10 per cent since 1812, then the make for South Wales would be 125,000 tons.

These three major districts alone account for 285,000 tons; and even if it is assumed that the rest of the country's capacity expanded by a mere 5 per cent from 1806, then the total production for Britain may be estimated as 344,000 tons approximately.[19] Hence the British iron industry had about quadrupled its productive capacity during the Napoleonic wars. That this was an excess capacity will be shown by the extent of the post-war depression.[20]

The question to consider next is the effect of the war on the development of the industry. Of actual production of war-like stores, such as cannon and shot, there is, of course, the outstanding example of Carron, which was the arsenal of Europe.[21] However, there is no

[18] As explained previously, Welsh furnaces were the largest in the country.

[19] It is probable that this is an underestimate, since both the years 1812 and 1815 included periods of depression in the industry. Thus, as we have seen, according to our estimate based on the year 1815 (September) the make of Shropshire was well below its 1806 level. At its peak the capacity of the iron industry in 1815 may well have been 375,000 tons.

[20] See below, pp. 54ff.

[21] See again R. H. Campbell, *Carron* (Edinburgh, 1960). Carron, exploiting the newly-invented carronade, had poured out munitions to supply the belligerents during the War of Independence. During the Napoleonic Wars, in 1793, Carron was writing to its agents at Greenock to ask them to seek permission to allow the shipping of 32 guns to Liverpool: 'The merchants in Liverpool are daily writing to us in the most pressing manner for the guns in order to enable them to arm their vessels for the defence of their private property, and we are anxious to give them the most speedy assistance in our power.' (*Carron MSS*, James and Co., Greenock, 18 February 1793). In 1805–6, when Philipp Andreas Nemnich visited the Carron works, he was told that in a good year over 5000 cannon were cast; some went to Russia and Germany but most were for the British Government. (*Neueste Reise durch England, Schottland, u.s.w.* (Tübingen, 1807), pp. 551–2.)

Campbell makes the following points about the incidence of the war on Carron's trade: the beginning of the war, 1793, was a year of panic and the sale of peace goods fell from £65,000 to £45,000. 'The one complete year of peace, 1802, was a peak of economic activity in the country generally, but not at Carron. For the first time, total sales fell below £10,000, particularly because of the fall in sales of shot. Afterwards, in 1803, the renewal of war led to the upward trend appearing again. In 1804, the large sales of shot and carronades meant a new record in total sales, while non-military sales were down . . . ' (p. 221).

mention among the *Cyfarthfa MSS* of the production of cannon which had earlier been Anthony Bacon's chief product there. Its neighbour, Penydarren, it is true, had a long experience of cannon-boring and, as we shall see, the Dowlais partners investigated the possibility of making cannon in 1806. This plan did not materialise, nevertheless the Navy Board became an important customer of Dowlais, which supplied it with much bar and bolt iron. The South Wales industry—though we do not know the extent—was engaged upon war work, as a correspondent of Taitt observed in 1815: 'The Welsh Pigs were only used for making cannon—now the war is over there is no demand. . . .'[22]

In North Wales, at Brymbo and at the other Wilkinson establishments in Shropshire and the Black Country, it is quite likely that the manufacture of munitions went on after John's death in 1808. In Lancashire, at the Haigh works near Wigan, the venture of the Earl of Balcarres into the casting of shot, as we shall see, was again technically and financially unsuccessful. The high proportion of rejects resulted in the works owing a large debt to the London banker who had financed this expansion of the business. But this was unusual, for the prosperity of several major firms was founded upon the profits of munitions. Particularly was this so in Yorkshire and Derbyshire. In the latter county, the Butterley ironworks made cannon among other things (including rails for tramways);[23] in the neighbouring county, works at Leeds and Sheffield were prominent in the manufacture of armaments. From 1795 onwards, guns and cast iron shot were made at Low Moor[24]—the manufacture incidentally continued until as late as the Crimean War and until steel guns began to replace iron cannon. Further south, we have evidence in the shape of a token coin of the Phoenix ironworks, Sheffield,

---

Campbell gives certain figures or profit to confirm these trends. '. . . Carron Company's profits fell with the reduction in the sales of ordnance, from just over £20,000 in 1789 to over £13,000 in 1790 and 1791, rising again to over £15,000 in 1792. In the following year, 1793, war and increased ordnance sales increased profit to nearly £26,000' (pp. 221–2). Afterwards, as he says, it is more difficult to use the figures of profits because of the Carron bookkeepers' manipulations of profits and reserves. (On this point see also my essay in *Explorations in Entrepreneurial History*, Vol. 8, No. 2 1956, and subsequent discussion by Drs. A. R. Hall and R. H. Campbell. See also *Carron*, pp. 164–6.

The supply of cannon to the East India Company was kept up until 1822, and according to Cadell the business with the British Ordnance Department lasted until 1848 (H. M. Cadell, *The Story of the Forth* (Glasgow, 1913), p. 183).

[22] *Dowlais MSS*, W. Hood—W. Taitt, 7 July 1815.

[23] R. H. Mottram and C. Coote, *Through Five Generations—The History of the Butterley Company* (London, 1950), p. 50.

[24] *Souvenir of Low Moor, 1791–1906*, p. 38.

dated 1813, which shows a pyramid of balls and a long cannon before its iron mills.[25] The manufacture of cannon at the Walker concern near Rotherham is well known from the business records of the foundry company partnership edited by Dr. John.[26] This manufacture, begun in 1774, grew to such dimensions that the ironworks were one of the principal sights to be examined by the foreign observers to this country. Broling, the Swede, described the method of casting cannon there, as did Johann Conrad Fischer of Schaffhausen in 1814.[27]

It was the Birmingham district, however, which was the centre of the small arms industry, and a speciality of the Monway Ironworks was best gun barrel iron costing £33 per ton.[28] Much of the local Staffordshire iron was consumed by the gun foundries.[29]

The only detailed estimate of the consumption of pig iron compiled in the period is that of David Mushet, in Dr. Rees's *Cyclopaedia*. The year to which it refers is uncertain, but a contemporary writer puts it down as 1801. After an explanatory remark about the strict accuracy of this estimate, precluded by 'the endless detail into which the foundery trade has now branched itself, the almost universal fabrication which it embraces, and the extensive diffusion of the scites (sic) of manufactories themselves'. Mushet gives figures as follows:[30]

[25] W. J. Davis, *Nineteenth Century Token Coinage* (Birmingham, 1904), p. 183.

[26] *Minutes relating to Messrs. Samuel Walker and Co., Rotherham, etc.* (edited by A. H. John), London, 1951.

[27] J. C. Fischer, *Tagebücher*, pp. 163–4. See also a reference to an account written by W. A. Lampadius (in *Neue Journal für Fabriken*, Vol. III, p. 161 (1814), given by Beck, *op. cit.*, Vol. IV, p. 158.

[28] F. W. Hackwood, *Odd Chapters in the History of Wednesbury* (1920), p. 45.

[29] F. Crouzet, *L'économie britannique*, Vol. I, p. 83, gives these figures for the numbers of weapons made at Birmingham, the chief centre of the armaments industry:

	April 1804–September 1815
Armes complètes	1,743,000
Canons de fusils	3,038,000
Platines de fusils	2,879,000

To these requirements should be added more than a million muskets bought by the East India Company in the same period.

[30] A. Rees, *Cyclopaedia, or a New Universal Dictionary of Arts and Sciences* (London, 1819), art. 'Blast furnace'. Mushet pointed out that this estimate does not take into account of a considerable exportation to Ireland and the 2,000–5,000 tons of melting pig iron each used by the 'numerous and extensive' casting foundries of London, Liverpool, Manchester, Birmingham, Workington, Newcastle and Glasgow. Svedenstjerna in *Nagra underrättelser om Engelska Jernhandteringen* (Stockholm, 1813) gives 1801 as the date of the estimate. Cf. Rhys Jenkins, *Collected Papers* (London, 1936), 'about 1795 it is estimated that 26,000 tons of iron were cast into war material per annum'.

		Tons
It is reckoned that the bar iron forges in Britain manufactured annually from pig iron 40,000 tons of finished bars, which at the rate of 35 cwts. or pigs for every ton of iron bar produced will account for		90,000
Consumed yearly in the erection of new furnaces, forges, machinery, etc.		5,000
Purchased by the Board of Ordnance in the state of cannons, mortars, carronades, shot and shells and etc. on an average of 1794, 1795 and 1796	10,935 tons	
Waste in melting from the pig, boring, etc.	1,300 „	
		12,235
Purchased by the navy board in the state of ballast		2,664
India Coy's. annual supply in guns, shot, shells, carcases, etc.	5,000 „	
Waste in metling and boring etc.	700 „	
		5,700
Merchant guns, carronades, shot and etc. for arming trading vessels	10,000 „	
Waste in metling and boring etc.	1,000 „	
		11,000
Ballast for merchantment and Indiamen		5,000
		111,599

It would appear that just over 31,000 tons went directly to naval and military purposes. It is likely, of course, that this figure is underestimated. However, the fact remains that the demand for iron for armaments and ballast for the fleet amounted to less than one-quarter of the total production of the country.[31]

So much for these estimates of production and direct war-time consumption, which, it must be emphasised, can be used only as a general indication of the extent to which the industry participated in Britain's war effort. That war-time demand, however, had not assumed such dimensions as to gear the whole economy to its priorities can be seen from the careful analysis of Professor F. Crouzet in his *L'économie britannique et le blocus continental* (1806–13).[32] His conclusion, from the point of view of the vulnerability of the iron industry to the possible dislocations of the Continental Blockade, is that the iron industry was only to a limited extent dependent on foreign markets; it was protected against the fluctuations in the British economy directly attributable to the pressures of war. In fact,

[31] Crouzet, *L'économie britannique*, Vol. 1, p. 83, cites an estimate of N. Vansittart for 1804–5 that consumption of iron by the government did not exceed 15,000 to 20,000 tons. In the year 1804–5, the Board of Ordnance and the Royal Navy Board purchased 16,086 tons of British iron and 1458 of foreign (Swedish) iron.

[32] *Op. cit.* (Paris, 1958, 2 vols.).

this situation reflected the limited participation of the iron industry in the military and economic struggle.[33]

TABLE OF VALUES OF PRINCIPAL BRITISH EXPORTS (GT. BRITAIN, 1792–1808 (TABLE IX)
Iron comprising bar, cast, nail, pig, wrought and hardware

	Total	**Ordnance only**
1792	1,379,524	—
3	1,031,126	—
4	1,038,533	7,480
5	960,464	9,790
6	1,187,844	6,050
7	1,080,871	9,900
8	1,088,177	—
9	1,415,179	—
1800	1,483,601	1,100
1	1,330,341	—
2	1,486,455	4,950
3	1,075,522	14,850
4	1,083,575	678
5	992,840	6,164
6	1,248,290	22,886
7	1,382,510	28,640
8	1,178,575	14,621

The reader who wishes to trace the reactions of the metallurgical industries to the expansions and contractions of this war-time period is referred to Crouzet; however, it is clear that the industry in its expansion had run the risk of excess capacity. In 1808, the industry experienced a depression and the more far-seeing of the ironmasters, judging that the supply of iron was sufficient for the demand, were contemplating a system of limiting production.

The worst crisis came in 1812,[34] when there was a slackening of war-time demand, only partly compensated for by peace-time requirements, such as water-pipes and agricultural equipment. Attwood, the Midlands ironmaster, gives a general account of the situation:[35] 'Every branch of the trade is in a deplorable state; the first branch of it, which is the smelting of pig iron, till within these two years was in a tolerable condition, all other branches of the iron trade are now in a dreadful condition . . . even the smelting, that till within these two years, afforded some degree of profit, now affords no profit whatever.'[36] He goes on to give more gloomy details about the decline of the iron trade, 'which about seven years ago was a

[33] *Op. cit.*, p. 83. Mrs. E. B. Schumpeter's figures of British exports of iron products 1792–1808 (including ordnance) reveal that the export trade fluctuated considerably during this war-time period.

[34] Crouzet, *op. cit.*, p. 339. The war of 1812, in which Britain became involved, was also a potent dislocating factor, closing the important American market to the Birmingham and Sheffield trades.

[35] *Op. cit.*, pp. 746 ff.

[36] *Minutes of Evidence*, 1812.

flourishing concern'. The price of iron he had seen drop from £22 per ton to about £11 or £12; with ironmasters unable to produce iron without great loss. 'All men of capital and prudence have left the iron trade', he says with some exaggeration, but he meant that others were forced to go on to save some of the money invested in their concerns.

This gloomy account was substantiated by the evidence of the Midlands ironmasters to the Board of Trade, reviewed by Crouzet.[37] Works were being closed, stocks were mounting, even if some ironworkers succeeded in reducing outputs by as much as a third. In November 1812, Samuel Fereday, the Bilston ironmaster, pleaded strongly to Lord Harrowby, the Lord Lieutenant of Staffordshire, seeking either financial aid from the government or else orders to give the 15,000 workers in the industry three or four days' work a week.[38] The East India Company, in fact, was persuaded by the Board of Ordnance to anticipate its orders of the next year for munitions in order to meet the demands of the ironmasters for orders for 4000 to 5000 tons of iron.[39]

The crisis was of course not confined to the Midlands, though it is probable that it was here felt the most acutely: in South Wales, in Scotland, the downward movement of prices of pig and bar iron put pressure on the ironmasters. This was aggravated by keen competition between ironmasters and districts, the results of which will be seen in the accounts which follow of individual ironworks.

The iron industry did not recover from this crisis when the coming of peace in 1815 precipitated a period of stagnation from which the industry did not recover until the 1820s. And although wartime demand was not over-large in itself, the whole complex of war-time circumstances—including inflation and periods of bad harvests and famine conditions[40]—worked up to a final paroxysm in the first months of peace.

According to Butler's figures of furnaces in blast in 1815 and 1816, in Staffordshire in September 71 furnaces were working. In the next year, 1816, by June less than half of the industry's capacity was being used, only 35 were in blast.[41] The position in Shropshire was equally acute. From the *Dowlais MSS*[42] we learn that there, in the worst

[37] Crouzet, *op. cit.*, p. 747. [38] *Ibid.* [39] *Op. cit.*, p. 748.

[40] There is much evidence to suggest, as Professor Ashton has pointed out, that these natural events were much more potent in their economic repercussions than the artificial stresses of war in the eighteenth century.

[41] Butler's Journal.

[42] *Dowlais MSS*, G. Gilpin—J. J. Guest, 12 October 1817. Cf. *Annual Register* 1816: 'It appears that in Shropshire at this moment (August 1816) there are 24 furnaces out of blast and only 10 in blast.'

period of the depression, only 15 furnaces out of a total of 37 were in operation. The effect of the slump as it hit Shropshire will be discussed when the history of each major iron-producing district is surveyed, nor can an overall and precise picture of the slump be indicated here, as figures of output for the whole country are not available. But even if the profits of the Carron Company, which one would have expected to be the one most badly hit, only fell from 8·6 to 7·2 per cent in 1815–16 and 6·6 per cent in 1818,[43] it cannot be doubted that this slump was the severest of any which had yet hit the iron industry. Ashton mentions that in Shropshire the price of bar iron fell from £18 to £7; 'ironworks everywhere were being shut down: many works were sold at a sixth or a seventh of their original cost . . . and in some cases whole ironworks were abandoned and allowed to revert to the landowner'.[44] Elsewhere, he points out that the firm of Newton Chambers at Thorncliffe made no profits in 1815–16 and returned a loss of 7·7 per cent in 1816–17.[45]

The incidence of the slump fell particularly upon those ironworks working to satisfy the home demand. The authors of *The Growth and Fluctuation of the British Economy* come to this conclusion;[46] they observe that the figures of exports for these years show no decline. This inference is confirmed by the fact that the slump was badly felt by the Midlands ironworks supplying local demands for nail-rods and for cast iron pipes for the water and gas-lighting schemes of enlightened towns. Ketley, as we shall see, was itself closed down by Joseph Reynolds, who retired to Bristol. As for Coalbrookdale, it appears that the partners continued to pay themselves dividends, but only at the expense of the reserves of the company. In 1813 the valuation was £157,313, and ten years later it had fallen to £137,367.[47] It is evident that some losses were made.

Even the great ironworks in South Wales, like Cyfarthfa and Dowlais, felt the effect of the crisis. It was at this time that the young Crawshay was instructed to 'learn the manufacture of every article without delay. . . . This is our last and only salvation.' The reductions

[43] Campbell, *op. cit.*, p. 232. Table of Half-Yearly Profit.

[44] Ashton, *op. cit.*, p. 153, quoting *V. C. H. Shropshire*, p. 469 and 1833 *S. Committee on Mfres., Commerce and Shipping*, evidence of William Matthews, Q. 9637.

[45] Ashton, *op. cit.*, p. 160. Newton in his diary (22 September 1816) comments upon the necessity 'of contracting our business and of reducing wages . . . at the close of a long, tedious and expensive war' (*Thorncliffe MSS*) in terms of the gloomy 'commercial atmosphere'.

[46] A. D. Gayer, W. W. Rostow and A. J. Schwartz, *op. cit.* (London, 1953), p. 127.

[47] Raistrick, *op. cit.*, p. 297.

in the workers' wages during this depression, which were a part of the economy campaign, provoked on this occasion one of those periodic bloody encounters of the iron men with the military. The letters sent to Dowlais, too, during this period reflect the concern of the ironmasters to weather the storm. In these the failure of Fereday, the Bilston ironmaster, was reported, and there are frequent references to bankruptcies in the Midlands. However, it appears that Dowlais itself did not begin 'to feel any difficulty' (whatever that may mean) until the month of May in 1816. Nevertheless, there were alarming accounts of cut-throat competition between the Welsh ironmasters.[48] The depression was aggravated to some extent by the fact that immediately after the war there was a brief speculative demand for exports of iron to America which soon collapsed. In January 1816, Hood, an agent of Dowlais in London, reported that America had been 'completely glutted with iron and British goods'.[49] Moreover, in the export market itself, the East India Company had shipped Swedish iron to India on account of the bad quality of the native product.[50] Sweden, indeed, despite the competition of British iron in the home market, on account of the higher price of the foreign metal, was a strong competitor overseas. In 1815 it was reported that 27,000 tons of Swedish iron had been exported to America from Gothenburg alone.[51]

This severe depression was felt in all districts, but naturally, the inland areas suffered worse than the others on account of their relative disadvantage in transport costs. The Midlands ironmasters could not meet the increased competition resulting from the slump. Indeed, it was asserted by one ironmaster that the iron industry would be 'centred in South Wales'. This was not to be so, but already the harsh realities of competition, regional and individual, were exposed by this sudden stress put upon the industry. And, so far as we can judge, whilst the demands of war—economic or technical—were not primarily responsible for the expansion of the industry, it cannot be doubted that the cessation of hostilities set off the various forces of contraction in the economy. So far as the iron industry is concerned, the result was a long depression which lasted until the early 1820s.

[48] *Dowlais MSS*, Wyndham Lewis—J. J. Guest, 26 May 1816.
[49] *Ibid.*, W. Hood—J. J. Guest, 4 January 1816.
[50] *Ibid.*
[51] *Ibid.*

CHAPTER V

Ironworks and Ironmasters during the Industrial Revolution

This chapter contains brief histories of five ironworks—Coalbrookdale, Dowlais, Cyfarthfa, Silverdale in North Staffordshire, and Haigh in Lancashire. Since they have been based upon the business records available[1] there is naturally a difference in the emphasis given to the several aspects of industrial development. The examples of Haigh and Silverdale show the establishment of small-scale enterprise and some of the many factors accounting for the failure of these ironworks to grow into industrial giants like Cyfarthfa and Carron.

A further explanation ought to be offered concerning the treatment of ironmasters in this chapter. This again is not an explicit statement of these entrepreneurs' characteristics and activities, but their attitudes are revealed in the decisions they took in controlling the development of their ironworks. During the period under review in this present chapter, which roughly covers the period 1760–1830 conventionally assigned to the Industrial Revolution, the captains and lieutenants of this industry were more personally involved, than ever since, in the struggle of their enterprises for financial success. Here they will be met in the counting houses of their ironworks.

1 THE COALBROOKDALE COMPANY

Shropshire has a special place in the history of the iron industry. On either side of the Severn Gorge, which cuts across a not very extensive coal and iron field, lie Broseley and Ironbridge. In this valley formed by the Wenlock Edge, on the one hand, and the Wellington–Wrekin foothills on the other, the Industrial Revolution had its beginnings.

Whatever one's views of the origin of economic developments,

[1] In the case of Coalbrookdale, whilst Dr. Raistrick's book has been extensively used here, many of the MSS have been examined by myself for the present purpose.

whether they tend to be evolutionary or more drastic conceptions of innovation, Abraham Darby's innovation is a significant one. The indissoluble alliance of coal and iron, which the Darbys had achieved in the small Coalbrookdale furnace and forge, was henceforth to determine the superstructure of the modern world. The results, in many ways, have been unfortunate, as artists and art-critics have so eloquently observed;[2] yet this new industrial process was the key to the construction of large-scale mechanised economies.

In the eighteenth century the excitement and wonder of this achievement was new and unsullied by aesthetic disgust. It did not require, indeed, a poetic imagination to be fired by the red glow of the furnace fires suffusing the night sky. This narrow valley of the Severn was endowed with a romantic glamour as county-bred squires, foreign-born travellers and wondering ladies were drawn by its strange attraction to gaze at the blast furnaces, symbolising a new age of industry. The following is a particularly colourful description, written in 1787:

> 'The approach to Coalbrookdale appeared to be a veritable descent to the infernal regions. A dense column of smoke rose from the earth; volumes of steam were ejected from fire engines; a blacker cloud issued from a tower in which was a forge; and smoke arose from a mountain of burning coals which burst out into turbid flames. In the midst of this gloom I descended towards the Severn, which runs slowly between two high mountains and after leaving which passes under a bridge constructed entirely of iron. It appeared as a gate of mystery and night already falling, added to the impressiveness of the scene, which could only be compared to the regions so powerfully described by Virgil. . . . The noise was so great in the long tube which was joined to the furnace that it could easily be imagined the hissing of the irate furies. The steam of which hot liquid appeared as the lava of Vesuvius.'[3]

This was at Coalbrookdale; on the other side of the river, at Broseley, lived that other ironmaster who personified the new order. No modern publicity agent could have conceived a better advertisement than this iron-willed John Wilkinson, with his paraphernalia of iron—cannon balls for his living inspiration, and an iron coffin ready for his expiry. Moreover, as a link between the two worlds of the hard individualism of Wilkinson and the ethic-inspired industry of the Darbys was the iron bridge.

[2] R. Furneaux Jordan, 'A Man-made World—Philistine Victory', *The Listener*, 16 October 1952.

[3] *Viaggio in Inghilterra di Carlo Castone della Torre di Rezzionico, Comasco* (Venice, 1824), quoted in *Salopian Shreds and Patches* (1890), p. 337.

Today a buckled and rather frail structure, and exposing the limitations of a massive heavy construction of cast and wrought iron, it is difficult to see in this the first bold leap of that metal across space. However, in the hands of the engineers of the Industrial Revolution, iron was to span yet other worlds. In the future there lay a world of smooth and swift communications, of steel wheels upon steel rails; left behind was the slow plodding world held back by the grip of its boggy roads.

When the author of the Torrington diaries, the Hon. John Byng, visited Ironbridge in the 1780s he exclaimed: 'Judge then of the flourishing state of this branch of trade. . . . Every cart belonging to this trade is made of iron, and even the ruts of the road are shod with iron!'[4] It was here in Coalbrookdale where these innovations of momentous consequence were conceived.

Finally, in this 'brave new world' there was the steam engine. This the Darbys or their engineers had no part in inventing; nevertheless, first the Newcomen atmospheric engine was brought to greater utility by their skill, and secondly, the Watt engines were built and adapted to new industrial uses by the Shropshire ironmasters.

Following the development of Abraham Darby's initial success, the Coalbrookdale partners applied coke-smelted pig iron, which was particularly suitable for casting and founding, to making iron cylinders for engines. Thus the expensive use of brass was avoided.[5] The result was a chain of reactions which Raistrick[6] well sets out: 'The effect of the use of cast iron was to cheapen cylinders considerably, so that it was now an economic proposition to apply a large cylindered engine to mine pumping. This in turn made much deeper and more extensive mining possible and a cycle of development by interaction—foundry—engine parts—deeper and better pumping—easier and cheaper ore and fuel—larger furnaces and foundries—larger engines—was soon established.'[7]

The distinctive quality of the Watt engine depended upon one thing more than anything else—accuracy of measurement. Indeed, it may be said that since that time this has characterised modern engineering. Here it was the contribution of John Wilkinson who, with an improved boring mill, made possible that precision of fit desperately needed by Watt.

The second achievement of James Watt—that of rotary motion—had been anticipated by an improvisation of Smeaton, and Coal-

[4] Ed. C. B. Andrews, *The Torrington Diaries*, Vol 1, pp. 184–5.

[5] The iron rail had been laid on to Coalbrookdale railways in 1767.

[6] A. Raistrick, *Dynasty of Ironfounders* (London, 1933), p. 129.

[7] Raistrick, *op. cit.*, p. 148.

brookdale made engines for this purpose (that is to turn a water-wheel) from 1760 onwards.[8] However, the Watt engine possessed obvious advantages, and it is not surprising to find the Coalbrookdale Company setting up a boring mill upon Wilkinson's principles in 1780 in order to begin constructing these engines.[9] By 1796 they were boasting that their cylinders were being bored true to a ten-thousandth of an inch.[10] These first engines, however, were built for their own use at Coalbrookdale, Ketley and Horsehay (the associated works in the concern) and were to supersede old machinery. The ironworks were modernised and their productive capacity increased. Thus, as Raistrick says, 'during the twenty years of co-operation with Boulton and Watt, the Coalbrookdale works acted in many ways as a stimulant to the development of the steam engine for new uses'.[11]

Finally, whilst considering the development of the steam engine, which provided both the stimulus and the means for the expansion of the iron industry, it may be briefly mentioned that Coalbrookdale was associated with the invention of both the high-pressure and compound steam engines.[12] These were to be of considerable importance to the development of the railway locomotive and marine engineering.

Against this background of inventive and engineering achievement the Coalbrookdale Company evolved. Our chief concern here is to trace the growth and progress of the company as a business unit.

First, the changes in ownership must be briefly indicated, and this, indeed, is no easy matter, because of the complications of family interests. However, throughout that part of its history which we are discussing here the ownership and control of the company was vested in the hands of the various members of the Darby and the Reynolds families.

The early history of the company will be found in detail in Dr. Raistrick's very valuable book, *Dynasty of Ironfounders*. Briefly, the principal control of the partners' concern in the eighteenth century was in the hands of Richard Ford (1717–45), Abraham Darby II

[8] Raistrick, *op. cit.*, p. 150.

[9] *Ibid.*, p. 152.

[10] *Boulton and Watt MSS*, Dale Company—Boulton and Watt, 23 January, 1796: '. . . they say the diameter of the opening at the bottom is nearly $6\frac{1}{2}$ inches or $6\frac{4125}{10000}$ . . .'.

[11] Raistrick, *op. cit.*, p. 155.

[12] It has been discovered that the drawings of the Trevithick locomotive, thought to be those of the famous Penydarren trial model, refer in fact to a locomotive made at Coalbrookdale some time previously. (See *The Engineer*, 22 February 1952, pp. 266–8.)

(1738–63), Richard Reynolds (1763–72), Abraham Darby III (1768–89), and Richard Dearman (1789–1803).

In 1708[13] when Darby moved to Coalbrookdale the concern was valued at £2804 4s. From this small beginning, to Abraham Darby's death in 1717, the capital grew but slowly to £3200. At that time Darby's sons only held three-sixteenths of the shares, the majority being in the hands of the Bristol merchant and banker Thomas Goldney. When the new company was formed in 1718 it was Richard Ford, with a two-sixteenths share, who directed the fortunes of the partnership. They evidently prospered, for in 1740 the total value of the concern (including a dividend which Raistrick says was paid out) was £21,323.[14] It was during this period that the Willey and Bersham furnaces were run as a part of the concern, although a separate reckoning of profits and losses was maintained.

The period 1745–70 saw the great expansion of the concern and the consolidation of the Darbys' interests in the partnership. The Horsehay and Ketley furnaces were built in 1755 and 1756[15] and the heirs of Thomas Goldney were eventually bought out. The position up to about 1770 may be summarised in Dr. Raistrick's own words: 'We thus see the business dominated by Thomas Goldney and his family from 1718 to about 1770 so far as the actual shares and profits were concerned, but the Darby family, with the assistance of Ford and Reynolds, controlled all policy and the actual conduct of the works.'[16]

Naturally, new capital had to be put into the concern during this period, although we do not know exactly how much was contributed. In 1756 Abraham Goldney and Abraham Darby each put in £1350 extra capital to develop the collieries. However, much of the extension of the property and repurchasing of shares seems to have been done by means of loans. The chief source of this loan capital was Richard Reynolds who, coming to the Dale in 1756 from Bristol, had married into the family and was a partner in the Ketley and Horsehay concerns. By 1780, having bought the manor of Madeley from Abraham Darby for £30,673—the culmination of a series of important purchases of land and shares—Reynolds became the

[13] Raistrick, *op. cit.*, p. 6.

[14] It is not clear whether this included the value of buildings.

[15] *Forester MSS*, Wm. Ferriday to W. Forester, M.P., 11 February 1755. 'Mr. Darby hath lately had a large supply of money from Mr. Goldney who I hear is resolved to make the Horsehay furnace answer, if men and money can. They are mustering up all that can be met of any sort or size old of young; none escapes them. It will be a monstrous expense and after all I think to no purpose.' (I am grateful to Dr. Chaloner for this reference.)

[16] Raistrick, *op. cit.*, p. 13.

principal landlord of the company.[17] Nine years later, upon the death of Abraham Darby III, he vested his interests in the Ketley and Horsehay works in his sons, William and Joseph, and assumed control of Coalbrookdale until 1804, when he retired to Bristol.

As already indicated above, the extension of the Coalbrookdale concern by Abraham Darby III had only been achieved by borrowing money for the expansion. Thus, some of the Darby family shares had been mortgaged to William and Joseph Rathbone and other members of the company, and at Ketley the Dale Company only held half the capital (the rest being in the hands of the Reynoldses). Richard Reynolds, too, had lent the company £20,000 at 5 per cent interest.[18] As a measure of recovery, it was resolved in October 1789 to continue the policy of returning profits to the partnership (at an interest rate of 5 per cent) until all the debts were repaid. Further, even the profits of the associated Ketley Company were to be paid into the Coalbrookdale's coffers, although it was not until October 1790 that some measure of consolidation of the two companies—Coalbrookdale and William Reynolds and Co.—was formally agreed upon.

There were several transactions at this period to discharge the debt of the Darby brothers to the rest of the partners—a debt amounting to more than £50,000—and the capital of the concern was reduced by this amount. The allocation of this reduction of capital reveals that, at that time, the shares were held in the following proportions: Mary Rathbone one-fifth; Sarah Darby one-fifth; Abraham Darby two-fifths; William Reynolds one-tenth; the trustees of Hannah Mary Rathbone one-tenth. As Raistrick says, 'the Coalbrookdale Company was essentially the Darby family concern'.[19] Yet it should be noted that in the 1790s it was Ketley which was the predominant concern with William Reynolds at its head; for it was a new works with new equipment (including Boulton and Watt engines for the blast) which was the main factor in the company's struggle to pay its way. As it was, Richard Reynolds, as the mineral landlord, agreed in 1794 to abate the company's rent by £1000 a year. Moreover, the company sold its dependent Madeley Wood furnaces[20] to William Reynolds and reduced the extent of its leases to an area sufficient for the supply of one furnace.

The inventories drawn up for these transactions show that the net

[17] Raistrick, *op. cit.*, p. 89.

[18] *Ibid.*, p. 209.

[19] *Ibid.*, p. 213. The Reynolds family only held one-fifth; at Ketley, as mentioned above, the Dale Company held one half of the stock and the Reynolds brothers the other half.

[20] *Ibid.*, p. 214. The Madeley Wood furnaces were built in 1766.

value of the Coalbrookdale concern was only £39,858. This included, it is interesting to note, a forge at Bridgnorth and a warehouse at Chepstow upon the River Severn. William Reynolds at the same time valued the Ketley concern at £138,067; however, this high figure was objected to by the rest of the Coalbrookdale partners.

The outcome of all these transactions was that the Reynolds's[21] interest in the parent concern was bought out at a price of £8000 and William Reynolds purchased the independence of Ketley by buying the Coalbrookdale half share.[22]

However, if the finances of the two companies were at last distinct, the plants were still interdependent. Horsehay,[23] Coalbrookdale and Ketley each specialised in different processes and supplied each other with semi-finished products. The group's output at this time consisted of boiler plates and boilers, foundry work, timber and bricks, as well as the stock product, pig and bar iron. The level of output of the associated ironworks can be judged from figures of pig iron made in the year 1805. In the list compiled during the ironmasters' campaign against the threat of taxation the Dale Company's two furnaces at Coalbrookdale produced 2962 tons and at Horsehay 3834 tons; whilst Reynolds and Co. at Ketley, Madeley Wood and Queenswood produced 7510, 2951 and 2605 tons respectively. Six of their seven blast furnaces were in blast.[24]

The general position of the district and of the company's prosperity will be indicated in another chapter. In the early years of the century it was involved in the price war with the Welsh barmakers and, as Raistrick points out, measures of economy (e.g. in the consumption of fuel) had to be adopted. Depression caused the blowing out of a furnace in 1812.[25] The amount of cash paid out to the partners had declined from £30,252 in 1807 to £3571 in 1813 and the works were in a state of decline when Thomas Butler visited them in 1815. He gives a very black picture of them—indeed, he seems to have exaggerated the description of their outdated structure. He says the furnaces at Coalbrookdale were blown by a water-wheel and all the machinery was 'very old and clumsy'. The two furnaces only made 48 tons of iron between them in a week. The Horsehay works too,

[21] Raistrick, *op. cit.*, p. 216. The figure agreed upon was £105,279.

[22] *Shrewsbury Chronicle*, 10 May 1797.

[23] For greater detail on the Horsehay works the reader should, of course, be referred to Dr. Mott's articles in *T.N.S.* cited above.

[24] *Boulton and Watt MSS* (Assay Office), List of Furnaces in Great Britain, 1806.

[25] An analysis of the company's furnaces follows, showing the effects of boom and slump upon the valuation of the company and the amount of money paid out to the partners.

'a very considerable work', was only half employed, according to Butler. His summing up was: 'They gave us a melancholy picture of the iron trade. . . . To make these works do equally to the Staffordshire and Welsh people, they must be entirely new built from the very foundations.'[26]

The concerns of the Reynoldses, Ketley and Madeley Wood were not in much better shape. Here once more it was a case of past glory: 'These have been perhaps the most profitable iron concerns in the Kingdom. . . .'[27] At Ketley, which Butler notes had 'been for a very long time conducted in the most respectable manner', reconstruction was in hand and they were 'altering their works from the Old to New School'. He was, incidentally, most struck with the 'noble office for the principals' and those for the minor officials. 'The whole', he says, 'may be compared to the Commons, Lords and King'.

As we have seen, Shropshire was hit a very hard blow by the post-war slump; only 15 furnaces out of a total of 37 were in operation[28] in 1816. The result was that Joseph Reynolds closed down and dismantled the Ketley works. However, the Coalbrookdale concern, whose dividends to its partners dropped from £8621 to £6610 between 1815 and 1816, succeeded in surviving under the managership of Barnard Dickinson and in 1819 showed a good recovery from the slump by increasing its payments to the partners. The products of the company at this time were boilers, as previously, and sugar pans and furnaces for the West Indian sugar trade.

But the works had been neglected during this unpropitious period and in 1821 Joshua Field, the civil engineer, was not favourably impressed by the deserted look of the place. (True, it was a Sunday visit, as Raistrick points out.) The furnaces had been blown out and the work was confined to founding 'and a little bad mill work'.[29]

The chief clue to the reason for this great change in the appearance of the parent concern at Coalbrookdale had been seen by the keen eye of Butler in 1815. He had noted the fact that their minerals were at a distance from the original furnaces and, six years later, Field

[26] Alan Birch, 'The Midlands iron industry during the Napoleonic wars', *Edgar Allan News*, August and September 1952. The main body of this article is reprinted in R. Butler's *The History of Kirkstall Forge* (York, 1954), Appendix C, pp. 242–9.

[27] Cf. Raistrick, *op. cit.*, p. 239.

[28] *Dowlais MSS*. The *Annual Register* for 1816 puts the position in an even worse light: only 10 were in blast, according to this authority.

[29] Hall, J. W., 'Joshua Field's diary of a tour in 1821 through the Midlands', *Trans. Newcomen Society*, Vol. VI, 1925–6, pp. 1–41. The relevant extracts are quoted in Raistrick, pp. 242–3.

indicated that 'the Furnaces are all removed to where the ironstone coal, etc., are found'. This was, indeed, part of the migration of the Shropshire iron industry eastwards. Henceforth, the first smelting was to be concentrated upon the furnaces nearer Wellington.

Finally, there was uncertainty about the renewal of the lease. It is not surprising that Raistrick says: 'During these ten years 1818–1828 Coalbrookdale was at its lowest ebb in its history with most of the Company's activity concentrated at Horsehay and Lightmoor.'[30] Thereafter, the Coalbrookdale works were to develop as foundry and engineering shops.[31] For Coalbrookdale the wheel had come full circle; in 1699 in Bristol Abraham Darby I had started his industrial career as a pot-founder. By 1851 the art of founding was still the company's mainstay and Coalbrookdale was famous for its ornamental ironwork.

Raistrick gives a table showing changes in the capital accounts and dividends paid out for the period 1798 to 1810. Up to and including the year 1804 these refer to the Coalbrookdale concern only; in 1805 the Horsehay works entered into the reckoning, and in 1809 the Dawley Castle furnace was added.[32]

In 1798 the 'total capital value' (which was the value of tangible assets, such as buildings and plant, plus debts owing to the concern—a total reduced by provision for bad debts, changes in the stock position and debts owing by the Coalbrookdale Company) was £95,424. In 1805, the year of financial amalgamation, the Coalbrookdale concern had a net value of £109,786 and the Horsehay concern brought the total to £134,665.

These terms will, of course, make it quite obvious that the finances of the company were not recorded in the modern way with a balance sheet and with the notion of a fixed capital, as, for example, were the Carron Company's affairs. Indeed, it is extremely difficult to interpret the Coalbrookdale finances in this light at all. Dr. Raistrick's table shows: total capital value, capital increase, capital withdrawn and finally the amount of dividends paid out.[33] On occasions, though not always, the capital increase plus capital withdrawn equals the amount of dividend. This seems to indicate that the term dividend should mean 'gross profit'. However, this must, unfortunately, remain a mystery unsolved; particularly as further figures of cash paid to the partners, which the historian of the company gives, and which overlap the period just mentioned, do not correspond with any computations of capital withdrawn and dividends paid.

[30] Raistrick, *op. cit.*, p. 249. [31] *Ibid.* [32] *Ibid.*, pp. 244–5.

[33] In the text, Dr. Raistrick refers to the interest paid out on a quarter share. Surely, this should have little to do with dividend paid out?

The nearest we can get to seeing the actual position of the company is in 1810, when Raistrick reproduces a summary of the value of the concerns for that year. The net total value[34] of the three concerns —Coalbrookdale, Horsehay, Dawley Castle—is £164,594 13s. 1d. The profit and loss account shows a gross profit of £14,417 12s. 8d. This was appropriated by interest payments of £8096 6s. 5d., by £1680 11s. 1d. of bad debts written off, and £6043 0s. 10d. 'Miscellaneous Inventory' (a reduction of stock). Thus there was a deficiency of £1402 5s. 8d. which was met by reducing the capital accounts of the seven members of the Darby family and two other relatives who had interests in the concern. Thus, whilst the 'value' of the concern cannot strictly be called its capital, it was the record of the value of the investments and undistributed profits which had been sunk into the firm.

From Dr. Raistrick's second table, showing the 'Profit and Loss Account' from 1805 to 1852,[35] it is possible to trace the progress of the company. The valuation mounted, we can see, up to the year 1809 (£165,996); it slumped slightly to £157,313 in the year 1813. Thereafter, in the depression of the post-war years, the valuation fell until in 1823 it stood at £137,367. The boom year 1825 showed a rise to £149,434. Again depression followed and the valuation figure stood stationary at £120,052 during the lean years of 1831 to 1833. In fact the fortunes of the company show a fairly close correlation to the generally recognised periods of boom and slump. In 1836 the buoyancy of the iron market made itself felt in a rising valuation figure of £162,261. The upward movement continued until 1843; thus Coalbrookdale had missed some of the bad times of 1839 and the following years, but eventually at the tail end of this slump the concern's valuation dropped from £252,628 in 1842 to £246,347 in the succeeding year. 1845 again was a boom year (with £301,814 on the books), but so were the following years until 1851. Then the last year, 1852, shows a drop to £313,461. Of course, too much cannot be made of these figures. We do know, nevertheless, that exceptionally large sums were paid to the partners in the years 1806–9, in 1819 and 1825, in 1836, 1838, 1839, 1840 and in 1846 and 1847.[36] Thus there was prosperity and it was passed on to the

1806	£25,178	1819	£12,491	1839	£50,042
1807	£30,252	1825	£14,753	1840	£21,252
1808	£26,631	1836	£13,785	1846	£75,454
1809	£23,397	1838	£18,718	1847	£77,832

[34] As reduced in the manner indicated above.
[35] Raistrick, *op. cit.*, p. 297 (Appendix 7).
[36] *Ibid.*, p. 297

proprietors of the business. But there are longer-term trends evident, too; dividends were on a fairly high level, in the few years we know about, up to 1809. Then for almost thirty years the average level was barely one-quarter of that of the former period. This period came to an end in 1836 and the following years up to 1852 show a much higher level of dividends—even if they were more fluctuating. It is true that this last period also shows a higher rate of capitalisation; indeed, the foregoing analysis could easily be vitiated by decisions to depreciate, which we know nothing about. Perhaps the most that can be deduced from these imperfect figures is that in the second half of the 1830s the Coalbrookdale Company entered into a new period of expansion.

2 THE DOWLAIS IRONWORKS[37]

The Dowlais ironworks, like its rival and neighbour Cyfarthfa, had small beginnings. But unlike Cyfarthfa it was not the enterprise of a London merchant; it was the outcome of prospecting in the locality by Thomas Lewis, who was the owner of the Pentyrch furnace and other small forges in the district. As the articles of partnership show, in detailing the assets brought into the concern by Lewis, he had obtained leases of coal and minerals at Gelligaer and other places around Dowlais upon that large tract of heathy upland which had been originally let by Lord Windsor as a hunting chase. So it came about that the mineral wealth of these heathy slopes on the north-eastern fringe of the South Wales coalfields came into the hands of the Dowlais partners at an absurdly low rent of £31 a year.[38] This circumstance, incidentally, one of the chief reasons behind the ultimate success of the company, was, as the lease drew near to its date of expiry, a very powerful weapon in the hands of the Earl of Bute and was the cause of great uncertainty about the great ironworks' future.

But in 1758 the immense income which was to be at stake not even being a dream for the future, the coming of industry to Dowlais was opposed by the local men who pastured their sheep on the moor-

[37] The selection of Dowlais MSS given in *Iron in the Making*, the Dowlais Iron Company, 1782–1960 (Cardiff, 1960), and in particular, the vast collection of MSS to which this book is a guide, should also be referred to. Unfortunately, Miss Elsas's introduction does not include a detailed epitome of the company's history.

[38] The original lease was a 99-years term from 1784 at an annual rent of £26. In 1763, the lease being reviewed, the rent was increased to £31. No royalties upon coal and iron were payable.

land there. A 'Combination Bond' drawn up these men[39] tells of their protest: '. . . for the new furnas is builded by Mr. Thomas Lewis and Company and they do come to break the incroquement next monday down and gather money amongst the supscribers to take advice counsels before they go to break the new furnas down and turn the watter from her . . .'. Nevertheless, the building of the furnace went on and a partnership bond was drawn up in September 1759. This took in eight men and the £4000 capital was to be divided into sixteen £250 shares; the company was to come into existence on 1 January 1760.

The company was not at first a success. There was difficulty in raising the capital and the concern had to depend upon hand-to-mouth loans, making new mortgages to cover expiring ones. One reason for this shortage of capital was that, although this partnership possessed interests in the iron industry in Shropshire, Bristol and elsewhere in South Wales, because the partners had literally so many irons in the fire they could not afford the investment.

Thomas Lewis, who headed the partnership (whose family interest in the firm was not bought out until 1848 at a cost of £200,000),[40] besides running the Pentyrch ironworks as mentioned above, had also absorbed the Melingriffith tinplate works into his integrated concern. Isaac Wilkinson, too, whose patent 'of blowing Furnaces, Forges and other Iron works by Valves, Cyphons or other Engine(s)' was taken up by the Dowlais Company,[41] also had competing interests in the Bersham ironworks as well as some brief connection with the

[39] *Bute MSS* (National Library of Wales), bond dated 8 July 1758. Lloyd, *op. cit.*, p. 37, quoting a case submitted to Counsel in 1795, says the furnace was begun in 1757.

[40] Bessborough (ed.), *The Diaries of Lady Charlotte Guest* (London, 1950), p. 251.

[41] For an account of Isaac Wilkinson's career see W. H. Chaloner's article 'Isaac Wilkinson, Pot-Founder' in *Studies in the Industrial Revolution* (ed. L. S. Pressnell), London.

In his petition for a patent, dated July 1757, Wilkinson claimed to have invented 'a method or invention for moving or blowing a bellows for a furnace or forge with one, two, three, four or more cranks or elbows made of cast iron, wrought iron . . . to be moved or worked by a water wheel . . . so as to blow two . . . or more bellows . . . through one or more pipes into the furnace or forge to melt iron, iron stone or iron mine'. He also claimed to have invented a method of blowing a furnace harnessing the bellows to a 'fire engine' 'by the going down of the beam in condensing whereby the bellows are lifted up or depressed . . . '. Isaac Wilkinson took out a patent No. 713 in March 1757 (H. W. Dickinson, *John Wilkinson*, Ulverston, 1914, pp. 56–7). Wilkinson at Dowlais was to be paid a generous bonus of £50 for every ton if the furnace exceeded more than 20 tons a week during the first blast. It is certain that Dowlais never produced 1000 tons of iron a year at this period.

Cyfarthfa works at Merthyr Tydfil. Edward Blakeway, a Shrewsbury draper, one of the original partners in the New Willey Company set up by John Wilkinson and other Bristol merchants in 1757, had stayed only two years in that company. His connection with Dowlais was also a brief one. He was discharged from a bankruptcy in January 1768[42] and his £250 share in the Dowlais Company was bought up by another partner, Thomas Harris of Bristol, for £60. Another partner defaulted on his payments altogether. Others were connected with the Cardiff Forge.[43]

It was through Wilkinson that the Guest family appeared on the scene in South Wales. John Guest, born in 1722 at Broseley (not far from Coalbrookdale), had worked in a small family furnace and also for John Wilkinson at the New Willey Furnace.[44] Yet it was Isaac, the father, whom he joined in an enterprise near to Dowlais in December 1763. This is the first reference to Guest's activities in South Wales.

We may discount Wilkins's account of Guest setting off for the benighted country of Wales, and prudently making his will before starting upon this pioneering journey on an old grey mare: Wilkins also likens him to Bunyan's pilgrim on his journey, 'guided by faith and guarded by prayer over his solitary way'. Yet his account of Guest may be true. He was, he says, a tall, strongly-built man, somewhat eccentric, yet reserved and kind.

However, instead of embarking upon this enterprise with Wilkinson, John Guest became manager of the Dowlais furnace in 1767. Actually, it is possible that Guest took up this post a little earlier, in July 1765, but this seems to have been only to the end of the year. From December 1765 to January 1767 Dowlais was managed by two partners, Webb and Price.[45]

The position of Guest was at first that of sub-contractor: he was at this time a mere subaltern of industry. The 1767 agreement being renewed for ten years in 1781, he was then to guarantee to make 700 tons of pig iron a year, selling it to the partnership at £2 10s. bearing all expenses.[46] Even if this arrangement was not very profitable for Guest, it is not surprising to learn that the partnership was prospering, for between 1770 and 1776 it sold its iron at Cardiff for £5 10s. a ton.[47]

[42] I am indebted to Dr. W. H. Chaloner for this information about Blakeway and for the details of the New Willey Furnace Company.

[43] John, *op. cit.*, p. 32. This was financed by the Bristol ironmongers' capital.

[44] *An Account of Dowlais*, by Lord Wimborne (in typescript).

[45] P.R.O. E.112/2096/115.

[46] *Dowlais MSS.*

[47] John, *op. cit.*, p. 99.

Not until 1782 did Guest become a partner. Thomas Harris, one of the original members, then sold him six shares for £2600.[48] Thus the concern was only valued at just less than £7000 at this period after twenty years of existence, although perhaps the sale of the shares was made at a concession rate.

John Guest died, aged 65, in 1787. His heirs included John, who stayed at Broseley; Thomas, who had taken over the management shortly before the elder Guest's death; a daughter, Sara, who had married a merchant, William Taitt. They succeeded to a large share of Dowlais; Thomas Guest continued in the management until 1792, and Taitt, who was the architect of the concern's commercial success during the Napoleonic wars, took over the merchanting side of the business.[49]

The incoming letters to the firm date from this period, and this correspondence shows the continued struggle to build up the Dowlais Company. By this time it possessed two blast furnaces, and their output was about 1800 tons a year. This supplied the neighbouring works of Cyfarthfa and Penydarren,[50] where Cort's puddling process had been adopted. Not until 1801 did Dowlais really take up the manufacture of what was to be its staple product—wrought iron.

The year 1792 was a year of unsettled relations at Dowlais. A new manager, Robert Thompson, was appointed (who stayed until 1798 when he took over the Tintern Abbey works), and the majority of the partners, including the Guests and Taitt, offered to sell out to William Lewis for £25,000 plus stock at valuation.[51] It seems likely that the partners had been looking after their own interests—Lewis, in particular, was more concerned about Pentyrch—and the management of Dowlais had suffered. In December 1792 it was agreed at a meeting of the partners to offer Dowlais to Crawshay at a price of £60,000.[52] However, it is certain that Richard Crawshay, the owner of Cyfarthfa, declined this proposal.

These letters and others printed by Miss Elsas are extremely valuable as showing the extremely onerous life led by an ironmaster at that time. Master, quartermaster (a necessary precaution to prevent

[48] Guest also bought one share for his son Thomas, who took over the management soon afterwards.

[49] *Dowlais MSS*, Articles of Agreement, 30 November 1787. Thomas Guest and William Taitt were each to receive a salary of £150 a year.

[50] *Dowlais MSS*, Minute Book of Proprietors, 1792. Dowlais had contracted to supply Jere. Homfray and Co., with 600 tons of pig iron for seven years from 25 June 1791; and Richard Crawshay 200 tons.

[51] *Dowlais MSS*, Minute Book.

[52] *Dowlais MSS*, Letter Book, William Lewis—R. Thompson, 11 December 1792.

hunger riots), house-builder, farmer, banker, sometimes publican—as well as being a technician and experimenting scientist—Guest Taitt, Crawshay and the other ironmasters had many demands upon their capabilities and patience. Rivalry between the ironmasters was intense, flaring up into disputes about 'the seduction' of skilled men, rights of way, mining drainage, 'scouring', and in particular, the canal. On one occasion, the enraged Crawshay was said to have found an outlet for his feelings by draining a stretch of the Glamorgan Canal—that vital artery of trade—bone dry.[53] Still, they co-operated in a railroad to attempt to improve the means of communication with Cardiff. This was the venture which saw the early Trevithick locomotive put to trial.

There was, too, the heavy social responsibility of the ironmaster who had the lives of his men and their families in his hands to shape as he could according to his conscience. The isolated yet teeming, brawling, ramshackle village of Merthyr desperately needed the magisterial care of the masters to introduce some elements of civilised life into the place. This was not the least of the worries of John Josiah Guest when he succeeded his father as manager in 1807. Moreover, at this time there was the threat of French invasion. In this connection, Thomas Guest, having dispelled the suspicions of the authorities about his devoted Methodism, became an officer in the Dowlais contingent of the Merthyr Sharpshooters. Instead of against Bonaparte, however, the pikes were used in the bloody battles with the regular troops in the riots which periodically swept the district.

But these were distractions from the main task of expanding Dowlais and meeting the competition of its neighbours, Cyfarthfa, Penydarren and Plymouth. There are many letters of Taitt urging the manager 'for God's sake' to increase production.[54] A third blast furnace was constructed in 1793,[55] and a few years later a Boulton and Watt engine was bought to replace the atmospheric one which had been there since the beginning of Dowlais as an ironworks. Having attended to the technical resources of the furnaces, the Dowlais partners had to see to a supply of labour. This was a perpetual thorny problem. A necessary condition of having a labour force at one's disposal was to provide food for man and beast, and

[53] *Dowlais MSS*, W. Taitt—R. Thompson, 2 September 1794: 'I understand that Mr. Crawshay has broke the locks and let off all the water from under the boats loaded with our iron. . . . '

[54] *Dowlais MSS*, W. Taitt—R. Thompson, 31 August 1792: 'For God sake make as much D(ark) and G(rey) iron as possible the present is our harvest for Pigs if we can but make them.'

[55] *Dowlais MSS*, Minute Book, 28 January 1793.

much of the haulage up the canal from Cardiff consisted of fodder. Taitt even ranged as far as Ireland for potatoes, but was refused on account of the dangerous consequences of exporting this staple food from that starving land. Dowlais too, in common with many other ironmasters, had to resort to a 'tommy shop'.

The problem of selling their products, too, was one giving many anxious moments to Taitt—this despite the contracts for a considerable part of their output.[56] In 1793 Taitt considered extending their activities to gunfounding in collaboration with Crawshay, but they could not agree upon terms.[57] The idea was taken up again in 1798, but it appears that the only part that armaments played in the Dowlais business was the unexpected one of being used as raw materials for the forges when they were brought into production. Good quality Spanish charcoal iron guns were bought up by Taitt and shipped to Dowlais.[58] By this time (1806) the forges were pressed to exceed their production 'target' of 80 tons a week and by now the position of Dowlais and Cyfarthfa had changed. Dowlais was now dependent upon Cyfarthfa, Beaufort and other neighbouring works for supplies of pig iron. Attention was not only fixed upon quantity, however, and the managers received many reprimands about the unsaleability of the Dowlais iron because of faults of workmanship.

Since the internal account books of the company have not survived, we do not know a great deal about the growth of capital invested in the concern, or whether it was fed by the profits of the merchant house. In 1793, when Taitt was complaining that times were bad, the net trading profits for the previous year were just over £1900. This was on a capital of approximately £50,000[59]—less than a modest 4 per cent profit. However, the capital value of the company rose from £38,000 in 1787 to £61,072 in 1798.[60] The letters, of course, reflect the impact of slump upon the trading of the company. 1792, as we have seen, was a good year; the next year was one of falling prices. So was the next year. Taitt was writing: 'Our own iron is not at present wanted—as I do not know where to send a Ton. I hope this stagnation will not long continue.'[61] Again, if we can believe

[56] In 1792 Lewis was trying to obtain a contract from the Anglesea Copper Company for 2000 tons a year. These were plates used in refining the crude copper.

[57] *Dowlais MSS*, W. Taitt—R. Thompson, 10 and 21 March 1793.

[58] W. Taitt—Thomas Guest, 21 November 1806.

[59] In 1787 Thomas Guest sold a one-sixteenth share for £2400; in 1798 Taitt bought a share for £3817.

[60] *Ibid.*

[61] *Dowlais MSS*, Letter Book, 12 June 1794.

Taitt, 1794 was a year of reduced demand and increased supply. Unfortunately we are not given many figures to gauge the extent of the depression. In 1797 William Lewis put the unprofitable situation very strikingly: 'Coak Furnaces will never make us rich, at present we are Pack Horses for the Lower Works at Merthyr'[62] (i.e. Penydarren); but this may only point to the desire of the company to get into the more profitable wrought iron trade. It is true, however, that a little later Lewis reckoned for the fourteen months ending 26 May 1798 his profits on all the pig iron made were only £183 on an outlay of £9121.

It was undoubtedly the lack of working capital which was not increased very much during these lean years that delayed the Dowlais venture into the wrought iron industry. When it had been decided in 1801 to risk this Taitt wrote to Thomas Guest: 'Mr. Lewis has been here and thinks we had best only go as far as the Bloom making[63] 'till we grow richer (I doubt it must be so) and that we must individually find our proportions of money. . . .'[64] This extension could not be financed out of reserves, and money difficulties were apparent again in October 1802. Taitt complained: 'I cannot find money to go on as we are doing', and the next month he was facing ruin because the furnaces would not yield enough pig iron to cover the expenses of the concern with a capacity of 76 tons of bar iron a week. It was alleged that even the 'shop keeping' was not very profitable at that time! So harassed was Taitt that his one concern on re-marrying in 1803 seems to have been his wife's connections with the Government and the East India Board which would secure preference for Dowlais iron.[65] The profits of Dowlais for that year 1802–3 were then said to be £4000.[66]

Taitt's marriage was perhaps advantageous to Dowlais, for soon orders came in from the Navy Board and the rebuilding of the furnaces was commenced in 1806. In this it is impossible not to see in the letters of Taitt at this time the straining to meet the competition of his neighbours—always it was, 'You must push on faster.'

In 1806 Dowlais had an output of 5432 tons of iron and the future ironmaster magnate, Sir John Josiah Guest, was introduced to the business and science of iron smelting, being put in charge of the furnace office;[67] a year later it appears that he assumed the manager-

[62] *Dowlais MSS*, Letter, W. Lewis—R. Taitt, 31 March 1797.
[63] That is, there was to be no rolling mill.
[64] *Dowlais MSS*, W. Taitt—Thomas Guest, 7 May 1801.
[65] *Dowlais MSS*, W. Taitt—Thomas Guest, 26 November 1803.
[66] *Ibid.*, 20 December 1803.
[67] *Dowlais MSS*, W. Taitt—Thomas Guest, 20 March 1803.

ship of the works. This year again was one of falling prices in the iron trade and one of Josiah Guest's first important actions was to negotiate an agreed measure of reduction of wages between the masters of Merthyr. The young ironmaster, however, had yet to be schooled in the grim necessities of trade and manufacture, for in 1809 he was sharply reminded by Taitt: '. . . you have had enough of Dancing at this time—you must now stick to business. . . .'.[68] The captains of industry of the first generation were hard driven by the need to run an enterprise at a profit sufficient to provide for the expansion of their cotton mill or ironworks. This attitude of mind is apparent throughout the early years of the history of Dowlais for which we have records.

Despite the severe threat of the slump of 1811–13 and competition from the other Welsh ironmasters and even from Shropshire, John Josiah Guest was evidently a wealthy man when his uncle, William Taitt, died in 1815 leaving him the largest shareholder in the Dowlais Iron Works, which was then the third largest in Merthyr Tydfil.[69] We may close this account of Dowlais up to the end of the Napoleonic wars by indicating the partnership as it stood in 1819: Josiah J. Guest, Wyndham Lewis of Green Meadow; William Price Lewis of the same place, and Thomas R. Guest of Cork, Ireland. The Guests held ten-sixteenths and the Lewises six-sixteenths. Dowlais was then shipping over 10,000 tons of iron down the Glamorgan Canal to Cardiff.

The later history of Dowlais, when Guest became a member of the aristocracy and the works and Dowlais came under the remarkable influence of Lady Guest, will be related in a later chapter.

3 THE CYFARTHFA IRONWORKS

A much fuller account of the rise of Cyfarthfa and of the part played by the Crawshay family has been given by J. P. Addis in his book *The Crawshay Dynasty*.[70] That account is based on a much closer examination of the Crawshay MSS in the National Library of Wales than was possible for me. The brief sketch which follows is intended to present the salient features of the business history of this famous South Wales family and ironworks. Addis's book is

[68] *Dowlais MSS*, Wm. Taitt—J. Guest, 20 January 1809.

[69] In 1813 the firm consisted of William Lewis of Cheltenham, Gloucester, and William Taitt of Cardiff, according to a lease granted to the Penydarren Iron Company (Lloyd, *op. cit.*, p. 41). Penydarren's partnership was Thomas Homfray of Hagley, Stourbridge; William Forman of the Tower of London, and Henry Forman of Greenwich.

[70] *Op. cit.* (University of Wales Press, Cardiff, 1957).

discussed briefly in my introductory chapter: neither that account nor my own penetrates sufficiently into the causes of the decline of Cyfarthfa.

* * *

The Cyfarthfa works were originally part of 'Bacon's Mineral Kingdom'.[71] This rich London merchant, a man of many activities,[72] had taken out a lease some time after 1755 which, according to Malkin, covered the mineral rights of an area eight miles by four at an annual rent of £200.[73] However, the first substantial piece of evidence of his entrepreneurial activities at Cyfarthfa is the partnership deed of 1765, leasing with William Brownrigg of Whitehaven the Cyfarthfa land for building a furnace.[74]

The extension of this furnace was the mill for boring cannon, the supply of which to the Board of Ordnance from 1773 to 1782 Professor Namier has described in his article on Bacon. It was in this business that Richard Crawshay of Lower Thames Street, London, became a partner in 1777.[75] In 1783 Bacon leased his mill to Frances Homfray, the owner of a furnace near Broseley and of a forge near Stourbridge. Nevertheless, Bacon seems to have kept on the furnace, as the lease stipulates that Homfray was not to purchase his iron from anywhere else, and was not to build a furnace of his own. Bacon was intent upon keeping a ready market for his pig iron close at hand. These restrictions, of course, were also meant to keep up the value of the furnace.

The history of Cyfarthfa during the next few years is not clear. According to Malkin again, about the year 1783 Bacon leased his estate and works to Crawshay for £5000 and other portions to Homfray, Hill and Lewis and Tate.[76] Wilkins also more or less confirms this account, except that he says that this took place in 1784 and that a Mr. Tanner took over in Cyfarthfa.[77] The reason for this short tenure of Cyfarthfa mill (not the furnace, it is important to note) was, according to Wilkins, because Bacon had let Homfray

[71] R. L. Galloway, *Annals of Coal Mining* (London, 1898), p. 360.

[72] *Vide* L. B. Namier, 'Anthony Bacon, M.P., an eighteenth-century merchant', *Journal of Economic and Business History*, Vol. II, pp. 20–70.

[73] B. H. Malkin, *The Scenery, Antiquities and Biography of South Wales* (London, 1804), p. 174.

[74] J. Lloyd, *The Old South Wales Ironworks* (London, 1906), p. 48; C. Wilkins, *The History of Merthyr Tydfil* (Merthyr, 1867).

[75] Crawshay was associated with Bacon in the deals in cannon some time earlier. Namier, *op. cit.*, p. 50.

[76] Malkin, *op. cit.*, p. 175. The Dowlais works mentioned in this were never leased from Anthony Bacon, thus much doubt is cast upon Malkin's account.

[77] Wood's *Rivers of Wales* (1813), p. 5.

down with a supply of pig iron. Professor Namier is inclined to place reliance upon Malkin and to dismiss Wilkins's information as 'diluted with a great mass of quasi-pleasant descriptions and would-be lively stories'. However, Lloyd, to whom Namier does not refer, says that in October 1784 the forge and cannon boring mill were leased to Mr. David Tanner.[78]

At all events, after Bacon's death in 1786 his will was contested in 1793 by his illegitimate son, Anthony Bushby; the estate having been handed over to trustees, Cyfarthfa Furnace had been leased from Christmas 1786, for a term of nine years, to Richard Crawshay, William Stevens and Joseph Cockshutt of Cyfarthfa Furnace at 'the clear yearly rent of £1000'.[79] Thus it seems that Crawshay, having been associated with Bacon since 1777, when Bacon's partnership with Brownrigg came to an end,[80] in the business of selling cannon to the Board of Ordnance, decided to venture out into iron manufacture. It is curious why he should have taken this risk with only a nine years' lease to run (until Anthony Bushby's majority), but perhaps he thought he could buy Bushby's interest. However, in 1793, Bushby, or Bacon as he was known later, wished to take possession of Cyfarthfa.

We know more about the results of Richard Crawshay's ownership of Cyfarthfa than we do of the way in which the business progressed and the works developed technically at this time. As is well known, this ironmaster, who had started life as a runaway lad from Normanton in Yorkshire to embark upon a Dick Whittington career in London, died a rich man.

The few Richard Crawshay letters which survive among the *Crawshay MSS* refer to his activities as a political agitator, protesting against 'the absurd Tool Act' in 1786 and appearing before the Committee of Trade on customs duties on iron.[81] In 1793 he wrote to the Earl of Liverpool for the repeal of 'a plaguey act which he cannot understand but which he hopes does not place impediment

[78] Lloyd, *op. cit.*, p. 52. It is fair to point out that Anthony Bacon's will dated 14 June 1785 still refers to the tenancy of Frances Homfray. Tanner's tenancy could, of course, have been a sub-tenancy. Cf. Addis, *op. cit.*, p. 8: '... After complaining that he was not getting enough Cyfarthfa pig iron, Homfray disposed of his lease of Cyfarthfa forge and other premises there to David Tanner of Monmouth.'

[79] Lloyd, *op. cit.*, p. 59. This most probably should read 'James Cockshutt', who had worked with Tanner. Other parts of the estate were leased at the same time, e.g. Plymouth to Richard Hill.

[80] Lloyd, *op. cit.*, p. 50. Brownrigg received £1500 for his share in the concern. Addis mentions that Crawshay was able to finance this venture by winning a state lottery.

[81] *Crawshay MSS*, *Cyfarthfa Papers*, 3, Box 1. Copy of Addl. MSS No. 38346.

at our Customs on fresh arrivals'.[82] In 1799 he had taken up the cause of the English copper trade.[83] There is only one letter, addressed to his son William, dated 11 November 1808, expressing concern about a lack of orders, which deals with domestic business matters.[84]

However, we have seen how Richard Crawshay boosted the expansion of both Cyfarthfa and the South Wales iron industry by developing and exploiting Cort's puddling process. Not only did he take the newly patented technique to Wales in 1787, and with the help of his partner, James Cockshutt,[85] perfect the practical application of the process, but he also went out of his way to introduce Cort's puddling process to the rest of the iron industry.

The wrought iron industry, exploiting the cheap coal resources of the district, was the foundation of the Crawshay fortunes. A. H. John has investigated the financial history of the concern during this period of exploiting the very profitable patents of Cort and he has been able to show the rapid increase in capital invested in Cyfarthfa. (Crawshay, incidentally, much to his own advantage, suggested to Cort that he should reduce his royalty to five shillings a ton.) The Cyfarthfa capital grew from £14,369 in 1790 to £103,908 in 1798. This might seem astonishing if we did not know that in 1792 the profits amounted to almost 82 per cent of the capital. This, of course, was exceptionally high, but in two other years—1795 and 1797–8—the profits approached a modest 20 per cent. Only in 1793–4 was no profit shown.[86] Thus the concern was built up by a ploughing-back of these large profits, and Crawshay extended his activities to the Hirwaun and Union furnaces. John suggests that these extensions of Cyfarthfa's establishment were at the expense of bankrupt works.

At Cyfarthfa itself, the original furnace was added to until in 1797 there were four. These in 1800 were making 200 tons of pig iron a week.[87] The greatest part of this was converted to bar iron in the forges. In 1793 the works was visited by Peter Ewart, one of Boulton and Watt's engineers. Then there were '2 plenishing forges, 4 melting fineries and a rolling mill with 2 pair (of) rolls, for Rolling bars out

[82] *Crawshay MSS*, *Cyfarthfa Papers*, 3, Box 1. Copy of Addl. MSS No. 38229.

[83] *Ibid.*, *Addl. MSS* Nos. 38233 and 38311.

[84] *Ibid.*, Richard Crawshay—William Crawshay, 11 November 1808.

[85] James Cockshutt was the son of John Cockshutt of Wortley Forge and Mousehold Forge, Owlerton, near Sheffield. The Cockshutt wealth had come from a small clerical landowning ancestor of the early eighteenth century. (C. R. Andrews, *The Story of Wortley Ironworks* (Mexboro', 1950).)

[86] A. H. John, *The Industrial Development of South Wales* (Cardiff, 1950), p. 41.

[87] *Boulton and Watt MSS* (Assay Office), Iron Trade Box. 'Observations upon the Coal and Iron Mines district of South Wales and the Forest of Dean: May and June, 1800.'

of ½ blooms turned by a water wheel about 20 feet diameter'.[88] 'At the upper works—one large shingling forge with 8 puddling furnaces where they can make about 60 tons of ½ blooms per week.'[89]

Cyfarthfa was visited several times by Gilbert Gilpin from the Shropshire ironworks in the course of the next four years. He confirms that Cyfarthfa was rented from Anthony Bacon according to the terms to be indicated later, but according to him, that lease had 36 years to run. The last two of the furnaces were then just erected, the four together to be blown by a large water-wheel 50 feet in diameter. (This water-wheel was the wonder of the place and was commented upon by many other travellers, including Bonnard and the Rev. Sir Thomas Cullum,[90] who visited it in 1811. According to this last-named writer, it was claimed as the largest in the world and cost £4000 to build.) He gives no details about outputs, but mentions that there were '6 fineries blown by Engines'.

Another visitor was the Duke of Rutland, and from him we have a long detailed account of Cyfarthfa and some description of Crawshay himself.[91] 'He was', Rutland tells us, 'an elderly man, and seemed a singular character; fully convinced of the great importance of the works which he had accomplished, and talking in a high style, which however was perfectly excusable in him. He struck us as resembling his present Majesty very much in his features and figure.' Already at this time the Crawshay family regarded itself at the head of the iron industry and the ironmaster confessed 'that he had bent his whole mind on being a perfect ironmaster'. In 1802, according to Bonnard, Crawshay made £36,000 from his iron making.

Finally, the Rev. Sir Thomas Cullum in 1811 added his contribution to the descriptions of Cyfarthfa: 'The building where the furnace is placed, is very elegant in its form, and is perhaps the only building in the world which has a compleat iron roof. Iron is here converted into every possible use, even iron barges, which are supposed will soon be very generally used. . . .'[92] Cullum remarked upon the cost

[88] *Boulton and Watt MSS* (Assay Office), Iron Trade Box. From Mr. Gilpin's Memo. Book, October 1796.

[89] *Ibid.*

[90] H. M. Vaughan 'A Synopsis of two tours made in Wales in 1775 and 1811 by Sir Thos. C. Cullum and his son Revd. Sir Thomas G. Cullum', *Y Cymmrodor*, Vol. XXXVIII (London, 1927). The water wheel, according to Gilpin, was baptised and put into operation on 24 October 1796.

[91] Duke of Rutland, *A Journal of a Tour through . . . Wales, 1787* (London, 1805), pp. 61–72. As this description is so vivid and informative on the business attitude of Richard Crawshay himself, some extracts are reproduced at the end of this account of Cyfarthfa.

[92] Cullum, *op. cit.* It was in 1787 that John Wilkinson first launched his iron barge on the river Severn.

at which the ironmasters earned their great fortunes: 'But a fortune is dearly earnt in this place, which to reside in must be one of the most disagreeable places in the Kingdom. It is very cold and exposed from its high situation to the western blast, without any neighbourhood but the masters of the works, and without any society from the frequent disputes they have among one another about the right of Water, or the supposed encroachments in their subterranean works. These things combined with disagreeable smoaky situation of the place must render a residence here a perfect Banishment. . . .'[93] Nevertheless, the Crawshays set themselves up as leading members of the industrial aristocracy by building Cyfarthfa castle overlooking Merthyr Tydfil.[94]

This smoke-laden atmosphere is well shown in the drawing of Cyfarthfa works done in 1811 to illustrate Wood's *Rivers of Wales*. This shows the four furnaces, set into the hillside, at work, with the casting houses standing in front of them. The bare banks of the river Taff and a forge pool are in the foreground of the picture. In the middle distance we see the forges and the bleak rows of industrial cottages, which Crawshay had had built at 30 guineas each. Over all this scene of activity the chimneys throw a black pall of smoke.

At this date, in 1811, Crawshay, the London merchant who had built this 'Temple of Cyclops', had been dead but one year. From his initial capital won, it is said, in a State lottery[95] he had amassed a great fortune. Wilkins most typically says 1½ million pounds. This statement we can immediately discount; Cullum was probably nearer the mark with his estimate of £200,000.[96] However, his will, reproduced in full in Lloyd,[97] shows that he expected to leave a large fortune. For besides bequests of annuities and estate he also left shares in the Brecon canal alone worth £30,000. Moreover, in his lifetime he had endowed his son-in-law, Sir Benjamin Hall, with a large estate, and according to Wilkins, although the authority is questionable, the Rhymney ironworks.

Cyfarthfa itself was left to his son William, with a ⅜ share, his son-in-law Hall, with another ⅜ share, and to his nephew Joseph Bailey, a quarter share valued £25,000.[98] Thus the works were

[93] Cullum, *op. cit.*

[94] Addis, *op. cit.*, p. 54. It was built in the boom of 1825, at a cost of £30,000, by William Crawshay II, and to the horror of his parsimonious father. As Addis says, Cyfarthfa castle was of greater significance than a mere dwelling-place—it was the symbol of the power, prestige and independence that he (W.C. II) longed to achieve.

[95] Ashton, *op. cit.*, p. 96. [96] Cullum, *op. cit.*

[97] Lloyd, *op. cit.*, p. 65.

[98] Cf. John, *op. cit.*, p. 57, quoting a remark of Richard Crawshay, 'I shall not

estimated to be worth £100,000 in 1809. Much of Crawshay's fortune, of course, must have accrued from his merchant house in London—the partnership of Crawshay, Routh and Moser—which his son, already a rich West Indian merchant, had the option of buying. This merchant house was to be an important source of capital in the future and was, in fact, the centre of operations for William Crawshay's direction of Cyfarthfa.

It is William Crawshay I whom J. D. Evans has called 'The Uncrowned Iron King'.[99] He was the architect of the Cyfarthfa works' fortune during the difficult years of cut-throat competition and of post-war depression; yet, as Evans points out, he has been most erroneously treated as of little importance by the local historian of Merthyr Tydfil.[100]

William Crawshay made Cyfarthfa a single Crawshay concern and removed the dangerous restrictions upon its development which his father, Richard, had left untouched at his death in 1810. The first step to securing the independence of the Crawshay control of Cyfarthfa was the buying out of the other two partners. Fortunately the strong-willed and ambitious Joseph Bailey in 1813 sold out his interest in order to launch out as an ironmaster with his brother, Crawshay Bailey, at the Nantyglo works.

Hall was a more difficult proposition. He was the owner of the Rhymney works—a wedding present from Richard Crawshay—and this iron works was dependent upon Cyfarthfa for taking its pig iron in order to make it pay its way. Hence Hall wished Crawshay to continue this preferential treatment to Rhymney and, moreover, he wished pig iron to be taken as payment for a debt owing to the Cyfarthfa concern. Another circumstance leading to friction was that William Crawshay's son,[101] the manager of Cyfarthfa, a man of fiery temper and disposition, hated Hall.

Hall was offered sums of up to £30,000 for his interest in Cyfarthfa in 1814, and pressure was put upon him to repay the Rhymney debt and money on account of the Glamorgan canal. At last Hall gave way, and some time between 1817 and 1819 William Crawshay became sole owner of his father's works.[102]

take in any partners so long as I live', at the time of his giving a one-sixteenth share to Joshua Bailey! (Llanover Estate papers.)

[99] J. D. Evans, 'The Uncrowned Iron King', *Jnl. of Nat. Lib. Wales*, Summer 1951, Vol. VII, pp. 12–32.

[100] C. Wilkins, *History of Iron, Steel and Tinplate* (Merthyr, 1906), pp. 79–80. In his *History of Merthyr Tydfil* he even pre-dates his death by about twenty years.

[101] William Crawshay II.

[102] Evans thinks it likely that the crisis of 1816 and the labour troubles of the time actuated his retirement.

The second great threat to Crawshay's independence and any plan to expand the works was the mineral lease held from Anthony Bacon.[103] This had been obtained, obviously at a high price, by Richard Crawshay after 1795; there was a fixed rental of £5000 a year and a royalty payment of 15s. per ton of iron made.[104] Naturally, Bacon was unwilling to sell such a lucrative source of income and it took all the guile of Crawshay to get Bacon to consider his offer. His main plan of campaign was to frighten Bacon by stories of losses and the blowing out of furnaces. In January 1814 Bacon was told that Cyfarthfa was losing £10,000 a year.[105] This plan, as Evans points out, was comprehensively aimed at the solution of the troubles, for it was a time of labour troubles, overstocked markets and a bitter price-cutting war.

'I will blow out two or three furnaces', he wrote to his son, 'refuse the Romney metal, and thus reduce the refractory spirit of the Men, the too great make of Iron for the sale, the expenditure beyond the real funds of the concern, Mr. Bacon's rental, and my own disquiet as to finances';[106] and being threatened by a bluff threat to remove the Cyfarthfa works to Clydach, Bacon sold out in February 1814 for £95,000.[107] As Evans says, 'He sold out to a master manipulator.'[108]

Almost immediately afterwards Crawshay gave orders for Cyfarthfa to work at full capacity in order to take advantage of a rise in the iron market. To quote Evans again, 'The tactics were perfectly timed and brilliantly successful.' William Crawshay's policy of production and sales, however, cannot be separated from his conception of the place and importance of Cyfarthfa at the head of the iron trade.

As far as that attitude affected the history of Cyfarthfa we may note the consequences of the 1816 depression upon the ironworks. At that date, the capital was much more than £160,000 (it was this in 1813), for the resources of the concern now included the absolute ownership of the lease until its renewal in 1855 and the wealth of the merchanting partnership in London. This, however, was only the result of Crawshay coercing his unwilling London partners into supporting the Cyfarthfa works in times of distress. As Evans points out, it was a difficult task, since 'The policy of securing the maximum balance of advantages for himself as the head of one integrated

[103] Viz. Anthony Bacon II (Anthony Bushby).

[104] Cf. Addis, p. 56. The £5000 was a royalty payment for iron made up to 6400 tons a year and the 15s. per ton as surplus rent on outputs up to 10,700 tons. Bacon then initiated a Chancery suit to recoup on amounts in excess of 10,700 tons.

[105] *Crawshay MSS.* [106] *Crawshay MSS.* [107] *Ibid.*

[108] Evans, *op. cit.*, p. 16.

enterprise could, in fact, only be achieved by making himself sole owner of both the works and the London House.'[109]

The 1816 depression had already been preceded by a minor slump and a price war in 1813. Cyfarthfa at this time was producing about 15,000 tons of pig iron from its five furnaces, but it was not being converted into bar iron on account of the lowness of price.[110] It was a period of great economy at Cyfarthfa. The keen eye of Crawshay looked at expenses which should be drastically pruned: 'Postage is enormous, save it here and at Cyfarthfa', he said in advising short letters. It was at this time that he indulged in a typically mercantilist metaphor in lamenting his position: 'A man with Wealth is a piece of luscious fruit for Drones, Wasps, Flies and every Vermin to fasten on.' So William Crawshay I described his relationship with his son, who, in the father's eyes, was extravagant and improvident.

The economy campaign also included, of course, wage reductions and the abolition of a P.A.Y.E. tax system for his clerks. The letters from the elder William to his son are eloquent with the expression of his feelings of financial embarrassment; and much could be written about this crisis in the South Wales iron industry, which provoked riots and Luddism on the coalfield.

Some of Crawshay's concern and his actions during this time of undoubted crisis in the industry were, of course, a 'blind' to put off Bacon. In 1814, as we know, he bought the lease. Yet there was probably much truth in Crawshay's assertion: 'This purchase shows the good result of early economy.' The result was, as Crawshay put it: 'Thus Cyfarthfa is emancipated, if Mr. Hall will conduct himself as I expect and hope he will. The Power now is mine.'[111]

But the pressure of events was not yet over. In June 1815, when trade was again bad, Crawshay reported to his son that finances of the House were 'so reduced and in such a most unprecedented state of exhaustion that new and vigorous measures must be resorted to'. The measures were cashing money put into Funds and the abandonment of Crawshay's policy of holding stock.[112] At Cyfarthfa, Crawshay ordered the building of a mill and the learning of how to make every article without delay'. This crisis was made worse by the failure of Bruce and Company, the Crawshay's bankers, and the London

[109] Evans, *op. cit.*, p. 16.

[110] *Crawshay MSS*, Letter book, Vol. 1, 16 January 1813–5 February 1817, W. Crawshay—Benjamin Hall, 30 January 1813.

[111] *Crawshay MSS*, W. Crawshay I—W. Crawshay II, 18 February 1814.

[112] *Ibid.* 'We must no longer consider ourselves one Pound richer than others, we must sell and do as they do, and pride of capital and strength of any kind must no longer falsely buoy us up.'

House had to come to the rescue.[113] In all between the years 1813–19, between £250,000 and £300,000 was spent from the profits of the merchant house in London in carrying out Crawshay's plan for securing independence and for bolstering up the concern when it was tottering under the strain of this depression.[114]

Even Crawshay himself was in desperation at this time. In February 1817 he wrote to his son: 'I should not feel one minute's regret on my own or my family's account if the works were for ever destroyed and blown into Attoms or Air—they are no source of pleasure in the best of times.'[115] However, both Crawshay and Cyfarthfa left their depression behind them as trading prospects brightened in the 1820s and, once more, the Crawshays felt the urge to wear the crown of the Iron Kingdom. The history of Cyfarthfa henceforth is of that struggle.[116]

AN ACCOUNT OF MERTHYR TYDFIL, RICHARD CRAWSHAY AND THE CYFARTHFA IRONWORKS

by the Duke of Rutland[117]

August 7, 1797

'A little after one o'clock, we set off on our road to Merthyr Tydfil, 18 miles (from Brecon) . . . we soon got amongst wild, steep, and rocky hills, where not the least mark, not the least trace of cultivation, was visible as far as the eye could reach . . . all was dreariness, desolation, and wildness.'

'. . . We now at least were cheered with the sight of Merthyr, and the iron forges (of which there are three about the town) sending forth large columes of smoke. . . . Merthyr lies in the middle of these

[113] Evans, *op. cit.*, p. 18.

[114] John, *op. cit.*, p. 42.

[115] *Crawshay MSS*, W. Crawshay I—W. Crawshay II, 5 February 1817.

[116] In 1834 William Crawshay I died, leaving his son, William II, the sole owner of the Welsh estates. However, the London merchant house and agency of the firm, 'Richard, William & George Crawshay & Company' (in which Routh and Moser, Richard Crawshay's original London partners, had a one-eighth share), presented problems, both from the proprietary and sales points of view. As late as February 1839 it looked as if William Crawshay II was prepared to sell out his interest in George Yard, the firm's London premises. When the transactions had reached a climax and only awaited completion, the agreement was scrapped and William Crawshay II emerged 'as the purchaser of the entire right, title and interest of Richard & George Crawshay, Routh & Moser in the premises and goodwill of the London House' (Addis, p. 88). This was the last stage in the integration of the Crawshay concern.

[117] *Journal of a Tour through North and South Wales, etc.* (London, 1805), pp. 61–72.

desolate hills, rich indeed in their productions of ore; it is a large place, chiefly occupied by the families of the workmen belonging to the forges. Travellers do not often go there but it is a place well worth their notice, as any in Wales.'

'We dined heartily, and at dusk in the evening, the rain ceasing for half an hour, walked towards the forges. We wandered about for some time, and then went immediately to them, guided by the streams of fire which were bursting forth from the chimneys. The distance of them from the town is about three-quarters of a mile, and the railroad along which we walked ran by the side of a canal; as we approached them, the effect was grand and sublime beyond all description. The fires from the furnaces were bursting forth in the darkness of the night, and every moment we saw, as it appeared, a red-hot bar of iron walking towards us, the man who carried it not being visible. In the perspective, we could see numbers of Vulcans dragging about pigs of iron just taken from the furnaces (the fires of which would dazzle the strongest eye) and pursuing their different operations, while their grimy figures, and gloomy visages, were visible by the light of the forges. We saw them running about in all directions through the door-ways of the buildings, some of them hammering, others rolling the iron, while the regular thumps of an immense hammer, which we heard far off, before we came near the works, and gradually increased to a thundering noise as we approached, completed the grandeur of the scene. I never saw anything that gave me more the idea of the infernal regions. . . .'

'We walked in amongst the men, who were all very civil; and staid till our eyes were weakened by the powerful blaze of the furnaces, and our ears deafened by the beats of the hammer. . . .'

Wednesday, August 8th, 1797

'This morning we sent a note to Mr. Crawshay, the proprietor of the works, requesting his permission to see them. He returned a very polite answer, saying, he would be ready to attend us whenever we chose. . . .'

'He (Richard Crawshay) was an elderly man, and seemed a singular character, fully convinced of the great importance of the works which he had accomplished; and talking in a high style, which however was perfectly excusable in him. He struck us as resembling his present Majesty very much in his features and figure. He told us, that when he originally came to the place, about 10 years ago, there was only one furnace, and that all the other extensive and magnificent works and improvements were wholly his own. He said he could not form any idea of the number of men that he employed, as he had captains under him, who had each agreed to furnish him with a certain number; however there could not be fewer than 1000 able-bodied men employed, and after adding the women, out-labourers, & etc. & etc. the total number of souls who depended upon him, must amount to about 4000.'

'In the last expedition of the French on the Western coast of Wales, he proposed to his men to enrol themselves, and in two days 1000 stout and resolute fellows, armed with long poles, at the end of which were sharp spikes, made by themselves were prepared; and had not the expedition terminated so soon as it did, he intended, he said, to have marched at their head to the scene of action.

'We first saw, and entered one or two of the workmen's houses, which he had himself built for them at the rate of 30 guineas per house; they are extremely neat and clean. The works themselves consist of two divisions, one of them below, the other above a hill. He first took us to the summit, and explained to us the nature of the mines. . . .

'Very fortunately, iron ore, coal and lime stone, are all found in the same hills, so that Mr. Crawshay has every requisite for his works close at home. He pointed out to us one shaft of coal, which would yield daily 200 tons. This is the quantity expended every day amongst the works. . . . Before the end of the summer, he lays up a provision of 15,000 tons of iron ore for the consumption of winter, when the mine cannot be so easily worked. Most of this is generally consumed before the return of spring. . . .'

Next Rutland describes the operations of ironmaking. There were nine puddling furnaces, he tells us. He was particularly impressed with the rolling mill:

'. . . In the first cylinder, the ore (the puddled balls) becomes merely flattened pigs; in the second, these pigs are lengthened out into bars, three times their original length. It is astonishing to see the ease with which the workmen run about with the hot pigs of iron between large tongs, and with which they lift them without difficulty between the rollers. When thus lengthened into bars, they are taken to an immense hammer, which continually acts upon them, and gives the finishing to them, by straightening the bars. The same engine which works the hammer, moves an immense instrument like a pair of scissars (*sic*), which cuts off the end of the bar, generally the worst part of it. While red-hot under the hammer, a boy stamps the initials R.C. on the ends of the bar, (Mr. Crawshay's name) and the operations are thus completely finished.'

'The active and plodding mind of Mr. C. is continually engaged in some new piece of machinery, and expense appears to be the most trifling consideration. He talked of having spent £20,000 in improving some parts of his machinery. . . .'

'When we were there, three furnaces were at work, but there are five, all of which this single wheel is sufficient to blow, iron tubes connecting the whole, and joining them. . . .'

'Mr. C. said at present he made more iron than probably any person in the *world*, that he had bent his whole mind on being a perfect ironmaster, and that should he live long enough, he had yet great plans in view. He hoped to see the day when the quality and quantity of English iron would preclude the necessity of importing that metal from Russian and Sweden. The best revenue was that which arises from the labour of the people, and this ought to be the case in England. . . . He wishes to see a spirit of emulation in every man. . . .'

'In the meantime, while so much engaged in the iron trade, Mr. C. is by no means negligent of other concerns. He has cultivated the country around him, which on his arrival at the place was as barren as the bare rock. When his works were at a stand a short-time since, he employed all his men on half-pay to clear the country of stones, several thousand tons of which he threw into the river, and then cultivated the ground thus cleared. . . . Mr. C. has erected fresh bridges of his own construction, and placed iron buttresses along the banks to turn the torrent. In the last flood it carried away to the amount of £5000 of his property. He is now engaged in a very expensive work, in making another level for the water, which turns the large wheel, considerably above the former one.'

'. . . The difference between these and the Carron works is, that the former are solely meant for the purpose of making bar iron. . . .'

'Mr. C. showed us a turnpike he was going to make, and a canal is finished from the works as far as Cardiff, which is to communicate with the sea. . . .'

4 THE HAIGH IRONWORKS, 1789–1856: A NOBLEMAN'S ENTERPRISE DURING THE INDUSTRIAL REVOLUTION[118]

The iron and steel industry of Lancashire today is largely the product of a renewed burst of activity (following the adoption of the Bessemer converter process) in the eighteen sixties. In the Furness district in north Lancashire, it is true that the charcoal industry

[118] The following section is an abbreviated version of the article which appeared in full under the same title in the *Bulletin of the John Rylands Library*, Vol. 35, No. 2, pp. 316–33.

survived after passing its peak of production in the eighteenth century, but here, as in south Lancashire, the iron industry was to make a new beginning. On the Wigan coalfield itself, it was in 1858 that the large Kirklees Hall ironworks were erected to smelt iron ore from Ulverston, coke being obtained from the local source, although it had to be expensively and slowly washed to remove excessive sulphur. By 1867 there were ten furnaces in operation with an output of over 70,000 tons of iron a year.[119]

However, during the intervening period, when Lancashire was more important as a source of iron ore to the other smelting districts,[120] the manufacturing and finishing side of the industry flourished. Among the several examples which could be cited of this development[121] was the Haigh ironworks itself. It was, in fact, as a foundry and forge that Haigh first comes to our notice. The real mineral wealth of Wigan and district was the famous 'cannel' coal, remarked upon by travellers from the time of Leland in the sixteenth century onwards.[122] There was some iron ore to be found, but it was not of sufficient quality to persuade any industrialist to build a blast furnace in the seventeenth and early eighteenth centuries. However, pig iron (obtainable perhaps from Shropshire and Cheshire via Liverpool, or the Furness area) was to be had at an economic cost to supply the neighbouring coal-pits with castings and forgings. In 1766 the sale of a forge on the River Douglas one mile from Wigan was advertised in a Liverpool paper.[123] This almost certainly was the Brook Mill Forge. A few years later, in 1776, another sale was advertised by the assignee of a Mr. James of Wigan, a bankrupt.[124] As we shall see, it was an existing foundry which was taken over by the Earl of Balcarres.

The Earl of Balcarres (1752–1825) was all that one expected an

[119] F. Kohn, *Iron and Steel Manufacture* (London, 1869), pp. 20–1.

[120] In *The Torrington Diaries*, ed. C. B. Andrews, i. 24, it was noted in the 1781 Tour to the West, when speaking of Tintern Abbey, 'All the iron ore is brought from Lancs . . . '.

[121] Others were Hannah Lees and Sons at Ashton-under-Lyne, and the Dallam Forge established in 1840 at Warrington.

[122] See A. J. Hawkes, *Sir Roger Bradshaigh of Haigh* (Manchester, 1943). Hawkes, to whom I am indebted for much valuable information about Haigh, gives a detailed account, based upon the Haigh Colliery orders, 1636–90, of the industrial activities of the Earl of Balcarres' predecessor. In 1678 there was a glass-making enterprise, but most important was the construction of the great sough, 1652–70 (chap. iii, pp. 13–20). See also the Earl of Crawford's paper to the Manchester Statistical Society, 1933, 'Haigh Cannel'.

[123] *Williamsons' Liverpool Advertiser*, 28 November 1766.

[124] *Gore's General Advertiser*, 5 May 1776. Dr. J. R. Harris kindly pointed out these references to me.

eighteenth-century peer to have been. His descent, it goes without saying, was of an illustrious line. He himself, after a Continental education—not the usual Grand Tour, it is true, but two years at Göttingen University—made the Army his career. By 1789 he had risen to the rank of Colonel of the 63rd Foot, and during the threat of Napoleonic invasion of this country was in command of the forces in Jersey with the rank of Major General.[125] This episode and his subsequent appointment as Governor of Jamaica from 1794 to 1801,[126] it has to be admitted, interrupted his career as an industrial entrepreneur. However, finally incapacitated from further military service on account of a lame leg, he was forced by economic necessity to recognise the need for 'the improvement of his property', as the official history of the family succinctly puts it.[127] This property, which came to Balcarres as his wife's inheritance in Lancashire, obliged him to become an 'improving landlord' of necessity, 'the mansion house being in ruins, the furniture sold—the mines of coal and cannel forsaken, the lands undrained and every farmhouse and fence in the last stage of decay'.[128] Thus it was that, as a beginning to the exploitation of the great potential wealth of the estate, he entered into a partnership in 1788 (20 August) with one James Corbett, an ironfounder in Wigan.[129] The foundry was sold to the partnership for £400 and the new company, which was also to include Robert Lindsay, the Earl of Balcarres' brother (at that time in India), was to set up furnaces at 'Lalland Mill within Haigh', as part of an integrated concern together with the Brock Mill Forge, or, as it appears in the original, 'When the same should be compleated they and the said iron foundry should be thrown together and carried on to such an extent as the scheme would admitt of.' The ownership of the concern was to be divided in the following manner: The Earl of Balcarres, 5 parts; James Corbett, 3 parts; Robert Lindsay, 4 parts.[130]

[125] *D.N.B.*, Alexander, 6th Earl of Balcarres (1752–1825).

[126] Cf. G. H. Tupling, 'The Early Metal Trades and the Beginnings of Engineering in Lancashire', *Trans. Lancs. and Ches. Antiquarian Society*, Vol. LXI, 1949. He says the Earl left Haigh in 1798, and the furnace was abandoned and the works turned over to the manufacture of shot and shell. See below for a correction of this statement.

[127] Lord Lindsay, *Lives of the Lindsays*, II (London, 1849), p. 360.

[128] *Ibid.*, p. 361.

[129] The Earl also developed the coalmines again, selling coal by means of the Liverpool-Leeds canal in Liverpool and Lancaster. This side of the business activities of the estate is not dealt with in this essay, although it is certain they were by far the most remunerative of his activities. The deed of partnership unfortunately could not be traced among the Haigh muniments. It is recited, however, in substance in the contract dissolving the partnership, 1791.

[130] Robert Lindsay, it appears, at first withdrew; and then later consented to

The partnership, then, represented a typical alliance between landed, technical and commercial capital. For Robert Lindsay was, by experience and training, the vital man to the success of the concern. As a younger son, he had been put into trade; in fact, the service of the East India Company. 'He went out as a writer to the East Indies, but without any help to forward his views [other] than his own genius.'

In India he put his genius to practical use, building up a fortune for himself in the course of an eventful and versatile career as 'soldier, magistrate, political agent, farmer, ornamental gardener, elephant-catcher, tiger-hunter, shipbuilder, lime manufacturer, physician and surgeon!' Thus 'having in his line the quickness of a projector', as the official history describes this business instinct, the success of the Haigh concern would seem to have been assured if Lindsay had played an active part in the building up of the ironworks.

In fact, Lindsay, on returning to this country to enjoy his retirement, was reluctant to live at Haigh and take upon his shoulders the responsibility of managing the works, but, all the same, he was dragged in to get the works under way. Corbett, the original owner of the foundry, it appears was full of grandiose schemes for expansion, including setting up a boring mill to manufacture cannon. In June 1789 he wrote to the Earl: 'I am now making the pit, building [a] large receiving furnace and getting forward with the Boring Mill, which I will now venture to say shall not be equalled by any in the Kingdom. It will be constructed [so] as to bore seven cylinders of different diameters of six pieces of cannon at one time.'[131] However, he died in 1790, leaving the plant designed to outstrip John Wilkinson, the great ironmaster of the day, unfinished. What is more, the contract previously quoted reveals that 'the said business having been carried on for some time under the management of the said James Corbett, it was found upon his decease that he had conducted them with great want of economy and that they had cost a much more considerable sum than they ought to have done'. This state of affairs was to be endemic with several managers.

In contrast to this neglect of the appointed managers, however, several letters in the Balcarres collection from Lindsay to his brother show that despite Lindsay's protestations that he knew nothing of the business, his natural energy and practical ability were devoted to reorganising the concern. Finding that most of the men had left the sinking ship, he 'engaged an overseer for one twelve-month, his

become a partner. There was no mention, it seems, of the amount of capital to be paid into the concern by the partners.

[131] James Corbett—Earl of Balcarres, June 1789.

name—John Coward, who at present has the management of the Duckinfield works . . . he is perfectly master of the Manchester business'.[132] Also, he got his brother to recruit moulders from Scotland. The recruiting and handling of labour—in fact, disciplining hands into an efficient working force—was one of the chief concerns of the industrialist of this period, and there is some evidence in the correspondence to support this statement. On 20 July 1790, Robert Lindsay was writing: 'Yesterday warned off Sam Furnace for getting drunk upon his post, which had nearly played the devil with the machinery. In a few days I expect to have all our best people bound down for three years; this has required vast manoeuvring.'[133] Other essential acts of reconstruction were not neglected. He repaired the weir, got the building of the casting house under way, as well as tenements for the men and stables for the horses. He showed an intelligent interest in the metallurgical side of the business. He experimented with the different varieties of coal and ironstone available locally in the two blast furnaces (built 1789–90) and tried the method of casting direct from the furnace, as the following extracts from his reports show:

1. 'Our metal is again excellent, but very much in a state of fluctuation from some unknown cause–last week gave us nearly 15 tons–the furnace without doubt works best with about 2/3rd portion Bath and 1/3rd Brock miln stone.' (20 July 1790.)
2. 'The Black Brock Mill stone, hitherto universally condemned and with which they talked of repairing the roads, proves by much the best stone we have—the metal came down this morning, of an excellent quality and much kished, which has not been the case for two or three months . . . several other things that I have lately tried have succeeded well, particularly in the casting of goods from the blast instead of running the metal a considerable distance as hitherto. It goes into hole at the mouth of the furnace by a gentle stream and from thence (is) carried bit and bit as it comes out to the different moulds and a proper arrangement of carriers fixt beforehand to prevent confusion—this answers prodigiously—for many days we have not run a pig of metal.' (16 July 1790.)
3. ' . . . I have engaged as many hands as I think we can employ for the present . . . by degrees we shall weed until we can get a set of good hands . . . I shall push on the building rapidly to be ready when the

[132] Robert Lindsay—Earl of Balcarres, 5 June 1790.

[133] Robert Lindsay—Earl of Balcarres, 20 July 1790. A later manager complained not of drunkenness, which perhaps was all too common, but of the harm caused by the wife of one of the keepers with her intriguing, profligacy and sabotage (by putting sulphur in the furnace). A. Haliburton—Earl of Balcarres, 21 August 1806.

people come. If the loam moulder does not answer my expectations I will let you know. Our furnace is doing well. If it is now evident that she requires to have the dust drawn out frequently, we begin likewise to be much better acquainted with the materials.' (14 July (?) 1790.)

The final extract, however, makes it clear that the plan for moulding directly from the furnace had to be abandoned:

4. 'After some experience and much observation upon the propriety of using the Blast metal I perceive clearly that we must soon drop the plan altogether except in cases of heavy goods (and) use only the metal from the air furnace . . . for we seldom fail to make a number of wasters owing to the coldness of the metal.
   I am quite in love with the cupula (*sic*) for working up our scraps, of which we had a shameful quantity. This alone keeps all the moulders agoing, nor can they keep pace with it it melts so fast—I have known it charged 20 times a day.' (15 June (? July) 1790.)

As a result of this attention, the concern, which at one time Balcarres was considering giving up, was firmly established and in 1790 (September) the following advertisement appeared in the Liverpool newspaper *Gore's General Advertiser*: 'The Ho. Robert Lindsay & Co. announce they have erected a Blast furnace and Boring mill at Haigh, offer services in Foundry business. Having engaged Engineers and able workmen, they are enabled to undertake the complete construction of Fire Engines of every kind.'

Nevertheless, in one respect Robert Lindsay's influence upon the works was to be almost disastrous. For, despite his work of reorganisation, he was not reconciled to the idea of being tied to Haigh (perhaps because the Earl also was not prepared to become a tradesman); and in 1790, after an extensive correspondence in which the two brothers haggled over the conditions under which Robert should be allowed to withdraw from the partnership, in September a formal deed of dissolution was drawn up and under this Lindsay appears to have paid heavily for his freedom. He was to forfeit £3000 of capital advanced in the concern and was to leave £7000 as a loan secured only by the works buildings and stock-in-trade.

This, however, it should be noted, did not clear up the matter, as negotiations dragged on until 1793. One complication was the claim of Corbett's widow (the wife of the deceased partner) upon the business for a share in its uncertain profits.[134] This state of affairs is,

[134] *Manchester Mercury*, 16 April 1793. 'Haigh Iron Works, Partnership Dissolved. Notice is hereby given that the Co-partnership in the Ironworks erected on the estate of the Earl of Balcarres at Haigh . . . heretofore carried on

of course, another example of the inconveniences of partnership, which (apart from the rare chartered company, e.g. the Carron Company, chartered in 1773) was the only accessible form of joint stock investment then available to business entrepreneurs.

Since neither Lindsay nor the Earl wished to take upon themselves the burden of active management,[135] the problem of securing an honest, efficient manager was of the first importance to the prosperity of the enterprise. Indeed, in this matter, the unreliability of the first manager (which, again, is a cause for reflection upon the development of industrial discipline and responsibility represented by the rise of the managerial class) nearly brought ruin to the concern.

An engineer to the court of Bavaria, Joseph von Baader, came to England in 1789 or 1790. Very little is known of his connections with this country, which brought him here intermittently for a period of approximately eight years.[136] At all events the Earl arranged for Baader to meet Robert Lindsay at Haigh. Now, not the least of Baader's assets was a very plausible tongue and an attractive personality. At that time he was a pleasant young man of twenty-five, not a grumpy old German, as Lindsay expected. He obviously impressed Lindsay with his ability to expound engineering matters and with his accounts of the reorganisation he had carried out at the famous ironworks of Coalbrookdale as well as those of John Wilkinson. During his visit to Haigh, a letter tells us, he drew up a history of the works which, unfortunately, is not with the papers which survive. We know his opinion of the Haigh ironworks, for

---

by the Honourable Robert Lindsay, James Corbett, and the said Earl under the name of Lindsay, Corbett & Co., and which since the death of the said James Corbett hath been carried on by the said Robert Lindsay and Alice Corbett, widow, and the said Earl hath been this day . . . dissolved. Accounts to Mr. Alexander Halliburton, manager at Haigh. 21 March, 1793.

Robert Lindsay,
Alice Corbett
Balcarres.

The concern of Haigh Ironworks will be carried on in future under the firm of James Lindsay & Company.'

[135] In January 1790, Lindsay made his point of view quite clear: ' . . . Now that the value of the metal and manufacture is so far established that the works can be disposed of in a manner so that you will reap a benefit from the concern stead of losing, it is high time to think of myself. I say disposed of with much reason, for unless the Merchants bestow their whole time and attention to a manufacture of this kind, where there is so much competition, it never can be productive but to the proprietor of the estate . . . you cannot suppose that I can make so great a sacrifice as to take up residence at Haigh nor can your rank of life admit your doing of it.'

[136] W. O. Henderson, *Britain and Industrial Europe* (Liverpool, 1954).

what it is worth—'upon the whole he likes the situation of our works and pronounces that at a future day they will make a Figure—Hitherto he says everything has been carried on in a style infinitely too expensive'.[137] As this judgment confirmed the Earl's own ideas, particularly about the latter tendency, the young engineer made a favourable impression upon the two brothers. Baader did not take up the management of Haigh immediately—he came some time in 1791—but when he did, the result was the complete undoing of Lindsay's work of reorganisation; moreover he did nothing to alter the state of affairs which he had criticised earlier. Having caused offence to the English and Scottish workers, he introduced German carpenters and miners into the concern. According to the steward of the Earl, Donald McDonald, these underlings of Baader did nothing but smash the moulding patterns to make furniture for themselves, in the case of the carpenters, and the miners showed none of that industry which Baader had praised, but started the day's work as late and finished as early as possible. As for Baader himself, 'he looks after nothing, nor cares not, as far as his deportment can be read, whether matters sink or swim'.[138]

It took some time, apparently, for the Earl to recognise what the true state of affairs was, but at last it became clear to him. 'Contagion and destruction seems to follow his (Baader's) steps everywhere', he lamented; not without cause, for Reichenbach (Baader's associate) had dismantled the steam engine for the blast and left it in pieces. Thus Baader had Balcarres in his power. Unfortunately, we can only surmise exact details of Baader's dismissal and departure from Haigh.

Another manager, Alexander Haliburton of Inverkeithing, Scotland, was appointed in January 1793, but of the progress of the works under his rule we have little evidence. From the manuscripts

[137] Robert Lindsay—Earl of Balcarres, n.d. There is an interesting sidelight upon the Earl of Balcarres' lack of business acumen in the Boulton and Watt papers (in letters, Bundle E.MIV) by P. Ewart, who describes several visits to the works: 'On our arrival at the Ironworks (in 1791) we found Lord Balcarres to whom we introduced ourselves . . . he informed us—he had laid out a great deal of money there, he believed to very little purpose, for he had been beset constantly by a set of rascally swindling schemers . . . His Lordship, however, appeared to us to be rather a weak brother and was probably an easy prey to such gentlemen as he described. I remember Cooper [Ewart's companion on the visit] observed when he came away that he was certainly like a lord. His lordship informed us, however, that he was then getting into a good train for he had engaged the best and most ingenious mechanic in the world, a person who had invented great improvements in the Steam Engine, Cotton Machine and everything else. This man was the great Doctor Baader.' Letter to James Watt, jun., dated 27 July 1800 (a reference kindly supplied by Miss P. M. Giles).

[138] Donald McDonald—Earl of Balcarres, n.d.

which survive it appears that his association with the Haigh ironworks lasted until 1823 or 1824, when the Earl compiled confidential notes upon his manager's incompetence. This was again a belated realisation upon the part of the Earl of the consequences of a misplaced confidence. It was the anticlimax of a period of high hopes of an extension of the foundry's trade to the West Indies, supplying the plantations with sugar mills and steam engines. Haliburton had apparently weathered one crisis in 1808, a year of depression in the iron trade, and had managed to console His Lordship that 'if these works are getting nothing—Smalley and others must be losing, and that consequently there will soon be less competition'.[139] The works foundry was at that time reported to be 'now complete in all patterns and other utensils, and adapted to all sorts, large and small, and the style of the business, though not extensive, is established on pretty good grounds'. More will be said of the financial state of the business later; however, the firm launched out into the new market in 1811. A circular giving the 'Descriptive advantages of the Haigh ironworks', after informing the reader that they were favourably sited on the Liverpool Canal 'as to admit delivery of goods in 24 hours' to Liverpool, indicated that the works 'are of considerable magnitude, and the proprietors have particularly and pointedly adapted them to an extensive intercourse with America'. The circular also tells us that the works had long been in the habit of making 'Fire Engines, every description of machinery and all manner of castings'.[140] This new venture, for which Haliburton assured the Earl of Balcarres 'every exertion shall be made to ensure success', was, it is fair to note, begun on a note of cautious modesty. Haliburton's advice to his lordship was surely exemplary: 'Two conditions however, seem absolutely necessary—the assistance and guarantee of a substantial House or Houses in England, who are well acquainted with the Trade and stability of the merchants and proprietors of those Countries and a capital at the works here to warrant an increase of Trade and such extended credit as a Foreign Connexion generally requires.' But despite this initial statement of sound business sense there is among the papers of the ironworks for 1824 a list of bad debts amounting to £17,293 due from the sugar dealers of New Orleans, and £6575 of Isle of France (Mauritius) debts. Finally, as to the consequences of Haliburton's managership, although there is little evidence to substantiate the charge, it does appear that he had financial interests in mines and a copperas works which he fostered at the expense of the Haigh works.

[139] A. Haliburton—Earl of Balcarres, 12 July 1808.
[140] Circular dated 1811.

Thus it appears that again the Earl was unfortunate in his business associates. However, it must be recognised that there were deeper seated causes for the decline of the enterprise. The first and most important of these was the unsuitability of the Haigh coals.

Even in the early days of promise, under the keen supervision of Robert Lindsay, his experiments with the different kinds of coal and ironstone had revealed some of the disadvantages of the resources available locally. Even then modest results had only been achieved because they were 'using much [iron] stone which had been picked up in the rivulets and long exposed to the air and had thrown off all the nasty stuff that adheres to the newly got'. The output then per furnace was a mere 17 tons; and the average was nearer 14 tons per week. This prompted the Earl to compute that the concern would lose £1000 a year at that rate, when iron was selling for £6 per ton.[141]

In this connection, one cannot avoid observing, never was an enterprise so bedevilled by continual estimates of loss upon the working of the ironworks. The Earl of Balcarres devoted a great deal of energy to this aspect of the business. Indeed, there are at least twenty such estimates of losses impending. The worst of the matter is that it appears that the Earl was all too correct in his calculations in so far as the accounts will allow us to estimate the profitability of the firm. These, like the rest of the correspondence, are fragmentary, and no extensive analysis of the financial state of the firm is possible. The figures of profit and loss, indeed, relate only to the early period of the firms' history. A letter dated 29 January 1790[142] estimated a loss of £417 for the period of the blast, and the writer goes on to say: 'I am very clear that in making 14 Ton per week, supposing it all sold at £6 per ton we lost £1000 a year.' Perhaps this estimate was not sanguine enough of the ability of the furnaces to pay their way; at all events a few years later, between 1793 and 1795 a loss was sustained of £920 13s. 6d. over the three years.[143] The next set of

141 Estimates of 29 January 1790. Balcarres added: 'I am convinced that our expenditure, before our outlays are at an end, in completing the one furnace and the Boring Mill and the Stock on hand, will be £15,000 to £16,000.'

142 It is essential to point out that although the figures are of losses, the letters fail to make it clear how long a time and therefore no strict comparisons of progress can be made.

143 Perhaps it is not remarkable that another ironmaster, Thomas Butler of Kirkstall Forge, near Leeds, during a visit to Haigh in 1797, also commented upon the inefficiency of its management (28 February 1797): 'After Breakfast Harris Junr. accompanied me to Lord Bellcarras' (*sic*) Iron Works—about two miles from Wigan; they are most excellently situated and every material very convenient: but they are very improperly conducted: His Lordship knows nothing of the Business and his Agents almost as Little. Many a Thousand Pound has been expended to no purpose.—If those works were in the hands of people who

accounts is the most precise of all. Between 8 November 1806 and 31 March 1807, when 435 tons of iron were made, on the working of No. 1 furnace there was a gross profit of £828 3s. 3d., but on No. 2 furnace there was a loss of £517 16s. 5d. Of the balance some £245 had been spent upon a coal-pit, leaving a net profit of £65 or so.

In the period March–July 1808 the furnaces were even more promising in their prospects, a net profit of £221 10s. being registered. However, it appears that Balcarres was disappointed because he had expected a profit of £520 for the quarter, with a gain of £2 per ton of pig iron. Once more, this was an occasion for him to declare: 'Should the answer to this [query] prove unsatisfactory, I have done with my blast iron concern.' One furnace had been out of blast during this period and Haliburton in reply could only propose continuing this policy. 'I think, in the present state of the country and of the Funds of the works, it would not by any means be advisable to re-attune the blast.'

This was a period of widespread slump in the industry, and, in fact, Balcarres not having carried out his threat, the works turned the tide, and final figure is one of gross profit for the year ending March 1810. Sales amounted to £9723 9s. 1½d., and the total expenditure in the works was £9133 4s. If we forget about the change in the figure for stock, which Balcarres' queries to the manager reveal to have been in a chaotic state, there was thus a profit of £590-odd.

This rather unsatisfactory statement of the ironworks affairs goes, unfortunately, as far as the surviving accounts will allow; however, a word may be added about the period when the works were producing armaments for the Napoleonic wars. This stimulus was, as has been shown, not turned to good account as it was with so many other firms. The subsequent inquest, with family acrimony aroused because of attempts to impute responsibility, is of interest because the chief reasons for the relative failure of the concern were recognised and stated by the disputants. Once more Robert Lindsay was brought in. His letter to Balcarres indicates the course of events. The control of the business had been left in the hands of a relative, but his hands had been tied because he only had responsibility in time of emergency. 'The Ordnance concern was then going on briskly—but advances from Carstairs were up by £14,000.' The average product, however, was very small; only 11½ tons weekly. Contributing to the failure of the concern, some of the coal seams had been drowned out —the coal which was obtainable was 'fit for nothing but country

could conduct them with propriety, they might make a fortune.' (*Diary of Thomas Butler, 1795–99*), edited by A. E., B. F., and H. M. Butler, privately printed, 1906, pp. 118–19.)

sale—it was therefore time to blow out. . . .'.[144] The works, nevertheless, were not blown out and in the blast department seem to have been fairly prosperous. A note records a profit of £2000 and odd pounds in 1793 and 1794. It was on the newly-expanded part of the business to manufacture shot and shell (which entailed advances from Carstairs of up to £10,000—£8082 7s. 1d. of which was spent on the buildings) where the business sustained its most serious loss. According to Carstairs,[145] about one-fifth of the total output of shell and shot had been rejected. A rough sort of balance sheet drawn up in 1796 gives some idea of the position of the works during this crisis.

BALANCE SHEET

Liabilities	£	s.	d.	Assets	£	s.	d.
Debts due to works on January 1793 (Less bad debts, and debts due by works	2,612	9	7	Goods to London	1,400	0	0
				2 Bills	650	0	0
				Book debts	1,500	0	0
				Goods in hand	5,000	0	0
Balance of Mr. Carstairs when shot and shell concern is realised (for buildings and improvements since 1793)	7,175	4	4½				
Balance to Lord Balcarres for cannel and rent	825	0	0				
Timber sold	2,228	0	0				
	£12,840	13	11½		£8,550	0	0

Thus there was a deficiency of at least £4190. It is small wonder that the works were reported by Lindsay in 1796 as being at a stand and the shot and shell concern wound up. Yet Balcarres was assured that 'you have as complete a set of works in Great Britain with every requisite to start afresh to advantage'; all that was required was a new engine costing £1200 and 'fresh capital must be produced of at least £3000 for one furnace and double that sum for two'.[146] The last part of Lindsay's advice was by far the most realistic, when he gave his opinion: 'that the proptr. of these works by residing on the spot and being up to the business in all its subordinate detail may make a lively-hood, but that it would never, I say never, answer conducted by a Manager'.

[144] Robert Lindsay—Earl of Balcarres, 13 August 1796.

[145] There is no precise indication in the manuscript as to who Carstairs was, but it seems likely that he was a London banker.

[146] Robert Lindsay—Earl of Balcarres, 13 August 1796.

Nevertheless, there was associated with the firm one engineer who must have brought some profit to the concern. This was Robert Daglish (1777–1865), who settled with a relative at Wigan in 1804[147] as engineer to the works. Perhaps under his management the engineering side of the business flourished. Certainly, the elder Daglish[148] showed considerable skill as a mechanic. It was at Haigh foundry that the first Lancashire locomotive was constructed, this being 'The Walking Horse', made on the model of Blenkinsop's 'Yorkshire Horse', which worked at the Middleton Colliery near Leeds.[149] This was the beginning, it appears, of a successful venture in locomotive building, which continued until 1856, when the firm concentrated on the manufacture of mining machinery. Mention has been made earlier of the financial losses resulting from the export of sugar-mill machinery to Mauritius. However, technically the firm had mastered the production of steam engines and rolling mills, for in 1818 a sugar-mill unit consisting of 'boiler, condensing engine and mill made by James Lindsay of Haigh Ironworks, Liverpool, was imported by Lt. Hannibal Price, R.N.' into Haiti.[150] In the following year there was drawn up a price list of engines, ranging from £300 for a 4 h.p. to a 50 h.p. engine costing £1870, delivered in Liverpool.

The later history of the firm, as an engineering concern, may be briefly summarised here. The furnaces, finally blown out some time before 1815, were demolished in 1828.[151] Thus the Haigh ironworks developed as a foundry and engineering firm making such products as paddle shafts for steam boats, wrought iron work for churches

[147] I am indebted to Professor T. C. Barker for this information.

[148] His son, also Robert, left Wigan about 1830 for St. Helens, where he was prominent in industry and engineering. (See G. H. Tupling, *op. cit.*, p. 28 and n.)

[149] J. H. M. Bankes, 'Records of mining in Winstanley and Orrell, near Wigan', *Trans. Lancs. and Ches. Antiquarian Society*, Vol. LIV, pp. 57–60.

[150] *Trans., Newcomen Society*, Vol. XXI, p. 16. A photograph of this plant, as it stood in 1930, was reproduced as an Appendix (Pl. VIII).

[151] *Wigan Examiner*, 2 May 1856. Alexander, 6th Earl of Balcarres, had died by this time (27 March 1825) and had been succeeded by his son, James Lindsay (1783–1869). Haliburton left the partnership in 1829 (*Haigh MSS*). E. Baines, *History, Directory and Gazetteer of the County Palatine of Lancaster*, etc. (Liverpool, 1825), Vol. II, p. 611, states: 'The late noble proprietor of Haigh rendered productive the iron and coal strata of his estate by commencing a foundry upon it in the year 1787, and after contending with many difficulties, he at length succeeded in establishing works for the construction of steam and other mechanical engines upon an extensive scale. . . . About 16 years ago the smelting of iron ore at these works was discontinued on account of the low price of metal, but from the recent advance in the price of iron, the intention is said to exist of re-erecting smelting furnaces at Haigh.' In the *Directory of Wigan* the name of the ironworks is given as 'Jas. Lindsay & Co.' (steam engines, millworks, etc.).

and, of greatest importance, locomotives.[152] It is said that the partnership, which took over from the Earl in 1836 (consisting of Messrs. Evans, Ryley and Burrows) on a twenty-one years' lease, made 114 locomotives at Haigh in that period. When the lease expired in 1856 the decline from its period of greatest prosperity set in. The works, then in the hands of another partnership, Birley and Thompson, were concentrated upon the production of mining machinery, the locomotive business being given up. Here, there was more competition to be met, which Haigh could not face, and in 1884 the works were closed down.[153]

5 THE SILVERDALE IRONWORKS

As B. L. C. Johnson has shown in his account of the Staffordshire iron industry, the smelting and working of that metal by charcoal furnaces and forges extended from its concentration in the south through Cannock Chase to North Staffordshire.[154] It was here that the Madeley furnace[155] described by Dr. Plot and, less well known, the Oakamoor furnace (upon or near the site of the Cheadle Brass Wire Company's mill) had an insignificant part to play in the wide-ranging activities of the Foley family of ironmasters.

With the adoption of coke-smelting it might have been expected that the iron industry in the north of the country would have expanded. There were at hand the raw materials—ore, coal and limestone; there was a market near at hand in the machine shops of Manchester and the warehouses of Liverpool; and there was adequate means of communication.

As early as 1776, a local landowner had projected and constructed a canal at Apedale to connect the mines there and the ironworks to Newcastle-under-Lyme.[156] The Grand Trunk Canal, constructed

[152] The famous Laxey Wheel, on the Isle of Man, was also built at Haigh (*Manchester Guardian*, 10 October 1946).

[153] J. A. Melling in *Wigan Examiner*, 28 January and 4 February 1949. This writer, who had intimate connections with Haigh, gives a divergent account of its origin and development. He suggests that the Earl of Balcarres' estate workshop outgrew its functions and there was continued expansion from that time. The surviving manuscripts, however, do not substantiate this view, particularly for the early period of the firm's history.

[154] B. L. C. Johnson, 'The iron industry of Cheshire and North Staffordshire, 1688–1712', *Trans. North Staffs. Field Club 1953–4.*

[155] In the *Sneyd MSS* referred to below, there are carriage accounts for the cartage of ironstone from Mr. Sneyd's mines, 3 October 1656 and 22 December 1657.

[156] Under an Act (15 Geo. III Cap. 16): W. H. Chaloner, *The Social and Economic Development of Crewe* (Manchester, 1950), p. 14.

under the guidance of James Brindley, whose home was near Tunstall,[157] provided access to the chief industrial centres.

Yet, in the first half of the nineteenth century, the district did not expand, rather the opposite. When Butler, the Yorkshire ironmaster, visited the district in 1815, he found the works declining. His observations are as follows: 'Near Talk o' the Hill, on the left, a small furnace belonging Mr. Layton—in bad condition, makes poor iron and little of it.' At Golden Hill furnaces Barker, the proprietor, Butler notes, 'has 150 tons on hand, will blow out directly, while times mend'.

'Silverdale Works have been discontinued about two years, they produced inferior metal. Golden Hill, only one furnace produces 18–20 tons per week, the iron very tender and weak and breaks close —the best sort melts pretty well—this iron in much demand for castings that require to be of a close texture, such as pipes, cylinders, etc. . . .'[158]

The reasons for this decline may make themselves apparent by an examination of the early history of the Silverdale Ironworks. For amongst the *Sneyd MSS* of the Raymond Richards' collection,[159] and in particular the few documents relating to this firm, there is the *Minute Book* for the period 1793–1815. The construction of the blast furnace and the exploitation of 'all the Ironstone and Coalmines on the N. and N.W. side of the Haying lane in the parish of Keele belonging to Mr. Sneyd' (together with a navigation to the Newcastle Road) was first mooted in August 1790. The draft agreement for a lease of that date provided for about 40 acres of land for a farm and a furnace for a term of 63 years or 21 or 42 years, as the lessees should choose. They undertook 'to erect on the premises a Farm House and other necessary Buildings, a Fire Engine Blast Iron Furnace with other necessary buildings, the same with good and well burnt Bricks to carry the same upon a true level and complete the same in a workmanlike manner within the space of two years'.[160]

This lease was agreed to and on 5 April 1792 a partnership of Thomas and John Breck, William Hyatt, Thomas Pool and Samuel Hopkins of Lea Forge in Cheshire, coming into existence upon the following day, took up the lease at an annual rent of £63 a year. No definite amount of capital to be called into the concern was fixed

[157] See the picture of Turnhurst, near Tunstall, with a model of the canal in the garden in P. W. L. Adams, *Wolstanton* (Tunstall, 1907).

[158] A. Birch, 'The Midlands iron industry during the Napoleonic wars', *Edgar Allen News*, August and September 1952.

[159] Now in Keele University.

[160] *Sneyd MSS*, Box 4, Proposed Agreement for Lease, August 1790.

upon, but there were to be monthly meetings for managing the concern.[161]

Some time after this Sneyd came into the partnership, probably as part payment for the grant of the lease. There had not been a precise provision for the investment of capital into the concern, and at first it appears that the concern was dependent upon loans from local people. In 1793, £500 was borrowed from William Daniel of Newport, a potter.[162] Sneyd, who later assumed the virtual ownership and direction of the company, was 'out of the country for the commencement of the concern' and 'the capital or Money advanced was irregular and unequal'.

The following extracts from the Minute Book[163] provide the best commentary upon the progress of the concern.

The opening resolution speaks for itself:

7 March 1793: 'Ordered by Mr. Hyatt that no Pig Iron shall be sent off without being marked in 4 different places on each Pig.'

By 23 January 1794 some determination about capital had been come to: each partner was 'to advance his share up to the amount of £1700'. A sixth partner had come into the partnership—George Wood—and this brought the capital of the Silverdale works up to £10,200. At the same meeting, it was resolved that a house was to be built and 'to have cast iron window frames if approved'.

Pool was a local owner of land and the next meeting (in April) allowed him £40 'for his attendance at Silverdale and for the use of his Road & Trespass upon his land'.

The pig iron was hauled up by a horse team at this time between Silverdale and the wharves at Stoke and Shelton. This iron, it should be added, had been the produce of two blast furnaces, but in February 1794 it was decided to blow out one of them 'as soon as it can with propriety'. The coal for the smelting came from the nearby Leycett pit, also owned by the company, and they had the prevalent trouble with the miners and ironstone getters, who took coals home for their open fires. The partners had to regulate this practice.

By 30 July 1795 times had mended and the Minute Book states: 'It is this day agreed that the Furnace No. 2 shall be put in blast immediately. That the Cupilou (*sic*) be put on and all the Scraps worked up.' Provision was also taken for replacing a hearth stone 'which was to be procured from the most approved place in Shropshire'.

[161] *Sneyd MSS*, Box 4. [162] *Ibid.*, Draft Case Statements. [163] *Ibid.*, Box 4.

There is a significant gap in the meetings of the partners between this last meeting and 13 April 1797. This evidently was a period of dislocation, from what cause we do not know. However, in 1797 a halt was called, and it having been found that 'the Company's Leger (*sic*) was in many Folios very imperfect and no account from Mr. Hopkins of his Transactions' having been received, a statement of the affairs of Silverdale was called for. The next meeting, which carries on this intention to put things in order, reveals for the first time the activity of Walter Sneyd in the concern. The capital shares were to be equalised to one-fifth each. (William Hyatt had in the meantime disappeared, being replaced by Sneyd; and there is no mention of Wood.)

The succeeding entries for that year show that a new start was being made; the company was taking up the casting of pipes and 'the customers on the Lancashire Journey' were to be visited by a representative of the firm. The furnaces were reconstructed—new hearths being again necessary—and a furnace constructor from Wednesbury (one Simcox) was to do this and to install a further tuyere to each furnace. It was evidently necessary to build a new boiler in order to increase the blast (and so to increase output), and this was done. The company also looked to its supply of coal from the neighbouring pits—new pits were required, and 'Whimsey Engines' were to be installed at Leycett colliery and 'the other upon Colonel Sneyd's royalty at Silverdale'. Most important of all, it was proposed to set up a slitting mill and forge 'to make 40 tons of Barrs or Rods weekly'. However, this plan was held in suspense for two years, and it was not until 27 November 1797 that it was decided to obtain the materials for a rolling and slitting mill to be worked upon Cort's principles. The entries during the year 1800 indicate that building was going on.

However, all did not go well. An accident to the blast engine during its modification into a 'Patent Engine' stopped production and the furnaces were out of blast until February 1802. During this time, too, there was trouble with the colliers and ironstone getters. The cause is not stated, but it may well have been in connection with the hunger riots which disturbed many parts of the country in that year. At all events, it was resolved that any men who neglected their work should 'be taken before a magistrate and punished as the Law directs'.

Something will be said of railway construction later, for when the North Staffordshire iron industry did get under way in the second half of the nineteenth century, it was very dependent upon railway communications with the markets for its pig iron. However, as early as 1803, the minutes quoted here show that the idea of a railway connection with Newcastle was being discussed. And in 1805 it was resolved to lay an 'Iron Railway' (that is, a tramway) from 'the Nantwich Turnpike Road at or near John Thomas's . . . through the Wood, called Wharton's Wood . . . to the Coal Pitts at the Sawney in Leycett'.

But again the works seem to have been out of action, or badly managed, and five years afterwards, in 1810, a new resolve was made to work both the furnaces. Meetings in 1813 were spent in settling accounts, and this, the only minute book, ends in March 1815 with some mention of another railway scheme from another pit to a point on the turnpike road.[164]

[164] All the above extracts have been taken from the Minute Book of the Proprietors of Silverdale Ironworks 1793–1815 (*Sneyd MSS*, Box 4).

CHAPTER VI

The Ironmasters' Associations in the Early Nineteenth Century

Integration of plant and finances among firms in the iron industry was paralleled with combinations among the ironmasters themselves. This procedure had its origins going back to the mid-seventeenth century, as when George Sitwell, the Derbyshire ironmaster, made an arrangement with two competitors to fix the selling price of bar iron.[1] Indeed, subsequent research has shown that local combinations of merchants fixing selling prices, and prices of raw materials, were being organised,[2] and that Ashton's statement that 'so long as charcoal remained the only fuel used in the making of iron, works were necessarily widely dispersed, combination was difficult . . .' requires modification.

The technical transformation of the industry nevertheless provided further necessity and opportunity for commercial combination. The expansion of the industry in the second half of the eighteenth century, and the recognition by the government of its potential value as a source of revenue made individual ironmasters aware of common interests for marketing and for political purposes. Moreover, it is likely that the various joint enterprises, such as turnpike trusts and canal companies, necessitated by industrial expansion brought together ironmasters who then, naturally, spent some of the time at these meetings coming to terms about prices and wages.[3] Indeed, it was asserted by an iron merchant in the nineteenth century that the association of ironmasters had 'prevailed so long, they were a

[1] Ashton, *op. cit.*, p. 162.

[2] John, *op. cit.*, p. 131, also *Econ. Hist. Rev.* 1st Ser. Vol. XIII, 'Coal and iron on a Glamorgan estate'.

[3] *Dowlais MSS*, Richard Hill—J. J. G. 20 February 1825: 'I presume Mr. Crawshay and Mr. Bailey will be at the Canal Meeting the 3rd February at the Bridgewater Arms when we will have some discussion about the Iron Trade, the system being adopted of taking the stint when we can get it. . . . '

matter of notoriety even in the times of Adam Smith, and were particularly commented on by that great master of Political economy'.[4] There is no direct reference by Adam Smith to the ironmasters' meetings, but it is possible that his famous remark, 'People of the same trade seldom meet together, even for merriment and diversion, but the conversation ends in a conspiracy against the public, or in some contrivance to raise prices',[5] had the ironmasters in mind. Be that as it may, certainly the quarterly meetings taking place at convenient hostelries were convivial occasions. Ashton has noted that, after 1790, a weekly dinner was held in the Ironmasters' Room of the Union Tavern in Birmingham;[6] in 1802, a Welsh Quarterly Meeting was instituted at Abergavenny, members agreeing to pay an annual subscription of a guinea and each new member had to provide a bowl of punch for the meeting,[7] and similar arrangements for providing hospitality were made by the Yorkshire ironmasters. The Shropshire ironmasters met at the Jerningham Arms, Shifnal,[8] and the Welsh ironmasters met at the Bush Hotel, Dowlais, and the New Inn, Merthyr. This close association with hotels was, of course, to be expected in the days of horse travel; the house of call, too, it will be remembered, was similarly associated with the friendly societies of the workmen. No one who has read Ashton's account of the meetings of these Yorkshire ironmasters, with their concerted policies, will underestimate the extent of this commercial combination; however, it should not be overlooked that in Yorkshire, if not elsewhere, there was an attempt to promote scientific and technological discussion, such as was to be achieved by the Iron and Steel Institute seventy years later. Through the ironmasters' meetings, too, some attempt was made by the industry to repay its debt to Henry Cort, the unfortunate inventor of the puddling process. It was

[4] John Hall, *The Iron Trade (with remarks pointing out the true cause and cure for its existing state of depression)* (London 1843).

[5] Adam Smith, *The Wealth of Nations* (Everyman Edn.), Vol. I, p. 177. However, it might be argued that the tradition of consultation had been strengthened by the religious affinities of the ironmasters—especially the predominant Quaker element. The business ethics of the members of the Society of Friends favoured fixed prices for goods; moreover, it is possible that the quarterly meetings, bringing the ironmaster and their families together for spiritual stocktaking, were the precedent for the commercial organisations also meeting quarterly.

[6] Ashton, *op. cit.*, p. 177.

[7] Ashton, *op. cit.*, p. 177; John, *op. cit.*, p. 133; C. Wilkins, *op. cit.*, p. 66.

[8] *Coalbrookdale MSS.* However, in 1819 Gilbert Gilpin explained to Guest of Dowlais, 'The Shiffnal ironmasters' meeting is never final, merely preparatory to the Stourbridge one' (Elsas, *Iron in the Making*, p. 5). The Stourbridge meeting, apparently, co-ordinated prices between the Staffordshire and the Shropshire ironmasters.

resolved in 1810 that R. Butler should 'apply to the iron trade for subscriptions to Mrs. Cort and family in order to act in conformity with the intention of the last meeting at Gloucester'.[9]

However, with the exception of the Midlands meetings which took place at local centres, Shifnal, Stourbridge, Wolverhampton, and Walsall, culminating in a final conference of the ironmasters from the several counties at Birmingham[10] regularly throughout this period, the first half of the nineteenth century does not appear to have witnessed any spectacular progress in the consolidation of these commercial combinations. Indeed, the tendency appears to have been for these organisations to disintegrate. So much so that Professor Ashton, viewing the disappearance of the district organisations in the 1820s, has suggested that 'local organisations were breaking up and giving way to larger bodies, of themselves expanding and assuming a form in which they could play a part on the larger stage of national politics'.[11]

This view cannot be altogether discounted, since the development of a national body of ironmasters has not yet been uncovered by historians. For even if the existence of such a body were continuous (as seems unlikely) it is submerged and only comes to the surface in the evidence afforded by the response of the local associations to the concerted policies of the national meetings. Nevertheless, Ashton's statement in the light of more recent research needs some qualification.

As we know, the industry had been organised on a national basis for short periods to agitate opposition to such political measures as the Irish Propositions of 1784, and the proposals to impose taxes on coal and iron in 1796 and in 1806. In that last agitation the interests of all the iron-producing districts were represented.[12] Again, in 1819, since the industry had been one of the few to escape the burden of war taxation on its raw materials or its products, it was still sensitive to government moves to tax coal, and in that year deputations and petitions were sent to the House of Commons on behalf of the ironmasters. Once more there was co-operation between the Scottish pig iron trade and the associations in England and Wales, as shown by the following extract from the minutes of a meeting of the iron

[9] *Jackson MSS*, Minutes of Meeting of Principal Ironmasters in counties of Yorkshire and Derbyshire, 11 April 1810.

[10] *Jackson MSS*, April and October 1826. According to the *Salopian Journal* of 1 July 1867, 'for more than a century (these meetings) have been held at the towns of Wolverhampton, Birmingham and Stourbridge'.

[11] Ashton, *op. cit.*, p. 185.

[12] *Boulton and Watt MSS* (Birmingham Public Library).

trade of South Wales held at the Castle Inn, Merthyr Tydvil, on 26 January 1819:

> 'a letter from Mr. Price, the chairman of the Staffordshire iron trade having been received and also one from James Christie, Esq., of Glasgow, stating than an application is intended to be made to Parliament to repeal the duty on coal carried coastwise and in lieu thereof to impose a duty upon all coal at the pit mouth. Resolved that it is the opinion of this meeting that a tax upon coal at the pit mouth will in its operation be particularly injurious to the iron trade. . . .'[13]

Understandably, political measures like this, producing a definite opposition and limited aims, could weld the ironmasters into a more or less unanimous body represented by a national committee in London. But when it came to a question of economic policy for the whole industry, it was another matter.

The first indication of tentative hopes of co-operation between regional organisations in the period under review appears in the minutes of the Yorkshire and Derbyshire ironmasters in 1808:

> 'Resolved that this meeting fully approve of every attention being paid to the communications that may be hereafter from the ironmasters of Wales, Salop and Staffordshire, and that a good understanding with these gentlemen be at all times recommended to the makers of iron in the counties of Yorkshire and Derbyshire.[14]

That a national association was set up in the same year is indicated by a minute of a South Wales ironmasters' associations meeting of the same year.[15] In 1810 the Yorkshire ironmasters decided to send a deputy to these meetings, and three years later a resolution was passed recommending this Gloucester meeting to resolve an increase in price 'such as may make the trade in general a beneficial one'.[16]

This meeting in 1813, which arose from the depression following the dislocation of trade by the American war, and a price war waged by some of the Welsh ironmasters, was the subject of several comments by William Crawshay, sen., whose attitude will be discussed in some detail later, at a time when he viewed associations of this sort with some favour. Hence it is possible to give a full account of the developments in this instance. On 13 January 1813, Crawshay wrote of the severity of the price war: 'Hill says it is now

[13] *Dowlais MSS*, 26 January 1819. See also H. Hamilton, *The Industrial Revolution in Scotland* (Oxford, 1932), pp. 205–6.

[14] *Jackson MSS* (Sheffield Public Library), Minutes of Meetings of the Principal Ironmasters in the counties of Yorkshire and Derbyshire, 1808.

[15] Ashton, *op. cit.*, p. 247 Appendix.

[16] *Jackson MSS*, 20 December 1813; cf. Ashton, p. 180, who says the Yorkshire ironmasters welcomed an increase in price fixed by the Gloucester meeting.

come to the point who can make it lowest and he appears desirous to try that not only with our Staffordshire opponents but with each other in Wales. What will be the end of his contest, who can or will sell lowest, time only can show. . . .'[17] After a second meeting in London, a reduction of 10s. to 20s. per ton was proposed, but a decision was postponed until a meeting at Gloucester. This meeting was apparently promoted by Richard Fothergill, of Tredegar, who made an effort to reconcile the conflicting opinions of the Welsh, Staffordshire and Shropshire ironmasters. Crawshay remarked that it 'is to be held fourteen days sooner than usual at the request of the Staffordshire people who say they will then attend and endeavour to arrange better for the whole trade than this destroying each other'.[18] The Shropshire ironmasters were even less interested than those of Staffordshire, with the exception of Joseph Reynolds of Coalbrookdale, who drafted a set of resolutions calling for a reduction of output to raise the price from £11 10s. per ton. These were not adopted, and Fothergills' reply is particularly suggestive in indicating the obstacle—the extreme individualism of the ironmasters—which lay in the way of restriction of output: 'I join in your ideas that reducing the make (if it could be adopted) will alone be effectual in getting up the price of bars; however, the difficulties are such as to induce me to think that it will not be submitted to till many have a bellyful of this £11 10s.'[19]

Following the failure of these negotiations, there ensued a bitter price war in Wales and elsewhere,[20] and such desperate measures as government intervention were contemplated by a Welsh ironmaster. Crawshay reports this, and his reaction to the suggestion: 'He [Samuel Homfray] called with several others one morning here in my absence, and [had] drawn out an advertisement to call a meeting of the whole trade to consider of the Plan of Regulation, Tax etc., by government and said Fereday [of Bilston] and others concurred with him. My reply in a moment was that I would not sanction or attend any such meeting; that we should be like the frogs who called for a king which was complied with and they were all destroyed, so it would be with us. . . .'[21] It is unfortunate that there are no other

[17] *Crawshay MSS*, Letter Book, Vol. 1.

[18] *Ibid.* He added: 'I do not flatter myself with any good result from that or any other meeting.'

[19] *Coalbrookdale MSS*, Nos. 75 and 76.

[20] Cf. A. D. Gayer, W. W. Rostow and A. J. Schwartz, *Growth and Fluctuation of the British Economy*, 1790–1850 (Oxford, 1953), Vol. II, p. 845. There the view is put forward that the ironmasters' associations were successful in keeping prices stable.

[21] *Crawshay MSS*, Wm. Crawshay—Wm. Taitt 18 March 1813.

details of this scheme of governmental control; however, Homfray thought better of his plan, though it was mooted again by him in 1821.

To resume the history of the Gloucester meetings, it appears that, following this failure, the association came near to disintegration. In 1814 the Yorkshire ironmasters resolved not to incur any expense through these meetings 'in consequence of the want of communication as well as from the consideration of the little utility the iron trade of this district derived from its connection with the meetings held at Gloucester'.[22] Competition between Staffordshire and Wales during this period was very keen in the Bristol market, and Thomas Botfield of the Old Park Ironworks, Shropshire, was this time the initiator of a scheme of reductions in each district, to be supervised by committees for each district,[23] although the Staffordshire ironmasters had already revived the idea of a meeting at Gloucester to consider a plan of restriction of output. However, Guest, speaking for the Welsh ironmasters in 1816, was disinclined to come to any agreement with the Staffordshire makers. 'I do not think under the present circumstances that a meeting of the Trade at Gloucester would be productive of any advantage, as I am confident that a reduction of make in this County . . . will not be agreed to by many considerable Makers here. . . .' It is clear he was under no delusions about the effectiveness of these agreements, for, speaking about the Staffordshire ironmakers who wished for the agreement in order to have a share of the market, Guest averred: '. . . they have always kept such bad Faith with us that we could not trust them. . .'.[24] However, two years later, in 1818, the chairman of the Staffordshire ironmasters wrote to Guest explaining why it had been necessary for the makers there to abandon a decision to advance prices. He gave details of prices and terms of trade, which, he said 'will correspond with the proportion as heretofore settled between the Staffordshire and Cardiff terms'.[25] However, in the following year, Botfield, the Midlands ironmaster, was against attempting to secure liaison between the two districts by correspondence upon the state of trade before quarter day.[26]

During the depression of 1831, there was again talk of a meeting at Gloucester, although it is obvious that there was not much chance of co-operation between Wales and Staffordshire, as the Midlands ironmasters complained that Hill of Blaenavon and others had

[22] *Jackson MSS*. The decision was modified by the next meeting.
[23] *Dowlais MSS*, Thomas Botfield—Wm. Taitt, 21 October 1815.
[24] Elsas, *Iron in the Making*, p. 3.
[25] *Ibid.* [26] *Dowlais MSS*, December 1819.

reduced the price pending the consideration of the measure. A Staffordshire spokesman promised that they would nevertheless go to Gloucester because they are pledged to do so, 'but,' he demurred, 'no good can be expected if the Welshman still hold the old notions of extermination'.[27] Moreover, during this slump, several ironmasters were doubtless of the opinion of Crawshay, who declared, '. . . I will not be bound by any resolution or fetters of any sort; to reduce make will most likely be also necessary but that I can do without Commissioners to direct me'.[28] However, in 1823, Crawshay was still finding it necessary to discourage a Staffordshire ironmaster from calling a Gloucester meeting, remarking, 'The trade cannot even agree in its own respective districts.'[29]

In 1825, a meeting at Gloucester was actually convened, with no less a person than William Crawshay, the younger, in the chair (though this aberration from independence was unauthorised by his father).[30] This, contrary to usual practice, was not during a depression, but during the first railway boom, and the object was to declare a rise in price. Nevertheless the iron industry was disturbed by the appearance of joint stock companies and a general levy upon the trade was being considered in order to institute parliamentary agitation against the granting of charters.

After the removal of protection in 1825, the British ironmasters' monopoly of the home market was in theory not so complete. Possibly because of this, and more certainly because the expansion of the industry, as stimulated by the demand for railway iron, had made it more susceptible to the commercial fluctuations of the period, further attempts to arrive at some kind of national co-operation in the industry were made. During the depression of 1831, the usual revival of the Gloucester meeting took place.[31] Before the Select

[27] *Crawshay MSS*, letter No. 111, 15 December 1821.

[28] *Ibid.* [29] *Ibid.*, No. 217b, 2 April 1823.

[30] *Ibid.*, 15 February 1825: 'I disapproved of that meeting being revived and lament that I did not say so when you proposed to go there. It was found useless before and was therefore discontinued. . . .' Cf. John, *op. cit.*, p. 134.

[31] John, *op. cit.*, p. 133, states an agreement for the restriction of output was reached; however, it was not carried into effect because of the independence of the Welsh ironmasters. *Dowlais MSS*, M. Grazebrook—J. J. Guest, 26 August 1831. 'The Staffs. ironmasters would (not) pledge themselves without (a) corresponding measure would be practised in Wales, which from the conduct of Messrs. Crawshay & Bailey appears to them impossible.' (See also Elsas, *Iron in the Making*, p. 9, Michael Grazebrook to J. J. Guest, 26 August 1831: 'The Committee of the Staffordshire Ironmasters met to consult and arrange the reduction in the make of iron, recommended by the 4th Resolution at the Meeting at Glo'ster, and they were unanimously of opinion that no plan could be adopted by which the Staffordshire Iron Makers would pledge themselves without a

Committee on Manufactures, Commerce and Shipping of 1833, Anthony Hill, the former leader of the price war, reported that there was 'a general disposition to reduce the quantity . . . and to call a meeting, not only of the ironmasters in South Wales, but of the ironmasters in Staffordshire and Shropshire on the subject, deeming it proper in consequence of the state of the trade, to reduce the quantity of produce'.[32] Because of conflicting interests the reduction was not adopted. We have already noted the Staffordshire ironmasters' opinion of the Welsh policy.[33] Thus the outcome of this national meeting was again a failure. There is no further mention of the Gloucester meetings among the records now available. The record, then, of these meetings during the period 1808–31 was one of failure; regional competition was so fierce that co-operation in a national price policy could not be maintained.

Nevertheless, the history of early developments towards a national organisation of the industry would not be complete without some mention of the successes achieved within the next few years. This surprising change, indeed, makes the paucity of evidence here the more regrettable.[34] In 1836 there was a move towards a concerted plan of restriction of output, but we do not know if this was carried out. In 1840, according to Scrivenor, an agreed reduction in output of 20 per cent took place in all districts.[35]

On the limitation of make in 1842 there is more information. Printed as an appendix to the *First Report of the Midlands Mining Commission* (1843) there is a document of the Yorkshire and Derbyshire ironmasters' association which gives the allocations of output to be made in the first six months of 1842, according to the agreement based on the previous year's figures. The areas which are included in the return are Yorkshire, Derbyshire, Staffordshire and South Wales; and with the exception of Wales, the reduction of 25 per cent in output was obtained. Scotland, also, was approached by the inspector specially appointed to superintend the scheme, but he reported, 'The trade here generally approve of the scheme, but

---

corresponding measure would be practiced in Wales. . . . They are willing at all times to meet the Welch Iron Makers when any benefit is likely to arise from it, but without unanimity nothing can be done. . . .')

[32] Evidence of Anthony Hill. Q. 10295 and 10433.

[33] Crawshay again opposed the measure, considering the basis of reduction not a fair one; those who built furnaces since 1821, he thought, should reduce first.

[34] Ashton, *op. cit.*, p. 178.

[35] Elsas, *op. cit.*, p. 10, quotes a letter from a midlands ironmaster to Guest in December 1835, suggesting that firms had agreed to share railway construction tenders among themselves.

are not sanguine enough to consider any benefit will result if not prolonged.'[36] However, Scotland did eventually join in, so that the measure was 'universally adopted in England, Wales and Scotland with the exception of Bailey (Nantyglo) and (in part) Crawshay in Wales, J. Dixon in Scotland'.[37] Thus this scheme was a close-knit affair; there was a system of inspection and monthly returns of outputs from each works, to ensure its success.[38]

It seems likely that the unprecedented severity of the slump of 1839–42, when in some districts nearly half the blast furnaces were blown out, at last, had shocked the ironmasters into co-operative action. But even that experience of a reduction in make, which acted as a temporary expedient in preventing over-large stocks from influencing expectations of further price-cutting, did not ensure the consolidation of the national organisation. The industry had not yet achieved that degree of control and unanimity which could prevent the vicissitudes of reckless expansion which were followed by demoralising depression. Indeed, there was no new national organisation of the industry until the British Iron Trade Association of the 1870s.[39]

To account for the failure to secure an agency to iron out the extreme fluctuations which beset the growth of the industry, it would be necessary to examine many factors. First, the working of the local associations reveals several fundamental reasons why success was usually confined to informal wage agreements. If these local bodies could not take common action in regulating prices in these districts,[40] it is not surprising that an *ad hoc* general body should

[36] H. Scrivenor, *The History of the Iron Trade* (1841 edn.), p. 314.

[37] *Ibid.* Harford Davies and Co.—Sir J. Guest 1st month 20, 1842. In view of this, they agreed to reduce but declined to pledge themselves to the arrangements (Elsas, *op. cit.*, p. 14).

[38] See the returns of furnaces in and out of blast from Scotland, North Wales, North Staffordshire, and Newcastle-upon-Tyne, in Elsas, *op. cit.*, pp. 15–16.

[39] However, this same period of depression brought to light the first positive plan for introducing order into the growth of the industry. This was the publication of Dr. F. P. Mackelcan's *Suggestions to Ironmasters on increasing the demand for iron also to the Ironmasters of Staffordshire on competing with those of Scotland and Wales* (London, 1843). Mackelcan's cure for the industry was the establishment of an ironmasters' institution: 'It would teach our manufacturers what to make; our merchants what to export; and to inhabitants of our own and other countries what to purchase for their own good.' Research would be conducted into the application of iron for new uses. Mackelcan also advocated greater efficiency of plant and of labour and the utilisation of by-products.

[40] Ashton states that the Welsh meetings at Newport came to an end in 1824 and the Yorkshire and Derbyshire association's records end in 1828 with the customary assertion of what was necessary but unrealisable as 'a diminished Make'. These meetings were revived during the depression of 1842. A Welsh Association

fail to reconcile the interests of the competing regions, all working under different conditions of 'geological, geographical and commercial advantages'.

Before considering the reasons for the lack of success even among the local associations, it may be as well to look briefly at their functions. Mention has been made earlier about informal wage agreements, and it is certain that these arrangements—made sometimes within the framework of the associations, but more often by secret understanding—occupied the greater part of the agenda of discussions among the ironmasters. Almost every set of business records for the industry, which survives, contains many letters about agreed reductions of wages.[41] By 1848, in the Midlands, the understanding that wages of ironworkers and the colliers should fluctuate according to the price of iron as fixed by the association of ironmasters resulted in a formal agreement by masters and men to what was known as the 'Thorneycroft scale'.[42]

Even this discussion of wage reductions, which one would suppose was a matter for little disagreement among ironmasters, sometimes found them at loggerheads. In the days of a shortage of skilled labour in districts like South Wales masters had to consider the need to retain the loyalty of their men. As late as 1853 Lady Charlotte Guest, standing out against a lock-out, encountered a conspiracy to trap her into losing her men.[43] It was a recognised move in the diplomacy of industrial competition.

The crux of this competition between ironmaster and ironmaster which manifested itself in all kinds of rivalry was the problem of selling the product of the blast furnace or forge. Fundamentally, the

---

reappeared as late as 1852. (*Crawshay MSS*, W. Crawshay—Bailey Brothers and Co; Elsas, *op. cit.*, pp. 16–17.) Another move for a general restriction was mooted by Crawshay, chairman of the South Wales meeting: 'I beg leave therefore to ask if you feel disposed to join in the measure of reduction of make or iron, to be arranged so as to be general in South Wales, and provided to Staffordshire, Shropshire, Northern & Scotch Ironmasters will also give in to measure.' Eventually, a month later, the adjourned meeting of the Welsh ironmasters resolved: 'That a general reduction of the make of iron, although in itself desirable, does not appear to this meeting under existing circumstances to be capable of being carried out by arrangement at the present time.'

[41] *Dowlais MSS*, Wm. Forman—Wm. Smith, Cardiff, 12 March 1813. 'In consequence of the further reduction of Bar Iron, I think something may be done by the Iron Masters to reduce the Wages, and the best time to effect it is when the trade is in that state that the Master is full as much benefitted by the Works standing still as going on. . . . '

[42] Royal Commission on Trade Unions (1867), *5th Report*, Q. 9496.

[43] Earl of Bessborough (ed.), *The Journals of Lady Charlotte Schreiber* (Guest) (London, 1952), pp. 12–13.

problem of the individual ironmaster was that of covering his costs and taking a gamble upon the swings of the market to snatch any profits that might arise. Even between ironmasters operating in the same district, with more or less the same physical advantages, it is evident the costs of production differed immensely. Here a major factor was the difference in the terms under which they mined their minerals.[44]

Again, closely connected with the problem of costs of production was the extent of the wealth and capital reserves of individual firms. Many ironmasters, particularly those in the Black Country, in business on their own in a very small way as an offshoot of coal or clay mining, or even in partnership (a common instance was that of the clerk of an ironworks and other skilled workers pooling resources and venturing into iron-smelting upon inadequate capital), were dependent upon loan capital from bankers. Hence there was a heavy burden of interest to be borne by these to be added to their costs. Few ironmasters were in the position of Crawshay, who was a rich man in his own right, besides commanding the profits of a merchant house in the London iron market, and who, with contempt, truthfully asserted: 'It is more the poverty of many of the ironmasters than the too great make that causes the bad state of trade.'[45]

It is significant that the successes in national co-operation were achieved, not with price control, but with the reduction of output. On the other hand, however, if necessity drove some ironmasters to blow out furnaces, poverty, as a contemporary commentator pointed out, kept others on the rack of vanishing profits and increasing outputs. (This is a striking reminder that the assumption of perfect competition, with a fall in price leading to a contraction of supply upon the market, is far removed even from the reality of competition in nineteenth-century industry and commerce.)

There were two basic weaknesses in the attempt to exercise monopolistic sales of iron even upon a local scale. The first lay in the nature of the product. At first sight this may seem to be sufficiently standardised and capable of simple measurement (viz., long tons of 2400 lb.) to facilitate control of production. And it is true that price structures were fairly stable around the price of Staffordshire marked bars. Yet the bars were branded, and the different reputations of each ironmaster's metal subdivided the market.[46]

[44] The Dowlais Company paid £31 per annum until 1848 and leased part of their mineral rights to Penydarren at £140 plus a royalty on output. Until 1814, when Crawshay bought the lease for £95,000, Cyfarthfa's mineral rent was £5000 plus a royalty payment of 15s. on each ton of iron made.

[45] *Crawshay MSS*, 19 January 1813.

[46] In a trade circular of 1861 there appear 46 prices for pig-iron alone, ranging

Then there were special prices for specific shapes and sizes. Moreover, as Crawshay shrewdly saw, so long as terms of credit and commission allowances were not fixed by a common scale, despite a show of observance of fixed prices, there was scope for underselling. Thus there was nothing like a standard product for which an agreed range of prices could be fixed. The meetings usually confined themselves to declaring an advance or reduction in the price of iron. This was supposed to adjust the prices of the different products, but it was left to the individual ironmasters to do this.

Secondly, there was the problem of maintaining discipline among members of the association. Fines for absenteeism or non-compliance with resolutions could be imposed; nevertheless, the most severe penalty which an association could exact was expulsion. In 1826, the Yorkshire and Derbyshire association passed a resolution against deviation from agreed prices, and by so doing weakend itself. In fact, it was destroying its own existence. Again, it is possible that the organisation of the meetings was too informal, because, it seems, they could be manipulated by ironmasters to their own purposes. A correspondent of Guest's, Gilbert Gilpin, who was well informed on the practice of the Midlands meetings, reported, 'Whenever a neighbouring county is short of orders, a meeting of the minor houses takes place immediately and resolutions are printed, and their travellers fly off with them in all directions, saying to their customers: "Although an advance has taken place we will execute any orders which you may favour us with in this quarter at the old price" . . . this was all the ironmasters expected from the meeting to procure orders sufficient to keep their works in motion, and money to pay their men.'[47] However, it is not likely that this sharp business practice undermined the authority of the regularly constituted meetings. The local associations disintegrated because of the recognition of the futility of indulging in self-commiseration and declarations of panaceas for the trade.

The difficulty in holding these associations together and the disappearance and reappearance of the South Wales meeting is sufficient evidence that local meetings of ironmasters were far from finding a unity of interest. The South Wales association, which Crawshay referred to as 'a rope of sand', was handicapped, it is true, by the super-independence of this ironmaster of the Cyfarthfa works and the Bailey brothers of Nantyglo. Yet despite this weakness, the far-reaching agreement of 1836, previously mentioned, for a reduction

from £9 7s. 6d. (a short ton) for Lorn charcoal pig to £2 10s. (a long ton) for South Staffordshire cinder pig.

[47] *Dowlais MSS*, 12 September, 1819.

of 20 per cent in the make of pig iron—proposed ironically enough by Crawshay's son—was the first successful scheme for a restriction of output. This scheme was all the more effective because of the unprecedented power of the ironmaster charged by the meeting to see that the reduction—made by roughly one-quarter of South Wales blast furnaces—was actually carried out. The output books were open to inspection, and the reduction made by roughly one-quarter of the South Wales blast furnaces was actually carried out.[48] However, the national scheme was far from commanding a majority support.

In the early nineteenth century, the British iron industry was at the high-water mark of uncontrolled Victorian industrial expansion. Given the technical foundations of large-scale output, the industry was possessed of the power to react freely to the stimulus of industrialisation and the growth of the British and continental economies in population and wealth. That it did so is attested by the phenomenal increase in output. But that development, arising especially from the projection of the railway systems, brought its own dangers of uncontrolled response to the stimulus—in over-production and excess capacity. The industry, already possessing the means to adjust itself to the situation, failed however to achieve a lasting and coherent means of control to rationalise production to actual demands. Instead, reliance was still placed upon the price-fixing of the quarterly meetings and this was by itself inadequate. The short-lived attempts to restrict output on a national scale, as in 1840 and in 1842, did not culminate in a continued system of control. That required the threat of foreign competition and at this period Britain was supreme in the metallurgical world. So the ironmasters plunged into delusions of boundless prosperity and fell victims of speculation. Depression was aggravated by competition: although, perhaps, it might be fairly asked if the monopoly, which had been attempted, would have weathered the depression better than the struggle of the individual ironmasters.

As has been seen, the associations of ironmasters, whether locally or nationally organised, at the best of times were insufficiently powerful to promote lasting stability in the iron trade. Nevertheless, as Professor Ashton has pointed out, the bonds of association were strong, and out-and-out individualists appear to have been few. In South Wales the two most prominent protagonists of self-sufficiency were Crawshay Bailey (of Nantyglo) and, of course, William Crawshay, the elder. Since the latter's attitude was based upon a statesmanlike disposition, and not like Bailey's, a 'devil take the

[48] *Dowlais MSS*, Samuel Homfray—Guest, Lewis and Co., 4 April 1837. The Dowlais furnaces kept well within the limit of their permitted output.

hindmost' scorn for his neighbours, Crawshay's opinions merit some discussion here.[49]

Crawshay pursued an aim of farsighted concern for the prospects of the iron trade and he refused to take advantage of immediate windfalls from price rises as a result of the swings of the market. In 1813 he offered to buy up a year's output of the Dowlais and Plymouth ironworks; again, in 1826, in order to maintain the market price, threatened by sales to realise money by banker-harassed ironmasters, he offered to buy two or three thousand tons of iron from Hill or any others.[50] The mainspring of 'The uncrowned iron king's'[51] action and policy was his recognition of the futility of the 'cardboard alliances' of the ironmasters' association of South Wales, and of the more irregular meetings of the entire trade at Gloucester already described. In line with this view was his belief that many producers had insufficient capital to follow a consistent long-term policy, and that from their poverty, and not from over-production itself, resulted the recurrent depressions in the industry. So that he forbade his son to take an active part in their proceedings and declined all invitations to join in the compacts of the ironmasters. Further, he refused to support agreements the object of which was to attempt to obtain higher prices. On this score, he made his attitude clear by stating he feared 'all the obloquy of the prodigous advance in the price of iron' (when the ironmasters took advantage of the boom associated with the first spell of railway promotion). This when the indiscreet William Crawshay the younger had taken the chair at a meeting of Welsh and Staffordshire ironmasters in 1825.[52] Doubtless on account of the immunity which his wealth and importance as a large producer and merchant brought, the elder Crawshay refused to acquiesce in the measures of the trade against the new threat of joint-stock companies of the same year, and earlier, in 1821, to reduce the number and output of firms by state intervention; '. . . though it would produce a fortune with our stock and make, it would be a scandalous oppression of those with small concerns, small capitals, a system quite incompatible with British principles of legislation'.[53] On the other hand, Crawshay, despite his distrust of the Welsh ironmasters, was content to fall in with decisions he

[49] Yet, at the outset, it must be pointed out that Crawshay's attitude was fundamentally inconsistent. Whilst he took an independent stand himself, he deprecated the results of individualism in the other ironmasters.

[50] *Crawshay MSS*, 21 December 1813 and 4 March 1826.

[51] For a full discussion of Crawshay, see J. D. Evans, 'The uncrowned iron king', *National Lib. Wales Journal*, 1951.

[52] *Crawshay MSS*, 17 February 1825.

[53] *Ibid.*, 22 November 1821.

approved of, though he was not a party to them, thus leaving him freedom of action if underselling ensued. Further, apart from influencing the association (through indirect advice and by himself refusing large orders), he also wished the trade to come to an agreement for allowing a common scale of commission on sales by agents and on extra charges for special sizes of iron.

If Crawshay thought little of the value of local associations where intercourse during social and business occasions could have been expected to secure adherence to common decisions, obviously he would have still less to do with the national meetings, especially those which planned to reduce output. In 1831 his and Bailey's refusal to co-operate appears to have brought down that plan. His son explained his father's attitude to Guest of Dowlais: 'I have used every argument in my power to persuade my father to accede to the proposed and partly agreed measure, and although he will not do this without a concurrence on Mr. Bailey's part, is quite willing to reduce his make as an individual measure. I shall recommend my father most earnestly to reduce our make and I am convinced he will do especially if he saw the same *wise movement* taking place elsewhere.[54] The son, ambitious and perhaps not so far-sighted as his father, was an independent supporter of the ironmasters' policies during his father's lifetime, and after his death, in 1834, proposed the reduction of 1836. Some of his father's independence and caution nevertheless remained with him and he withdrew from the agreement of 1842.

[54] *Crawshay MSS*, Wm. Crawshay II—J. J. Guest, 23 September 1831.

Part II

CHAPTER VII

The Expansion of the Industry, 1788–1880

The salient feature of the physical development of the British iron industry since the Industrial Revolution is undoubtedly the phenomenal expansion of its output of pig iron. We have already examined the rate of growth of the industry during the Napoleonic wars. In 1788 its 85 blast furnaces produced during the year some 68,000 tons; by 1815 the production of about 200 furnaces[1] was between 344,000 and 373,000 tons. In 1850 the industry's output was roughly 2,500,000 tons from some 450 of a total of over 600 furnaces. This, indeed, was a rapid rate of progress, but the second half of the century eclipsed this in a way which is readily discernible from the statistics of output. In 1865 the output of the British industry was not far off double the 1850 figure—4,825,254 tons—whilst the number of furnaces available and in blast had increased by some 300 and 200 respectively. This is striking evidence of the increased capacity of the blast furnaces resulting from technical innovations and the greater skill and knowledge of the ironmasters.

By 1880 the era of mass production had still far to run. Today, for instance, when the mounting yield to be run from the furnace has halted and is by no means as large as the maximum ever achieved, one British blast furnace will produce, on an average, in one year 175,000 tons. The total United Kingdom output of steel at the present time exceeds 20 million tons.[2] Nevertheless, the 1880 figures are impressive. Pig iron production amounted to 7¾ million tons, puddled iron to over 2¼ million tons and steel to over 1¼ million tons.[3]

[1] In 1805, the last date in this period for which we have definite figures, of a total of 216 furnaces only 161 were in blast. The iron industry always had a large excess capacity.

[2] In 1960 the total of steel production in the U.K. stood at 24,305,000 tons. (Carr and Taplin, *The British Steel Industry*, p. 596.)

[3] T. H. Burnham and G. O. Hoskins, *Iron and Steel in Britain, 1870–1930* (London, 1943), pp. 26–7.

However, this comparison between the outputs of the industry at the beginning and end of the period under review only sums up the process of development. The growth of the industry, unfortunately, was no regular and steady expansion to be plotted by a smooth curve upon a production chart. The iron and steel industry has been well named 'the boom and slump industry *par excellence*'. Therefore it is important to produce further statistics in order to analyse the rate of progress.

First, it is necessary to say something about the nature of these statistics. Today, the British Iron and Steel Federation produces admirable series of figures to indicate not only the production of different types of iron and steel, but also important statistics of employment, productivity and the eventual consumption of the product of the furnaces by the many other industries using iron and steel.[4] In the first half of the nineteenth century it was quite otherwise. It is true that the Government had been prevailed upon to establish, under the administration of the Department of Woods, a Mining Records Office in 1840.[5] But not until 1854 did its *Mineral Statistics* refer to the production of pig iron. Consequently, earlier statistics of furnaces and their outputs are far from official (not that the so-called official figures were very reliable at first, for the ironmasters did not readily collaborate with the interfering government official who required bothersome returns from them). As we have seen, some of the early output figures were the outcome of the ironmasters' defence against impending taxation and control; hence, these statistics, which were in fact weapons of propaganda, can only be accepted, in default of any better ones, with a certain reserve. The ironmasters' associations do not appear to have taken any steps to keep statistics of output until the 1870s, except in the 1840s when they co-operated in schemes for restricting output. Other collections of statistics were made by private persons, such as David Mushet, who devoted himself to investigating the history of the industry as well as to practical metallurgical experiments.

There are serious deficiencies, too, in the early statistics. Not until it became the concern of Robert Hunt, who was appointed to work

[4] See Robert M. Shone, 'Statistics relating to the U.K. iron and steel industry', *Journal of Royal Statistical Society*, Series A, Vol. CXIII, Part IV, 1950. More recently Mitchell, *Historical Statistics*, Chapter V, gives tables of outputs of ore, pig iron, puddled iron, steel ingots and castings; employment figures; imports of ore; imports and exports of iron and steel up to 1938.

[5] Cf. Shone, *op. cit.*, p. 464: 'Regular statistics to the iron industry began to be collected in 1839 when the Museum of Practical Geology was enjoined by the Treasury to preserve records of mining operations and collect statistics of the British mineral industry.'

under the direction of the Geological Survey, was the output of the not insignificant wrought iron industry measured. The most valuable indication of the extent of the production of wrought iron during the early period is that of the amount of iron shipped upon the Glamorgan and Monmouthshire canals.[6] Secondly, there is no series of the values of the output of the iron industry, except for the declared values of exports. It is true that, again, the *Mineral Statistics* give an estimate of the value of the total pig iron production based upon one price figure per ton of iron. This, of course, is a crude measure of the prosperity of the industry. Yet it is very desirable that there should be a more adequate series of statistics in this respect since, as we have seen, increased outputs did not always make for greater cash sales or increased profits. There is, too, the complex problem of prices current for the many products of the iron industry. To sum up, the statistics which are available relate only to the physical output of pig iron and no adequate information can be brought to bear upon the important question of the increased value of the product as a result of finishing processes (i.e., here the manufacture of wrought iron bars, profile iron and iron nails, and all varieties of steel).[7]

The series of statistics of output is not a complete one until the commencement of the official figures in 1854, from when it is possible to measure the rise and fall of the production of pig iron year by year. The eighteenth-century estimates have been dealt with; in the following century the first contemporary assessment is for 1818. In the 1820s one is fairly well provided with returns of outputs, but again in the 1830s figures are fewer. However, from the 1840s almost every year is covered until the annual *Mineral Statistics* take over.

As one might expect, contained in all these statistics there are elements of error. Sometimes the annual figure of output is derived from the rather bold assumption that the furnace was at work for fifty weeks in the year. Moreover, since the early furnaces were far from consistent in their weekly outputs, perhaps one may be permitted to wonder how the average of these figures was computed.

[6] See J. Lloyd, *Old South Wales Ironworks*, for details of this source.

[7] The most important deficiency is the lack of any figures of outputs of puddled iron until 1881, from which time the decline of that industry is dated. Hence, although we have a fairly precise idea of the trends in the blast furnace side of the iron industry, we have no indication at all of the distribution of the industry up to 1881 as between the founding branches of the industry and wrought iron manufacture. However, for an attempt to estimate the value of and to reduce the several products to a single figure, in order to work out the contribution of the iron industry to the national income see P. Deane and W. A. Cole, *British Economic Growth, 1688–1959* (Cambridge, 1962), p. 57.

Evidently, some estimates were made upon the principle of good round numbers. The desire for adequate statistics might well make one wish that the government schemes of supervision by the excise officers, mooted in 1806, had been adopted. As it is, the accuracy of these early sets of statistics depended much upon the thoroughness of the private enquirer. In the statistics of output given below it will be noted that in several instances alternative figures are given for a particular year. This arises, of course, where the early compilers give conflicting estimates. Unfortunately, it is not possible to resolve these differences, even if we so wished, since the statisticians seldom give the authority or source for their figures.

But in spite of these reservations, perhaps the margin of error is not sufficiently great to distort the picture of the general trend of the rise in production presented by the chart above.

The detailed statistics of pig iron production and of the blast furnaces from which these millions of tons issued forth are given in the table which follows:

OUTPUT OF PIG IRON, UNITED KINGDOM, 1806–80

	Year	Quantity (tons)		Furnaces In	Furnaces Out	Total
1	1806	(*a*) 243,851	(*b*) 258,206	161	55	216
2	1815	344–373,000 estimate				
3	1818	300,000	(*b*) 325,000			
4	1820	400,000	(*b*) 368,000			170
5	1823	(*a*) 455,166	(*b*) 452,000			266
6	1825	581,367		261	103	364
7	1826	(*a*) 600,000	(*b*) 728,000	280	25	305
8	1827	690,000	(*b*) 690,500			284
9	1828	(*a*) 702,584	(*b*) 703,184	277	90	367
10	1830	(*a*) 677,417	(*b*) 653,417	315	18	333
10a	1833	700,000				
11	1835	1,000,000				360
12	1836	1,200,000				
13	1839	(*a*) 1,248,781	(*b*) 1,343,000	396	142	538
14	1840	1,396,400		402	88	508
15	1841	1,387,550	(*b*) 1,500,000	350	177	527
16	1842	(*a*) 1,046,428	(*b*) 1,347,790	339	190	529
			(*c*) 1,099,608			
17	1843	(*a*) 1,215,350	(*b*) 1,214,550			
18	1844	1,575,240	(*b*) 1,999,608			
19	1845	2,200,000	(*b*) 1,512,500			
20	1846	2,214,000				
21	1847	(*a*) 1,999,608	(*b*) 1,998,808	433	190	623
22	1848	2,093,736		452	174	626
23	1850	2,500,000	(*b*) 2,249,000			
24	1852	2,701,000		497	158	655
25	1854	3,027,884		557		724

	Year	Quantity (tons)	Furnaces In	Furnaces Out	Furnaces Total
26	1855	3,200,580	590		761
27	1856	3,586,377	620		784
28	1857	3,658,147	625		812
29	1858	3,454,024	616		829*
30	1859	3,709,374	605		855*
31	1860	3,802,920	577		852*
32	1861	3,695,060	567		846*
33	1862	3,946,469	561		876
34	1863	4,510,760	597		945
35	1864	4,767,951	614		882
36	1865	4,825,254	659		919
37	1866	4,523,897	618		945
38	1867	4,761,023	563		913
39	1868	4,970,206	560		912
40	1869	5,445,757	600		901
41	1870	5,963,515	664		916
42	1871	6,627,179	673		897
43	1872	6,741,929	702		876
44	1873	6,566,451	683		892
45	1874	5,991,408	649		915
46	1875	6,365,462	629		899
47	1876	6,555,997	585		927
48	1877	6,608,664	541		940
49	1878	6,381,051	498		948
50	1879	5.995,337	496		929
51	1880	7,749,233	567		926

* These figures of totals of furnaces are an estimate, since the South Wales figures for 1857–61 and the Scottish total for 1860 are not given in the Mineral Statistics.

Sources of Statistics:

1. (*a*) Meade quoting Parlt. Return of that year.
   (*b*) 1871 *Coal Commission*, Scrivenor, p. 99.
2. Guest (1840).
3. R. C. Taylor, *Mining Statistics* (1848). (*b*) *B.I.S.F. Statistical Year Book*, *cit.* Mitchell p. 131.
4. *Edinburgh Phil. Magazine*, quoted by 1871 Coal Commission. (*b*) *B.I.S.F. Year Book*.
5. (*a*) Finch, *B. & W. Papers*, Birmingham Assay Office.
   (*b*) Guest.
6. *Boulton and Watt MSS* (Birmingham Assay Office), Iron Furnaces in Great Britain, 31 December, 1825.
7. (*a*) Taylor.
   (*b*) Dufrenoy and Beaumont (1827).
8. *Companion to British Almanac* (1829). *Edinburgh Phil. Magazine* (1828). (*b*) *B.I.S.F. Year Book*.
9. (*a*) 1871 Coal Commission.
   (*b*) G. R. Porter, *Progress of Nation* (1851).

10. (*a*) Finch (Scrivenor and Taylor give 678,417 tons), p. 268.
    (*b*) Porter, p. 378. (1836).
10a. *B.I.S.F. Year Book*
11. (*a*) Taylor.
    (*b*) Guest.
12. Taylor.
13. (*a*) D. Mushet, *Papers on Iron and Steel* (1840).
14. W. Jessop (1841).
15. Taylor. (*b*) *B.I.S.F. Year Book.*
16. (*a*) *1871 Coal Commission.* (*b*) Taylor. (*c*) *B.I.S.F. Year Book.*
17. (*a*) Buckley and Taylor. (*b*) Porter (*Glamorgan Gazette*).
18 Taylor. (*b*) *B.I.S.F. Year Book.*
19. *Ibid.* (*b*) do.
20. *Ibid.*
21. (*a*) Scrivenor, Porter. (*b*) Beck, *Geschichte des Eisens* (? figures for 1848).
22. Porter (1851), p. 269.
23. I. L. Bell (1886). (*b*) *B.I.S.F. Year Book.*
24. B. Poole, *Statistics of Commerce.*
25–38. *Mineral Statistics* (kindly supplied by the Iron and Steel Institute)
39–51. *Mineral Statistics.*

Thus the actual rate of output was much more than one hundred times as great as the 1784 level of production at the end of a century's development.

It is difficult to pick out any one landmark in the surging rise of the output figures. However, the real beginning to these high outputs dates from the 1830s, following the adoption of the Neilson hot blast—the first million ton mark was reached in 1835.

The most obvious feature in the production trend is that of the periodic recessions—in 1830, 1842, 1847, 1858, 1862, 1866, 1873, 1874, 1878 and 1879.[8] It will be noticed that some of these years of falling production correspond with periods of general commercial crisis and depression. In the first half of the century, the worst fall of this kind was in 1842, when production slumped over 300,000 tons below that of the previous year. In the second half of the century, however, as the drop in output suggests, it was, indeed, a great depression in 1873. Between 1872 and 1874 production fell by almost 800,000 tons, which was more than 10 per cent of the previous level. Of course, as the capacity of the industry increased through potentially greater yields from individual furnaces, so there was more leeway in which the output could swing from peak to trough and

[8] It is evident from this chronology of the troughs in production that it is difficult to establish a cyclic trend. It is to be noted that the authors of *The Growth and Fluctuation of the British Economy 1790–1850* do not attempt to measure the production of iron in this way.

vice versa. The most striking instance of this is the last year of the period under review. By 1879 output had fallen by over 600,000 tons from 1877, yet between 1879 and 1880 there was a sudden rise again from 5,995,337 to 7,749,233 tons. Such extreme fluctuations as this necessarily had severe effects upon the prosperity of the industry.

During this period under review the process of industrial concentration upon the coalfields had come to its culmination. The presence of coal and ironstone together had been the most important factor determining the location of the industry from at least 1750 onwards and for the next hundred years. From that date there had been a consolidation upon the coalfields; indeed, at one time the three districts, Shropshire, South Wales and the Black Country, had accounted for more than 80 per cent of total production. The revolutionary hot blast smelting, which had realised the potentialities of much of Scotland's hitherto useless coal and made that country the major producing area, had only increased the industry's dependence upon locally available coal and ironstone. Then the modern iron industry entered the second phase of its development and production came to depend upon the transportation of one of its vital raw materials. The South Wales ironmasters had for a long time relied upon cheap costs of transport to enable them to move increased quantities of Furness and Cumberland iron ore to supplement local resources. (Later on, they had to depend upon Spanish ore.) The South Staffordshire ironmasters, too, from the 1850s onwards found it necessary to import ore into the district from nearby Northamptonshire until, in the 1880s, approximately 90 per cent of the Black Country pig iron was made with raw material brought into the district.[9] Indeed, the problem of an adequate ore supply affected more than one district by this time. The establishment of the Bessemer steel industry had intensified the demand for haematite (the phosphorus-free iron ore) so that, according to Bell, in 1873 the margin between the price of suitable and non-suitable ores for steel had increased to 27s. 6d. a ton. Consequently, imports of foreign ores had increased within the space of ten years from less than a quarter of a million tons to more than three million tons. By 1882 more haematite pig iron was made in this country from imported ores than from the produce of its own iron mines. This reliance upon foreign iron ore, of course, applied particularly to the coastal regions of smelting—Cumberland and North Lancashire, the North-East and South Wales.[10]

[9] *R.C. on the Depression of Trade and Industry*, 1886 (C.4797 and C.4893), 2nd Report, Appendix A (1). Evidence of Lowthian Bell.

[10] For a full discussion of this problem, see M. W. Flinn, *British Overseas*

But internally, too, there was an important change in the factors of localisation. We shall see how the Black Country supplemented its ore supplies from Northamptonshire; soon afterwards the exhaustion of its coal reserves, which had been long feared, imposed a drastic scaling down of that district's output. But the supply of cheap metallurgical coke by rail was not always uneconomic and the Cleveland iron district is a notable example of the localisation of the industry at the sources of the ore. Coke came from the not too distant Durham coalfield. In the Midlands, the result of these economic pressures was the establishment of the iron districts of Northamptonshire and Lincolnshire based upon the lean ores of the Jurassic measures and the importation of coke from the South Yorkshire coalfield.[11] In this position the industry stands today, with its production concentrated upon these East Midlands districts. As for the Black Country itself, its smelting, which only just survives, depends upon both imported ore and scrap and coke from South Wales.

The detailed distribution of pig iron production from 1806 onwards to 1880, as between the nine important areas and other scattered places, such as Somerset and Wiltshire, is shown in the table below:

DISTRIBUTION OF PIG IRON PRODUCTION, 1806–80[12] (PERCENTAGES)

Year	Wales and Mons.	Staffs.	Shropshire	Scotland	Yorks. West Riding	Derby	Cumbs. and Lancs.	Northd. Durham and N. Riding	Lincs. and Northants	Sundries
1806	31·03	19·16	21·30	9·00	10·33	4·00	0·54	—	—	3·64
1823	42·93	29·36	12·73	5·38	6·00	3·08	—	0·52	—	—
1830	44·59	31·34	10·82	5·52	4·27	2·66	—	0·80	—	—
1839	38·80	28·99	6·46	15·70	4·20	2·73	—	0·04	—	2·08
1847	36·16	19·30	4·42	26·99	3·38	4·76	—	4·99	—	—
1852	25·77	30·18	4·44	28·69	2·22	3·33	—	5·37	—	—
1855	27·12	26·63	3·78	25·74	2·82	3·62	0·51	9·18	—	0·60
1860	26·63	16·11	3·79	24·50	2.58	3·28	4·42	17·22	0·19	1·28
1865	19·00	18·32	2·41	24·13	2·56	3·92	6·48	20·99	0·53	1·36
1870	17·15	14·97	1·81	20·25	1·38	3·01	11·31	27·31	1·25	1·56
1880	12·23	7·87	1·14	13·53	3·96	4·73	19·88	31·19	4·99	0·48

These percentages show the early concentration upon the three

Investment in Iron Ore Mining, 1870–1914 (unpublished M.A. thesis of Manchester University, 1952); also 'British steel and Spanish ore 1871–1914', *Econ. Hist. Rev.*, Vol. VIII, No. 1, and 'Scandinavian ore mining and the British steel industry, 1870–1914', *Scand. Econ. Hist. Rev.*, Vol. II, 1954.

[11] Here there had been a reciprocal exchange of raw materials, for these ore fields supplied and still supply the ironworks of Derbyshire and the Sheffield district with ironstone.

[12] Bell, *op. cit.* The sources of statistics for calculations are Scrivenor's *History of the Iron Trade* and the *Mineral Statistics*.

major districts; the rise of Scotland; and later, the predominance of the North of England at the expense of the other older districts. The increasing importance of Cumberland and Lancashire is also noticeable, as is that of Lincolnshire and Northamptonshire, although this had not yet gone far enough to look impressive at this stage. In view of the pattern of growth and decline which is discernible—particularly in the case of South Wales and Staffordshire—it is worth while pointing out, perhaps, that only in the case of Cleveland and the two other districts, mentioned above, which expanded in the second half of the century is there evidence of an unimpeded upward trend during the period.[13]

As for the rest, the figures, where the large fluctuations in output arising from depression do not obscure the long-run changes, show at some stage or other a decline in the relative importance of these districts. The most obvious case is that of Shropshire, which indicates the steady decline from its once predominant place as the chief centre and seat of the modern iron industry. Here, as in South Staffordshire, the exhaustion of local resources was a powerful factor.

Scotland exhibits an interesting movement in its relative importance. The effect of the Neilson process is most marked during the years 1830–50, when Scotland advanced into second place with a share of almost 30 per cent of the total output. However, its output still increased and it was at one time the largest producing district; yet in turn Scotland was overtaken by the rapidly expanding Cleveland district. Scotland shows most vividly the effect of technical and physical forces at work upon the changing location of industry.

Indeed, perhaps the most striking impression one receives from a study of the changing balance in the distribution of the industry is how rapidly this came about. It is true that South Wales, and to a lesser extent South Staffordshire, were constant factors during most of the century under review; but against this view of the industry one cannot help noticing the rapid rise of the two new districts, Scotland and Cleveland, within the space of forty years.

Nevertheless, the foregoing discussion, which has been concerned with the relative position of the various districts, understates these terrific spurts of activity in new districts, as well as masking the real state of progress of the districts which declined relatively to the new giants. For instance, the production of Shropshire continued to rise right up to the 1870s. In order to see the advances of these districts, it is necessary to look at their annual output figures.[14]

[13] That is, if we ignore the small amount of charcoal- and coke-smelting carried on in the early part of the century in Cumberland.

[14] B. R. Mitchell, *British Historical Statistics* (Cambridge, 1962), also gives

1 SOUTH WALES

OUTPUT AND NUMBER OF FURNACES 1806–80

		Furnaces		
Year	Output (tons)	Total	In	Out
1806	68,867	45	35	10
1820	154,000			
1823	182,325	72		
1827	272,000	90		
1828	279,812			
1830	277,643	113		
1839	453,880	127	122	5
1840	505,000	163	132	31
1842	317,430			
1843	457,350			
1847	706,680			
1848	631,280	196	139	57
1852	666,000	197	147	50
1854	750,000	162	121	41
1855	840,000	187	148	41
1856	877,150	200	162	38
1857	970,727	—	164	(36)
1858	886,478	—	148	(52)
1859	985,290	—	147	(53)
1860	969,025	—	139	(61)
1861	886,300	—	138	(62)
1862	894,309	197	125	72
1863	847,753	197	123	74
1864	937,621	197	123	64
1865	865,035	198	134	64
1866	927,454	193	132	61
1867	886,234	187	115	72
1868	894,255	184	108	81
1869	800,972	178	112	66
1870	979,193	184	114	70
1871	1,045,916	167	112	55
1872	1,002,623	154	114½	39½
1873	817,789	148	101	47
1874	714,724	139	88	51
1875	541,809	153	79	74
1876	756,121	134	69	65
1877	710,958	150	60½	89½
1878	741,136	145	57	88
1879	669,858	143	54	89
1880	889,738	141	69	72

In this chapter it is proposed only to comment upon the course of production in each district as it can be observed from the statistics given here; the particular character of each district's industry which emerges from its history will be discussed later.

---

figures of output, nationally and by district, but no furnace figures. See Table 2, Chapter V, pp. 131–3.

The figures for South Wales show both phases of remarkable expansion and of steady decline from a peak in the early 1870s. To confine our attention to the nineteenth century, output had risen from 68,867 tons of pig iron in 1806 to over one million tons in 1871 and 1872. This increase, it might be pointed out, is impressive evidence of the advance of technology in the iron industry, for where the early output was the product of a total of 45 blast furnaces, the make of 1871—which was approximately fifteen times as great—came from an addition of just over 100 more furnaces.

Apart from this striking growth of the capacity of the South Wales furnaces there is another feature which is most marked during the period of decline. The proportion of furnaces out of blast in any one year was always a generous one. This surplus capacity, indeed, which seemed an inevitable if unfortunate circumstance of the industry's development, was a problem becoming more oppressive as the district's resources became less plentiful and cost of operation increasingly higher. Thus, in South Wales, as the shortage of locally available raw materials and the growing pressure of unfavourable economic circumstances forced a reduction in the total number of furnaces, so the number of furnaces standing out of blast (some never to be relit) assumed greater proportions. This position is seen at its worst in the last years of the 1870s when the district's output had fallen by 400,000 tons from its zenith, this came from 54 furnaces only; at the same time there were 89 out of blast.

2 SCOTLAND

OUTPUT AND NUMBERS OF FURNACES 1806–80

Year	Output (tons)	Furnaces Total	In	Out
1806	22,840	27	18	9
1820	20,000			
1823	24,500	22		
1825	29,000	25	17	8
1827	36,500	18		
1828	37,700			
1830	39,500	27		
1839	196,560			
1840	241,000	60	54	6
1842	270,920			
1843	238,550			
1847	539,968			
1848	564,000	130	94	36
1852	775,000	144	113	31
1854	796,640	156	118	38
1855	827,640	160	122	38
1856	880,500	161	127	34
1857	918,000	165	124	41

OUTPUT AND NUMBERS OF FURNACES 1806–80

Year	Output (tons)	Furnaces Total	In	Out
1858	925,500	177	132	45
1859	960,550	175	125	50
1860	937,000	—	131	—
1861	950,000	169	124	45
1862	1,080,000	171	125	46
1863	1,160,000	169	134	35
1864	1,158,750	170	131	39
1865	1,163,478	180	141	39
1866	994,000	165	98	67
1867	1,031,000	167	112	55
1868	1,068,000	167	123	44
1869	1,150,000	165	132	32
1870	1,206,000	156	123	33
1871	1,160,000	156	127	29
1872	1,090,000	154	130	24
1873	993,000	156	126	30
1874	807,677	163	123	40
1875	1,050,000	159	119	40
1876	1,103,000	156	119	37
1877	982,000	152	109	43
1878	902,000	152	$94\frac{2}{3}$	$57\frac{1}{3}$
1879	932,000	151	$96\frac{3}{4}$	$54\frac{1}{4}$
1880	1,049,000	149	$112\frac{1}{2}$	$36\frac{1}{2}$

Whereas the expansion of the Welsh iron industry was a steady and gradual one, the output statistics for Scotland show that there large-scale production was the result of a sudden spurt after 1830. From 1806 to 1830 the industry made very slow progress, merely rising from 22,840 to 39,500 tons, a rate of increase at which not even her small contribution of 9 per cent of the total output could be maintained. In fact, by 1830 it had declined to just over 5 per cent; even Shropshire, which by this time had entered into that period of stagnation which foretold decline, had a share in the country's output almost twice as large as that of Scotland.

After being released from the cramping limitations of its hitherto unsuitable iron ore by a technical innovation, the Scottish industry based upon the blackband deposits of the Lanarkshire coalfield soon advanced in output to outstrip its English and Welsh competitors.

There is one noteworthy feature of this rapid development which, it will be seen, was almost unchecked by years of depression; when the million-ton mark had been attained a long period of almost twenty years formed a 'plateau' of a fairly stable level of output. This characteristic of the Scottish iron industry's physical growth, which was in contrast to its commercial fortunes, rocked as they were from booming prosperity to ruinous slump by the action of speculators in its markets,[15] suggests certain conditions under which its ironmasters were operating. However, this consideration of these limiting factors must await the discussion of the economic history of the Scottish iron smelting district.

3 THE BLACK COUNTRY: SOUTH STAFFORDSHIRE AND WORCESTERSHIRE

OUTPUT AND NUMBER OF FURNACES 1806–80

Year	Output (tons)	Furnaces Total	In	Out
1806	50,006	42	32	10
1823	133,590	84		
1827	216,030	95		
1828	219,492			
1830	211,604	123		
1839	346,213	120	106	14
1840	407,150	135	116	19
1842	346,840	129		
1843	300,250			
1846	500,760	146		
1847	320,320			
1852	725,000	159	127	32
1854	743,600	175	145	30
1855	754,000	178	146	32
1856	777,171	171	147	24
1857	657,295	180	153	27
1858	597,809	186	147	39
1859	475,300	184	123½	60½
1860	469,500	181	108½	72½
1861	395,650	182	114	68
1862	412,220	192	107	85
1863	691,157	200	110	90
1864	628,793	172	104½	67½
1865	692,627	172	114	58
1866	532,625	167	112	55
1867	515,638	177	91½	85½
1868	532,234	172	89	83
1869	569,562	164	95	69
1870	588,540	171	114	57
1871	725,716	163	118	45
1872	673,470	145	107½	37½
1873	673,397	142	99	43
1874	452,400	154	81	73
1875	470,540	155	76¼	78¾
1876	465,946	147	65	82
1877	428,276	146	57	89
1878	392,949	147	55	92
1879	325,780	140	44	96
1880	384,556	137	46	91

[15] Only in 1843, 1866 and 1873–4 did depression cause an appreciable fall in output.

The South Staffordshire district's output had exceeded that of Shropshire, its neighbour and forerunner in the establishment of modern industry, some time during the Napoleonic wars. During this period, too, it seems probable that output was expanding at a rate which exceeded that of South Wales. However, the Welsh iron district had left the Black Country behind in the output race, despite the fact that an increase in the make of the Midlands furnaces was maintained until 1840. As we shall see, exhaustion of fuel and ore supplies, together with competition from better favoured districts, henceforth limited output and the Black Country no longer played the role of a large-scale producer like South Wales and Scotland.

The period of decline, which ultimately resulted from the working out of its mineral resources, so that the district could only continue as a prime producer upon imported raw materials, does not show itself clearly until the 1870s. Once more, industrial and commercial depression hastened the impending decline of the smelting side of the iron industry,[16] despite new developments in the production of steel. The year 1871, which saw its greatest make, 725,716 tons, was actually the turning-point; by 1879 (which was another year of slump) the output had dropped down to 325,780 tons. What is possibly more expressive of the conditions prevailing in the Black Country are the figures of its furnaces. At this last date only 44 were in and 96 were standing idle. As between one-half and one-third of the capacity of the ironmasters had been unproductive for a very long time, it is obvious that this unemployment of its furnaces represented more than temporary recession.

4 THE NORTH-EAST: THE CLEVELAND DISTRICT

Unfortunately for the very important district of North-east England with furnaces in Northumberland, Durham and the North Riding of Yorkshire, it is not possible to arrange the statistics in order to give tables for each county or, alternatively, a consolidated table for this Cleveland district as a whole. For the period previous to the official statistics (which gave figures for each county) there are available sets of figures for Northumberland and Durham, whilst

[16] As early as 1854–61 there was a period of decline (with the exception of the year 1856) when output fell from 743,600 to 395,650 tons. There followed a short period of recovery between 1862 and 1865 when the output level approached the former peak of over 700,000 tons and possibly as the result of the importation of Northamptonshire ore. In 1866 and 1868 another spell of depression occurred, but strangely enough this did not prevent another short period of recovery.

the North Riding output is included in a general total for Yorkshire. The table for Northumberland and Durham, 1806–80, is as follows:[17]

Year	Output (tons)	Furnaces Total	Furnaces In	Furnaces Out
1806	2,500	2		
1823	2,379	2		
1830	5,327	4		
1839	13,000	6	5	1
1840	11,000			
1843	25,750			
1847	99,840			
1848	94,380	32	17	15
1852	145,000	39	25	14

The *Mineral Statistics* to complete this series are:

	Durham			Northumberland		
		Furnaces			Furnaces	
Year	Output (tons)	Total	In Blast	Output (tons)	Total	In Blast
1854	275,000	78	59	(combined figures for two		
1855	214,020	47	34	counties)	15	9
1856	269,250	46	40	62,120	18	8
1857	284,500	51	35	63,250	18	6
1858	265,184	53	44	45,312	18	6
1859	370,339	62	46	31,500	18	6
1860	340,921	61	37	69,096	18	7
1861	312,030	60	32	73,260	18	7
1862	337,218	61	33	46,586	19	6
1863	468,318	71	47	40,916	19	7
1864	466,980	62	42	55,467	17	7
1865	476,767	65	47	49,290	18	7
1866	298,867	70	47	50,456	18	6
1867	477,834	71	39	31,027	18	3
1868	499,592	72	35½	17,495	18	1
1869	658,506	73	40	15,942	18	1
1870	676,964	76	50¼	33,623	12	3
1871	759,244	61	47½	34,165	10	3
1872	760,172	56	49	38,766	4	4
1873	799,573	54	49¾	44,807	10	4
1874	829,235	55	49	33,142	5	2
1875	786,206	70	53	22,870	4	2
1876	823,172	69	50½	(included in Durham)	4	1
1877	734,438	70	40	„ „	4	1
1878	660,323	71	32	„ „	4	1
1879	557,255	71	31	„ „	4	2
1880	750,262	72	30	„ „	4	2

[17] This set of figures agrees with the statistics given for Northumberland only in an article: T. M. Hoskison, 'Northumberland blast furnace plants in the nineteenth century', *Trans. Newcomen Society*, Vol. XXV, 1945–7. There are two exceptions, for 1830 Hoskison's figure is 2247 tons and in 1852 he says the Northumberland output was 35,000 tons.

For the Cleveland area proper, the statistics from 1854 are as follows:

		Furnaces				Furnaces	
Year	**Output (tons)**	**Total**	**In Blast**	**Year**	**Output (tons)**	**Total**	**In Blast**
1854	73,444	28	21	1868	699,494	69	50
1855	84,500	23	21	1869	658,506	69	51
1856	179,400	31	23	1870	916,970	74	67
1857	179,838	32	23	1871	1,029,885	75	70
1858	189,320	30	20	1872	1,122,114	81	73¾
1859	216,127	30	23	1873	1,156,431	78	76
1860	248,665	33	25	1874	1,158,471	97	85
1861	234,656	33	26	1875	1,240,243	85	73
1862	283,398	33	28	1876	1,261,013	86	75
1863	315,917	42	33	1877	1,374,582	89	75
1864	409,106	47	41	1878	1,358,442	90	67
1865	486,421	65	53	1879	1,210,091	90	70
1866	546,091	67	55½	1880	1,666,156	91	80
1867	640,892	67	50½				

First we may note the minor importance of the Northumberland production in the composition of this district's iron smelting activity. It was the more southerly situated counties of Durham and the North Riding of Yorkshire, with their mineral resources of coal and iron ore respectively, which accounted for almost all of the output of this North-east region. By 1875, when the last figure for Northumberland is available, this county produced only 22,870 tons from two furnaces in blast as against more than two million tons from the 126 blast furnaces of its nieghbours.

The development of this area belonged to the second half of the nineteenth century, for in 1847 the North-east produced only 5 per cent of the total British output. The crucial period of expansion here was the decade 1850–60, when production rose with a leap from about 159,000 tons to 658,632 tons a year.

It was in the Cleveland district, the North Riding of Yorkshire, that the most rapid progress took place. During the quarter of a century following the discovery of the usefulness of the Cleveland ore the output in 1854—73,444 tons of pig iron—was expanded until in 1880 it stood at 1,666,156 tons a year. In the last decade alone of this unprecedented increase in the capacity of the Cleveland iron-masters output had been raised by 600,000 tons a year. The intensive rate of the exploitation of the recently found mineral wealth is also clearly indicated by the fact that, in 1880, only 11 out of its 91 blast furnaces were out of use. How very different from the situation in South Wales and the Black Country! Moreover, whereas depression elsewhere had either halted the advance or set in action the forces of decline, here, in the Cleveland district, output was little affected. The worst fall in output occurred in 1877–9, but during the so-called 'Great Depression' of 1873–96 the rate of production did not slacken.

Before passing on to the North-western iron district of Cumberland and Lancashire which also entered into a phase of rapid development in the second half of the century, we should note the more modest progress made by Durham. Although output had advanced by half a million tons from 1854 during this period, here the course of production was much more irregular. Between 1854 and 1867 output had fallen back five times; the most serious setback came in the slump period after 1876, when over 40 furnaces were out of commission and output dropped to the level of the 1860s. However, by 1880 the country's production was ¾ million tons—more than twice that of the Black Country.

5 LANCASHIRE AND WEST CUMBERLAND

	Cumberland			Lancashire		
		Furnaces			Furnaces	
Year	Output (tons)	Total	In Blast	Output (tons)	Total	In Blast
1854		2	2		3	3
1855		4	3		2	2
1856	25,530	8	3	(combined for two counties)		
1857	30,515	9	6	1,323	2	2
1858	26,264	9	5	2,840	4	2
1859	50,097	13	7	26,491	9	7
1860	87,950	13	8	81,250	10	8
1861	55,165	13	8	109,377	12	10
1862	103,455	13	7	138,563	14	11
1863	106,090	15	10	164,110	17	11
1864	141,033	16	11	195,460	15	13
1865	107,430	16	9	204,925	24	16
1866	136,343	18	12	268,680	25	19
1867	109,839	18	8	318,801	25	19
1868	116,864	18	8	325,367	28	22½
1869	129,107	18	9	436,662	30	27
1870	255,178	33	24	422,728	33	27
1871	336,569	34	28¾	520,359	41	34
1872	440,575	43	33¼	524,041	40	34¼
1873	456,877	48	33½	529,271	44	36
1874	390,840	51	30	488,672	47	33
1875	486,112	51	31¾	558,780	50	31
1876	436,887	49	27	552,984	47	30
1877	538,156	50	26¾	624,189	50	33
1878	542,904	53	27	616,255	50	30
1879	531,638	51	27½	631,343	49	37
1880	790,343	51	40	750,884	49	37

A district with a similar kind of retarded development was that of Lancashire and West Cumberland. The output figures for the first half of the century are merely an indication that the charcoal iron industry was lingering in the Furness area, in particular, at the Backbarrow furnace. In 1806 Cumberland produced 1955 tons and Lancashire 780 tons a year. Thereafter there is little or no record of any progress until the second half of the century, when in 1855 Cumberland was estimated to be making 16,574 tons of charcoal iron. In the following year the two counties together were making just over 25,000 tons of pig iron. From then onwards the development of the iron industry, as indicated by the statistics of production, is another matter. Lancashire, commencing in 1857 with only 1233 tons of iron from two furnaces, ended the period with an output similar to that of Durham—approximately ¾ million tons from 37 of its 50 furnaces. Cumberland, which for some years lagged behind its neighbour, did not catch up until 1880 when its output slightly exceeded that of Lancashire.

The development of the West Coast iron industry in the 1860s and 1870s was bound up with the adoption of the Bessemer converter which at this date depended upon phosphorus-free pig iron made from the haematite of this district. This, it seems, was the most important reason for the renewed importance of Cumberland and Lancashire at this period. Indeed, such were the potentialities of these districts which possessed a virtual monopoly in supplying the expanding Bessemer acid steel industry with the pig iron of the requisite composition that for the first decade Lancashire was the only district—the Cleveland district not excepted—to sustain an unbroken rise in output. However, in the 1870s these districts were not so favoured by their command of locally available iron ore deposits. Even Cumberland was becoming more dependent upon Spanish ore supplies, and the burden of higher costs of operation is shown in the susceptibility to the slumps which periodically beset the whole industry.

6 NORTHAMPTONSHIRE AND LINCOLNSHIRE

Northamptonshire

Year	Output (tons)	Furnaces Total	Furnaces In Blast
1857	11,500	3	3
1858	9,750	3	3
1859	12,800	4	3
1860	7,595	4	3
1861	7,730	4	3

	Northamptonshire			Lincolnshire		
		Furnaces			Furnaces	
Year	Output (tons)	Total	In Blast	Output (tons)	Total	In Blast
1862	13,471	4	3			
1863	14,590	5	3			
1864	22,823	15	9	(Lincolnshire total included)		
1865	25,728	10	7	”	”	”
1866	19,174	9	6	13,765	6	3
1867	25,184	9	5	25,579	6	5
1868	35,584	8	6	33,999	6	5
1869	41,500	8	7	33,786	6	5
1870	43,166	10	10	31,690	6	4
1871	60,512	12	9	30,122	7	4
1872	59,424	10	9	36,989	9	6¾
1873	58,480	16	10	52,076	13	9
1874	53,760	18	14	67,266	15	8
1875	80,689	18	12	111,683	21	14
1876	84,916	20	11	125,198	21	16
1877	106,948	20	13	116,857	21	10
1878	138,370	20	15	125,043	21	11
1879	165,317	23	17	131,678	21	14
1880	178,714	23	17	207,704	21	15

This is the area of 'lean-ore' production—with ore yielding on an average of 30 per cent iron—which the British iron and steel industry has belatedly come to exploit as its last reserve of native raw materials. And although iron and steel production in this area has the common characteristic that, like the Cleveland district, it is based upon ore resources only (i.e., coke has to be brought in), the two counties are not altogether alike in their geological and economic circumstances. The Lincolnshire ores are near the surface, can be quarried very cheaply, and are self-fluxing. According to D. L. Burn[18] its lowest lias ore was less phosphoric than that of Northamptonshire and offered difficulties to the basic steel makers. However, Lincolnshire was nearer to the South Yorkshire supplies of coke. The 'inferior oolites' of Northamptonshire were more suitable in respect of their greater content of phosphorus, yet still required mixing with other ores. This is not the place to consider the reasons for the unwillingness of the British steelmasters to smelt these ores. Here it is sufficient to point out that the statistics of output for this period reveal a very uncertain beginning to the exploitation of these ores and the re-orientation of the British steel industry to its position today. As to differences in the rates of development between the two counties, only the figures for 1880 show that Lincolnshire was to

[18] D. L. Burn, *The Economic History of Steelmaking* (Cambridge, 1940), p. 168.

draw ahead of Northamptonshire.[19] Thus in forty years the Northamptonshire industry's output had increased by only 100,000 tons. It may be noted that Stewart and Lloyds, who opened up large-scale production in Northamptonshire in the 1930s at the Corby establishment, at first desired to make steel in Lincolnshire, but according to Burn they could not obtain ore fields after actually buying an ironworks there.[20]

7 THE OTHER CENTRES OF THE INDUSTRY

Lastly, the following table gives the outputs for Shropshire, the West Riding of Yorkshire, Derbyshire, North Wales, North Staffordshire, Gloucestershire (the Forest of Dean), and indicates the scattered attempts to set up blast furnaces in Wiltshire and Somerset prompted by the shortage of iron ore in the major centres of smelting.

OUTPUTS (TONS) OF SHROPSHIRE, YORKSHIRE (WEST RIDING), DERBYSHIRE, NORTH WALES, NORTH STAFFORDSHIRE, AND GLOUCESTERSHIRE (FOREST OF DEAN) 1806–80

Year	Shropshire	Yorkshire	Derby	North Staffs.	North Wales	Glouces-tershire
1806	54,966	27,646	9,074	—	2,981	1,925
1823	57,923	27,311	14,038	Output	13,100	—
1827	78,000	40,000	20,500	figures	24,000	—
1828	81,224	32,968	22,360	included	25,168	—
1830	73,418	28,926	17,999	in St. total	25,000	—
1839	80,940	52,416	34,372	18,200	33,800	18,200
1840	82,750	56,000	31,000	20,500	26,500	15,500
1842	81,286	46,942	27,590	—	—	—
1843	76,200	42,000	25,750	21,750	19,750	8,000
1847	88,400	67,600	95,160	65,520	16,120	—
1848	111,020	59,800	78,100	—	22,256	—
1852	120,000	150,000*	—	90,000	30,000	—
1854	124,800	—	127,500	104,000	32,900	—
1855	121,680	90,840	116,550	101,500	31,420	19,500
1856	109,722	96,200	106,960	130,560	47,682	24,132
1857	117,141	117,000	112,160	134,057	37,049	23,882
1858	101,016	85,936	131,577	135,308	28,150	23,580
1859	149,480	84,950	139,250	148,500	26,950	31,750

[19] T. H. Burnham and G. O. Hoskins, *Iron and Steel in Britain 1870–1930* (London, 1943), p. 142. Burnham and Hoskins say Lincolnshire was the only district showing a continuous increase in pig iron production over the whole period (1870–1930); in 1920, Lincolnshire and Leicestershire together produced 650,000 tons of pig iron, whilst Northamptonshire had an output of 280,000 tons.

[20] Burn, *op. cit.*, p. 437.

Year	Shropshire	Yorkshire	Derby	North Staffs.	North Wales	Gloucestershire
1860	145,200	98,100	125,850	146,950	49,360	26,458
1861	140,791	142,865	129,715	187,706	46,658	23,163
1862	125,981	112,121	131,005	184,455	31,719	51,968†
1863	135,557	104,745	170,076	176,504	51,076	64,001†
1864	130,666	112,093	174,743	217,996	51,108	65,312†
1865	117,343	123,233	189,364	206,268	51,874	65,471†
1866	121,262	119,747	199,867	210,335	25,515	59,817†
1867	123,604	109,002	160,028	202,332	32,843	71,816†
1868	145,154	100,050	159,312	229,913	37,046	75,847‡
1869	197,443	105,765	188,353	231,913	38,530	81,306‡
1870	112,300	77,717	179,772	303,378	42,695	93,601‡
1871	129,467	114,549	270,485	268,300	41,893	99,997‡
1872	133,046	148,636	283,375	275,925	54,692	98,081‡
1873	135,149	151,511	296,468	241,166	67,463	93,093‡
1874	126,055	163,856	301,687	273,501	51,868	78,254‡
1875	120,996	267,153	272,065	241,398	55,099	59,819‡
1876	106,711	235,451	300,719	213,569	32,723	57,587‡
1877	102,180	229,027	328,203	255,383	26,715	50,752‡
1878	80,965	219,547	306,141	231,534	23,091	42,351‡
1879	60,790	218,805	291,455	210,374	18,953	40,000‡
1880	88,338	306,560	366,792	225,023	57,812	37,351‡

* Includes North Riding production.
† Includes Somerset and Wiltshire production.
‡ Includes Hampshire, Somerset and Wiltshire.

It is not easy to summarise the main trends exhibited in common by these districts. It is true that the iron smelting here can be regarded as a lingering relic of the charcoal-fired industry. However, all experienced the consequences of the eighteenth-century innovations —coke and steam-blast furnaces—which elsewhere expanded production in a revolutionary manner. Even this qualification might imply that these districts were backwaters of the industry which escaped the important developments of this period, and this evidently is not the case.

Shropshire, for instance, might be grouped with South Staffordshire, for its development runs parallel with that of its neighbour. Both iron-producing districts reached their climax in the years 1869 and 1871 and in both instances the decline which followed resulted from the exhaustion of local mineral resources.

In the case of the Forest of Dean there, too, was evident decline, but, as we have seen, it is by no means clear why this district did not expand upon the firm foundations of technical innovation in the eighteenth century. As it was, the Forest of Dean iron industry was finally put out of business by the competition of the new open-hearth mild steel.

With the exception of North Staffordshire and of North Wales these were the centres which concentrated upon the production of high quality wrought iron. In particular, both the West Riding (Yorkshire) iron district and that of Derbyshire were associated with the engineering and constructional trades of Leeds and Sheffield and Rotherham districts. Thus, as the local measures of coal and ironstone did not limit the modest expansion of output, these two districts were able to progress along the lines of development of the local wrought iron industry.

It remains to say something of North Wales and of North Staffordshire. North Wales, the scene of the activities of the Wilkinsons at Bersham and Brymbo, had seemed to stand on the threshold of a great industrial advance comparable with that of Shropshire during the latter half of the eighteenth century. However, although the output of this district was never of significant proportions in the national total, the smelting of iron did not altogether die out, even after a severe period of depression in the 1830s and 1840s. Instead there was a slight revival in the years between 1856 and 1865 and 1872–5. The chief centre in this district was again Brymbo, to which J. H. Darby had moved from Coalbrookdale, and was one of the first ironworks to experiment with the basic open-hearth process.[21] The modern development of this district is connected with the establishment of John Summers at Shotton on the River Dee near Chester in 1899.

Later, this firm was to acquire the Shelton ironworks in North Staffordshire established by Earl Granville, whose ancestor, Lord Gower, had been a prominent ironmaster in Shropshire. Previously, it had been in the hands of the Derbyshire Butterley Company for a few years. The Shelton ironworks had been operating as a basic steelworks and it is likely that the possibilities of this district, with locally available coal and ore and its central situation *vis-à-vis* markets at Manchester, Sheffield and the Black Country, invited favourable comparisons with the advantages of other areas. With improved means of communication to its markets, North Staffordshire entered upon a period of substantial expansion of output. However, once more, the commercial crisis of the early 1870s seems to have halted this advance. The peak year was 1870 when 303,378 tons of pig iron were made in its nine ironworks which had 30 furnaces in blast. Nevertheless, despite the subsequent decline, the production of North Staffordshire has survived the competition of other areas with advantages much more marked than this district can claim.

[21] P. W. S. Andrews and E. Brunner, *Capital Investment in Steel* (Oxford, 1951), p. 110.

It appears as an exception to the general pattern of development in the iron industry of the nineteenth century.

What conclusions can be reached from this brief examination of the trends of production in the individual districts? The most outstanding feature is without doubt the irregular path taken by the industry to its figures of production, which seem astronomical when compared with those at the beginning of the period. Almost every district reveals sudden slumps in output which were followed by years when output meteorically leaped upwards. Yet, in the first half of the century, only in the depression of 1840 was there a period when every district shared in a recession of output. (It must be remembered, of course, that expanding output did not always make for, or signify, prosperity.) For the second half of the century, or at least until 1880, we can observe the uneven pattern of recession. Strangely enough the statistics reveal no single year when there was a general drop in output. On the other hand, no district's prosperity was such that a fall in the yearly production was avoided. The worst years of reduction in the make of pig iron were 1858, 1861, 1867, 1873–4 and 1878–9. Although the reductions in output were not synchronised by economic circumstance or by deliberate restriction of output, it is evident that in nearly all districts the years 1870–80 were a trying time—only the Cleveland, Lancashire and West Riding districts managed to avoid the immediate effects of the 'Great Depression' of 1873 and even these suffered from the slump of 1878–9.

The statistics of output, of course, show the decline of the old established districts such as the Forest of Dean, the Black Country and Shropshire. This was to be absolute decline; for the process of concentration upon the coalfields had run its course by the 1870s—this was evident, too, in South Wales. However, it is difficult to conclude from these figures alone that these districts were more susceptible to industrial depression. Until 1870 this tendency for the newly-developed districts to escape the necessity to reduce output can be seen for Scotland, Lancashire and the Cleveland districts, which were almost unaffected. But from 1870 onwards only the production of the Cleveland district arose almost unchecked by depression.

There was a resulting shift in the balance of the distribution of production among the districts. In the first half of the century it was in accordance with the working out of the process of concentration of the furnaces upon the richer and larger coalfields. The three major coalfields, West Scotland, South Wales and South Staffordshire, in 1839 and 1852 accounted for 83 per cent of the production of pig iron. However, with the development of the new districts, more importance was attached to the location of the ore. Besides the special requirements of the Bessemer converter, economy in fuel consumption had pulled the scales the other way in favour of the ironstone. Thus it was considered economic to haul coke by rail to the ore-fields of Cleveland, Furness and Somerset. The development of the iron districts of Northamptonshire and Lincolnshire was to be the culmination of this process. Finally, there was one other important circumstance in determining the location of the blast furnaces. This was the import of foreign ore from Spain, Elba, North Africa and Sweden (in later years). In this respect the importance of the coastal areas, West Cumberland, Cleveland, was enhanced. The industry of South Wales, too, was rescued by this expedient. Upon the basis of imported ore and scrap (which was to be a new supply factor in the age of steel) this district was during the early 1950s the scene of unprecedented development at Margam Abbey. One of the reasons underlying this renewal of the importance of South Wales is, unexpectedly, the availability of local supplies of coking coal. Nevertheless, just as from 1750 to 1850 coal was the keystone of the industry's growth, so in this last hundred years ironstone itself has largely determined the location of iron and steel production.

CHAPTER VIII

The Regional Development of the Industry 1788–1880

This chapter will deal with the growth of the iron and steel industry in Wales, Shropshire, South Staffordshire ('The Black Country'), Yorkshire and Derbyshire, and finally Scotland. The history of the three districts, Cleveland, the North-West, Northamptonshire and Lincolnshire, where, since 1850 to the present day, the British steel industry has adjusted its centre of gravity in a new concentration of industry, are treated in the chapter 'The New Factors of Industrial Location'.

First, in order to provide a framework for these brief accounts of regional industrial history, and to keep in mind the relevant points which should be illuminated by local detail, it will be as well to consider the economic factors influencing the location of the iron and steel industry.[1]

The ideal location can be summed up in a sentence. It would be a site with good access to the primary raw materials—ore, coking coal and limestone; with access to suitable supplies of scrap;[2] and to the market to take its products. But, as a writer of an article on the location of the British steel industry has pointed out,[3] the ideal site is seldom met with, and in practice some compromise is almost always necessary: should the ore be taken to the coal, or the coal to the ore—or should both be transported nearer to the market for the finished product?[4] There are other problems too. Is there a good water supply? This was a factor important even after steam had eased the difficulty of high-and-dry water-wheels during summer

[1] For a more extended discussion of the distribution of the iron and steel industry in this period the reader should also consult Roepke, *op. cit.*

[2] Scrap, of course, is a new factor, only come into existence as a result of the Siemens-Martin process, although previously first the bloomery cinders and later the puddling furnace fettling—both waste products—were extensively used.

[3] *Monthly Statistical Bulletin*, Vol. 25, No. 11 (1950), p. 1 (issued by the British Iron and Steel Federation).

[4] As at Dagenham with the steelworks set up by the Ford Motor Company.

droughts, for engines need a regular supply of water too. Can adequate supplies of skilled labour be found? What are the relative advantages of the coastal as opposed to the inland site? The answer, as the article suggests, will depend, in theory, upon a nice balance of gains and losses—a balance which may shift with time. It will be effected by changes in the relative prices of the different raw materials, the level of transport costs, the progress of fuel economy and advances in iron and steel making practice.

To this list we can also add two other considerations: first, the availability of capital, invested capital and loan capital for the setting up and carrying on the works;[5] secondly, the social cost of transferring to a new location by uprooting a concern from an established centre. This factor was present, even if the entrepreneurs of the Industrial Revolution have been judged to have been ruthlessly individualistic in the pursuit of wealth. However, even if Guest's thought for the population of Dowlais, when the question of the renewal of the lease in 1848 threatened the abandonment of the works, was exceptional, it is, nevertheless, important to note that social considerations of this kind were not wholly absent from the mind of the nineteenth-century ironmaster.

1 SHROPSHIRE

We have already considered the geological, technological and commercial foundations of the establishment of the modern iron industry in Shropshire. Indeed the history of the Shropshire iron industry in the eighteenth century could be written around the works of the Darbys and the Reynoldses at Coalbrookdale, Horsehay and Ketley, on the one side of the river, and the Broseley ironworks of John Wilkinson on the other. We have already looked at the main outlines of the history of Coalbrookdale and have pointed out that by 1815 the small furnaces upon the right side of the river were being abandoned, whilst even at Coalbrookdale, primary production had moved further away from the Severn towards Wellington. There coal was more plentiful, but unfortunately it was more difficult to raise on account of its depth. However, by the nineteenth century the steam-engine, which had been developed by the engineering skill of this district, had facilitated the migration of the iron industry, and besides the Coalbrookdale concerns there were the large ironworks of the Botfield brothers at Old Park, Stirchley and Lightmoor; the Madeley Wood works, later to be taken over by the Stourbridge

[5] This might be available either locally or through the national financial system working through the banks based upon London.

ironmaster, James Foster; and the Lilleshall works, established by Lord Gower.

The post-war slump had hit the Shropshire trade very badly, as we have seen. The number of furnaces was reduced to perhaps not as many as fifteen. 'Ketley, the oldest, the most celebrated work in Shropshire,' reported Gilbert Gilpin, 'is becoming an appendage to the Wellington Workhouse; and is now advertised in the Shrewsbury papers in £20 shares, for the purpose of employing the paupers of that parish, who amount to more than a full complement for three Blast furnaces.'[6] Joseph Reynolds, the proprietor, had resolved to sell off at immense loss, and to remove to Bristol.[7] According to Dr. Raistrick, Barnard Dickinson at Coalbrookdale (where things were in a very bad way) was thinking of concentrating production at Horsehay.[8]

Then, 'in 1827, the Company was about to receive a double infusion of young blood, the energy of which, along with outside circumstances, was to carry the concern to its greatest level of prosperity'.[9] After a brief spell of apprenticeship, Abraham and Alfred Darby resumed the family direction of the Coalbrookdale Company's fortunes once more in 1830. To quote Raistrick again: 'The joint management . . . was initiated by almost revolutionary changes, not only in plant and technical processes, but in the regulation and relationships of labour and in working conditions.'[10] New methods were introduced to these moribund works and during the next twenty years the major inventions of the period were brought to Horsehay; all this was in strong contrast to 'a previous policy . . . of a strictly conservative character'.[11] From Staffordshire the important new method of 'wet puddling' was brought in 1832, and in 1838 the use of hot blast was begun. With these went measures of fuel economy, efforts to introduce discipline and efficiency into the works by new rules and premiums on output.[12] This period of efficient management which modernised the equipment and working of the Coalbrookdale concern, and which increased the output to 80–90 tons per furnace per week, however, came to an end with the death of Alfred Darby in 1852. Moreover, Abraham Darby had left the works in 1849 to take over the Ebbw Vale and, later, Brymbo iron and steel works in Wales.[13] Henceforth, Coalbrookdale concentrated upon the founding of the art ironwork for which it

[6] *Dowlais MSS*, G. Gilpin—J. J. Guest, 12 October 1817.

[7] J. Randall, *History of Madeley* (Madeley, 1880), p. 96.

[8] Raistrick, *op. cit.*, p. 248. [9] Raistrick, *op. cit.*, p. 256. [10] *Ibid.*

[11] *Coalbrookdale MSS*, 'History of the Horsehay Works' (Anon., *c.* 1876).

[12] For full details see Raistrick, *op. cit.*, pp. 256–65.

[13] *Ibid.*, p. 274.

became famous, and the Horsehay works and the Ketley concern were re-established by independent companies as bridge and structural engineering works.[14]

The eventual decline of the county as a smelting area was, as one might expect, connected with circumstances very similar to those in Staffordshire. Competition with the better favoured districts such as South Wales had always been severe, even at the beginning of the century; but then the Shropshire works could hold their own against the Welsh on account of a reputation for better quality wrought iron. However, as time went by, the disadvantages of Shropshire became more marked, particularly its inland position. The belated connection with the railways between 1848 and 1865, which had put an end to the expensive horse-teams as a mode of transport,[15] had long since made other regions, such as the Cleveland district, unassailable competitors. Moreover, with the rise in importance of the North-Western and Lancashire iron and steel districts, Liverpool, Shropshire's long-established market, was now supplied by these northern ironworks. Finally, the possibility that the River Severn, which once had served as the outlet to the coalfield to Stourport and beyond, could ever again serve this purpose and give Coalbrookdale the advantages of an inland port was removed in 1844. In that year, the Severn's navigation beyond Stourport, which had been badly neglected (and, indeed, never sufficiently improved), was again deferred for future action.

To conclude, as a writer has said of the development of the Coalbrookdale coalfield: 'Nothing was capable of withstanding the combined effect of adverse external and internal forces.'[16] Moreover, this smelting district had not a local industrial hinterland to encourage the gradual transition to engineering, as the Black Country had. Thus the decline in Shropshire was rapid and irrecoverable: by 1879 production had slumped to merely 60,790 tons from the furnaces of Dawley Castle, Lodge Wood, Lilleshall and Madeley Wood.

2 SOUTH STAFFORDSHIRE

The economic history of the metallurgical industries in the Black Country during the period when they, and the coal and iron industries

[14] Raistrick, *op. cit.*, p. 15.

[15] These operated within a 30-mile radius of Horsehay. In 1838 there had been a junction opened up with the Grand Junction Canal at Wappenstall. But this mode of transport was not so cheap as might be imagined and frequently inconvenient.

[16] T. W. Birch, 'The development and decline of the Coalbrookdale Coalfield', *Geography*, 1934.

in particular, were being established and were giving that district both its name and its reputation (as a producer of malleable bar iron) has yet to be written.[17] It is true that Professor Allen in his *Industrial Development of Birmingham and the Black Country, 1860–1927* necessarily has to lay the foundations for his study in an account of the early development of the iron and coal industries in the Black Country, but, so far, the course of the Industrial Revolution here has not been traced through the business records of the iron-masters of this district. This district, to which, incidentally, it is so difficult to set boundaries,[18] is meant for the purposes of this survey as the South Staffordshire, Worcestershire and Warwickshire coal-field of the West Midlands.

The beginnning of this period of rapid industrial development may be said to date from the introduction of coke-smelting at the Bradley furnace in 1757 by John Wilkinson. John Wilkinson, indeed, up to his death in 1808 was a central figure intimately associated with the expansion of industry in this area around Bilston, Dudley and Wednesbury, a district of a few square miles, 'bristling' (as French observers described the scene) with coal and iron mines, forges, slitting mills, foundries, naileries and other workshops. All these, of course, drew their power from the famous rich 10-yard coal seam of this district. To return to Wilkinson, Professor Ashton has already exposed the details of his important collaboration with Boulton and Watt,[19] which resulted, so far as the iron industry is concerned, in the application of steam power to the forge and the rolling mill and a powerful demand for engine castings for the newly-perfected steam engine. The best account of Wilkinson's family and his career yet available is Dr. Chaloner's 'John Wilkinson, Ironmaster'.[20] Unfortunately we cannot enter into the details of the life of this colourful personality here, but we may note that the iron-minded and iron-

[17] However, the following three contributions to *Birmingham and its Regional Setting. A Scientific Survey, 1950:*

(i) B. L. C. Johnson and M. J. Wise, 'The Black Country 1800–1950'.
(ii) Johnson and Wise, 'The Changing Regional Pattern during the Eighteenth Century'.
(iii) W. K. V. Gale, 'Development of Industrial Technology—the Black Country 1700–1900'.

together provide a sound résumé of the historical geography of the region.

[18] The Black Country is an area which grew out of the exploitation of the mineral resources. It had a local culture of its own with several dialects and many traditional sports and pastimes, but now with the decline of the metal industries its character is rapidly changing.

[19] Ashton, *op. cit.*, pp. 60–86.

[20] See his article in *History To-day*, May 1951. Dr. Chaloner has been working for many years on a biography of Wilkinson and it is to be hoped that this definitive study will not be too much longer denied to historians of this period.

willed Wilkinson had become a folk-hero in his lifetime, and Dr. Chaloner ends his account with the story of several thousand people assembling, in 1815, on Monmore Green, near Bradley, 'expecting his ghost to make an appearance, riding his grey horse'.

Wilkinson had died in 1808 and it is an interesting speculation to wonder if this innovator would have transformed the Staffordshire iron industry even further, had he not become embroiled in quarrels with the Soho engineers and also with his brother, William, of Bersham and Le Creusot in the last years of his life.[21] Actually, Wilkinson was not so conservative in his old age as has been hitherto believed;[22] in 1802–3, the Bradley works, as we have seen, were visited by the French metallurgists Bonnard and O'Reilly, and by Svedenstierna. They noted that he had taken up the puddling process of Cort and had adopted his massive oscillating rolls. This last-mentioned mechanism, incidentally, did not impress the Swede as much as it pleased Wilkinson, who was in love with this mighty conception. Nevertheless, Svedenstierna acknowledged the iron-master's ingenuity and ended his account of Bradley by describing Wilkinson in the following terms: 'He is an old man rich in new ideas, by which, however, he is believed to have enriched science more than himself.'[23] It is a pity that the history of Bradley, Bersham and the other Wilkinson works should have reached a hiatus before the end of this period of great development brought to a close by the Peace of Waterloo.

As a producer of pig and wrought iron (besides developing the

[21] Unfortunately, too, after his death the works at Bradley and Brymbo were allowed to stand idle for some time whilst family litigation for the spoils went on.

[22] However, in 1785 we see Wilkinson making a classic remark about his independence and his role in the industry. The occasion was the disturbance of the British ironmasters, alarmed by government encroachments on the industry; several of Wilkinson's colleagues had put pen to paper to demonstrate how necessary it was to promote the manufacture of iron in Britain. In the course of this battle of propaganda Wilkinson had been credited with being able to make 10,000 tons of bar iron annually. John Wilkinson, out of modesty or from a regard for the truth, denied this statement. 'I have not expectations in my Day of making Bars so low as to deliver them at Stourport for £14—or that the Quantity I may have to see made (by) myself at my own works will exceed one quarter of such annual amount.' Then comes this statement of the *laissez faire* philosophy of the eighteenth-century industrialist, 'I do not like these publications—I had much rather spend my time in making of iron, than in writing about it . . . it is to be hoped that Matters will rest where they are and that every Man will be permitted to pursue his business without further Regulation on the part of Government . . . my wish is that they may *let us alone*'. (John Wilkinson to Richard Reynolds, 28 May 1785 (*Rathbone MSS*).)

[23] Svedenstierna, *Resa igenom en del af England och Skottland, Aren 1802 och 1803* (Stockholm, 1804) p. 149.

many different metal-using hardware and tube trades) the Black Country industry had expanded enormously during the Napoleonic wars. The number of its furnaces had increased almost five-fold between 1796 and 1815 with a pace perhaps unequalled by South Wales at this time. It was claimed that in 1816 (a year of slump) 3000 tons of pig iron were converted in the immediate neighbourhood of Bilston and Dudley.[24]

In 1815 Thomas Butler visited the district, and one conclusion that one very soon comes to upon reading his observations is that this expansion was based upon a very flimsy foundation of capital.[25] Where the coal is readily accessible on account of its nearness to the surface, then very often there develops a rash of many small pits and furnaces operated by small coal and iron masters. This was the case in South Staffordshire and it was Butler's impression at the time that 'There is hardly an ironmaster in this neighbourhood worth a groat'. Of this poverty of the ironmasters, Gilpin had much to say a few years later. However, it was one of the largest owners in the district—Fereday, a partner in at least four works (including the former Wilkinson concern at Bradley) and a banker—who went bankrupt during the post-war crisis in the iron trade. It was his issue of notes and coin which brought him down, but he had been long embarrassed by being forced to buy pig iron under long-term contracts with the Shropshire works at higher prices than he could command when he sold. Butler had previously noted: 'He has extended his concerns to that magnitude that he is nearly crushed under the immense pressure of his engagements.'[26] Among these was a loan of £150,000 by the Bank of England to carry on the Bradley works.[27]

[24] *Kirkstall MSS.* List of furnaces in Staffordshire in 1815 and 1816 (attached to the Journal of the Tour in Staffordshire, etc., by Thomas Butler).

[25] Johann C. Fischer also visited the district in 1814 and his impression of the locality and in particular of the New Level Furnaces of the Gibbons's family are to be found in his *Tagebücher* (Schaffhausen, 1951), p. 102. Fischer described this district as the 'Homeland of Vulcan'; with its immeasurable treasure of coal, iron ore and limestone it was so overcrowded with houses, chemical works, iron works, blast furnaces and steam engines that 'one could count 500 manufacturers in the district'. See also W. O. Henderson, *J. C. Fischer and his Diary of Industrial England, 1814–1851* (London, 1966), p. 69.

[26] Butler's Journal. On 27 November 1815 W. Wood reported to J. J. Guest of Dowlais the failure of Fereday and the expected rise in the price of iron, in consequence (*Dowlais MSS*). Fereday was officially declared bankrupt in 1821. For further details of Fereday's career see W. H. B. Court's *Rise of the Midland Industries* (London, 1938), *passim*, F. W. Hackwood's *Sedgely Researches*, and Dr. W. H. Chaloner's article 'John Wilkinson as banker', *Seaby's Coin and Medal Bulletin*, December 1948.

[27] L. S. Pressnell, 'Public Monies and the Development of English Banking', *Econ. Hist. Rev.*, Vol. V, No. 3, 1953, p. 391.

In that depression only 35 out of 71 furnaces remained in blast.[28] A Midlands correspondent of Guest thought at this time that 'the iron trade would centre in South Wales'.[29] The Dowlais letters for this period have many references to bankruptcies among the Staffordshire ironmasters. A fierce price war with Wales ensued and the Staffordshire proprietors were reported as saying that they would rather give their coal and mines for nothing than be driven out of the market.[30] As a result of this, the state of the Staffordshire trade in the years following can only be described as convulsive—advances in prices prompting 'numerous erections of new works' which again increased the competition and resulted in reductions in prices and the blowing out of these furnaces.[31] In 1820 two works, the Deepfields and Gospel Oak, were suspended because they did not pay.[32] Yet despite these vicissitudes, in the first half of the nineteenth century there were appreciable increases in both the output and the number of furnaces. Undoubtedly, some of this increase in output resulted from two major developments in the iron industry in this area; the first of these, the 'Gibbonising' of the blast furnaces, increased outputs up to 100 tons per week for each furnace;[33] the second, by greatly increasing the capacity of the wrought iron side of the industry, stimulated the production of pig iron. This was Joseph Hall's system of 'pig boiling' or 'wet-puddling' which he perfected between 1825 and 1832. Unfortunately there is no extensive series of figures for the output of wrought iron at this period; however, in the year 1839 there were 581 puddling furnaces in South Staffordshire.[34] G. C. Allen estimates that in the 1850s the area was producing 500,000–600,000 tons of wrought iron a year, or a third of the total British output.[35] The Bloomfield works alone, which possessed no blast furnaces, and specialised in the production of the wrought iron puddled by Hall's method, was making upwards of 900 tons of iron a week.[36]

One of the most valuable sources of evidence of the organisation and development of the ironworks in the 'Black Country' and its

[28] List of furnaces in Staffordshire in 1815 and 1816.

[29] *Dowlais MSS*, T. Blakemore—J. J. Guest, 13 June 1816.

[30] *Ibid.*, W. Wood—J. J. Guest, 21 October 1816.

[31] *Ibid.*, G. Gilpin, 19 June 1819.

[32] *Ibid.*, G. Gilpin, 4 March 1820.

[33] Increasing output by the use of hot blast never found much favour with the Midlands ironmasters.

[34] H. Scrivenor, *History of the Iron Industry*, pp. 314–15.

[35] G. C. Allen, *op. cit.*, p. 40.

[36] *The Iron Question Vindicated, the Reviewers reviewed, by an Observer* (London, 1857), Inst. of Civil Engineers Tracts, Vol. 196.

surrounding districts, in particular the Worcestershire forges around Stourbridge, during the first half of the century, is the minutes of evidence of the *Select Committee upon the Oxford, Worcester and Wolverhampton Railway*, taken in 1845.[37]

The subject of enquiry itself—the problem of a railway connection with the Bristol market; the adequacy of the canal system and the river Severn; and the complications arising from the two competing broad and narrow gauges—throws a good deal of light upon the position of the Midlands iron district towards the middle of the century.

The outlines of this economic structure are given in an extensive table submitted by the agent of the Earl of Dudley.[38] The table, drawn up to give a computation of the mineral traffic upon the projected railway, gives:

(*a*) The names and location of the blast furnaces, mills, forges, tinplate works, foundries and other ironworks.
(*b*) Description of the raw materials and the products of these works and their source and destination.
(*c*) The volume of these goods.
(*d*) The present method of conveyance (viz. canal, River Severn and railway).
(*e*) The cost of transport per ton.

From these data a detailed picture of the structure of the industry can be got. Here some attempt is made to give a general analysis under the headings (1) integration of plant, (2) nature and sources of raw materials, (3) outlet for product.

First, the table shows, of course, an apparent degree of integration of plant. Foremost amongst the ironmasters with many ramifications to their enterprise was James Foster. He was the owner of the Shutt End furnaces, four of which were producing 400 tons of pig iron weekly. In the manufacturing side of the industry, the Brockmore Ironworks with a weekly capacity of 100 tons; the Brierley Ironworks with 150 tons; the Caponfield Ironworks with a mill and forge of 180 tons capacity (mostly rails); the new Bradley works, similar to Caponfield, producing 120 tons of iron; and finally, the Stonebridge Ironworks, making 250 tons a week—all these belonged to Foster.[39] Other ironmasters demonstrate the same integration, only on a

[37] *B.P.P.*, 1845, Vol XI (360).

[38] Evidence of R. Smith, pp. 186–99. Smith was later responsible for the planning of the Round Oak Ironworks, 1855–9.

[39] I have assumed that there is an error in the table and a confusion between James Foster and James Forster.

smaller scale, e.g. Messrs. Wheeley, owning Bromley Forge and the Brettel Lane furnaces. Moreover, there is in this connection another kind of integration which is not revealed by these details. That is, the actual ownership of plants. Here the most important example is Lord Ward himself, who was the owner of 22 blast furnaces, one foundry and one other ironworks.[40] Most of these were leased out to rent-paying occupiers.

The second aspect of the subject also shows if not actual integration, then, at least, an interdependence between the works. This is more a matter of inference than actual statement; however, from the details which Smith gives it is apparent that many of the ironworks which concentrated upon the manufacture of malleable and finished iron relied heavily upon local sources of supply. Again, there is an outstanding example of this. It is the Bloomfield Ironworks of Bramah, Barrow and Hall at Tipton, which was exploiting the 'wet-puddling' process. One of the partners of this firm also partly owned the two Woodside furnaces, giving 200 tons of pig iron a week; he also was on the point of negotiating for the New Level furnaces from the trustees of the late Earl of Dudley, when, in 1845, he died.[41] But the Bloomfield works also drew large supplies of pig and bar iron from the works around Dudley; Smith gives a figure of 29,108 tons a year. In fact, this works serves as a very good illustration of the wide field from which raw materials were obtained. Pig iron was also got from Shropshire, the Forest of Dean and Wales; scrap iron from Birmingham and the neighbouring manufacturers; charcoal from Birmingham and the Forest of Dean; cinders and hammer slag[42] from Golds Hill and other works. As to the basic raw materials, coal, of course, was obtained locally, from east of Dudley—from the famous 10-yard seam now nearing exhaustion—limestone from the same district, some of the iron ore locally, from the Walsall district. But even at this time supplies from outside were being brought into the district. Corbyn's Hall New Furnaces, belonging to Benjamin Gibbons, got their ironstone from North Staffordshire; other works obtained Lancashire ore via Ellesmere Port, in Cheshire, and from Coventry.[43] In all, Smith's table presents a picture of blast furnaces being dependent upon raw materials which had to be imported from some distance. In fact, it was said that only two or

[40] In 1860 he was created the Earl of Dudley.

[41] T. M. Hoskison, 'The Earl of Dudley's New Level Furnaces', *Trans. Newcomen Soc.*, Vol. XXVIII, 1956, pp. 153–61.

[42] A waste product used for making 'bulldog', with which the puddling furnaces were fettled or lined for the wet process.

[43] Whether this was the original source is uncertain. It was too early for Northamptonshire ore to be railed into the district.

three ironmasters owned coal pits adjoining their works.[44] By this time of course, subsidence and bad drainage had taken a toll of the local pits; yet the ironmasters were, apparently, content to bear the costs of the transport of minerals and fuel up to a weight seven times that of the 468,000 tons product of the one hundred furnaces in blast.

Of the third and last aspect—the market for the products of the Midlands furnaces—it is proposed to say very little here. Obviously the products of the industry were pig iron, bar and manufactured iron, rail rods, castings and other iron articles, such as screws and tinplates. Pig iron was consumed locally, at times there was even a deficiency on account of the demands of the puddling furnaces and the casting trade, but some was sent to the principal ports of London, Liverpool, Hull and Bristol; castings found their way to Birmingham. Bar iron, similarly, went to the ports, although some was sent to the tinplate works of Worcestershire. In all it was estimated that 98,800 tons of pig iron and 102,321 tons of bar iron, as well as 30,735 tons of tinplates, chains and fenders, would be carried by the London, Worcester and South Staffordshire railway. Not all of these goods, of course, were meant for the ports of London and Bristol, since the figures include the movement of the iron in the chain of processes of making 'finished iron'. However, it seems safe to say that over 100,000 tons of iron found an outlet through these two ports.[45]

The following decade, 1850–60, has been considered the period of greatest prosperity of the wrought iron industry. The district had been forced to concentrate on quality in this side of the industry, because of the acute competition in the pig iron trade and even in the common bar trade. Moreover, there was another fundamental factor at work (which has been referred to previously): supplies of ironstone were beginning to run out as early as 1845, when it had been reported that great numbers of furnaces were working to half capacity because of the lack of materials.[46] It was feared at this time that the iron industry was migrating northwards, just as earlier prophets saw the future of the iron industry centred in South Wales. However, this was not taking into account the development of the Northamptonshire ore field, which, as we have seen, came to be of considerable assistance to the struggling Staffordshire trade.

[44] Evidence of Benjamin Best, Qn. 15.

[45] The estimate gives 65,321 tons of bar iron and 28,770 tons of tinplate, etc., as traffic on the main line alone (p. 2).

[46] G. R. Porter, 'On the progress, present amount and probable condition of the iron manufacture in Great Britain', *Report of the 16th Meeting of the British Association of 1846* (1847).

Nevertheless the 1857 slump severely hit the smelting side of the industry; less than five-eighths of the furnaces were at work and several works including Birchills, Toll End and Netherton, were stopped completely.[47] It was during this period too that Blackwell, a Midlands ironmaster, is said to have failed owing £150,000 to the Birmingham Banking Company.[48] Other firms vanished in the 1860s also; for there was another depression in the early years of that decade and the Russell Hall and Oak Fair works were blown out for good by 1867.[49] The hitherto prosperous firm of Lloyd, Foster and Co. at Wednesbury, the first exploiters of hot-blast and, later, the Bessemer process (in 1862) in the district, had been brought down by a large contract to supply iron for the Blackfriars bridge in London. According to Lloyd, the contractor was unable to pay the debt and the Midlands firm incurred a loss of £250,000.[50] This necessitated the sale of the concern to the nearby Patent Shaft and Axletree Company of Charles Geach.[51]

Thus the situation of many of the Black Country ironworks was distinctly insecure. For besides the shortage of ore, now another bogy was looming upon the horizon—the prospect of the exhaustion of the seams of coal and, of course, increased difficulty in winning that vital raw material. In 1860 a Dudley ironmaster had forecast the duration of the coalfield to be another forty years.[52] It is true that others thought there was no fear of a coal shortage because of the availability of cheap transport facilities in the many canals in the district. However, circumstances conspired to combine to confound these optimists. The timing of the catastrophic decline of the production of pig iron in the Black Country coincided, as we have seen, with the Great Depression of the early 1870s. The number of furnaces in blast fell from 107 in 1872 to 44 in 1879.[53] The exhaustion of raw materials available locally, as a result of the reckless plundering of the strata and the consequent flooding out of the seams, went on. It is true, as Evans points out, that the decline in coal production was not nearly so catastrophic as was at first thought. Production only declined by one-tenth between 1870 and 1890.[54] Yet not all of this

[47] D. B. Evans, 'The Iron and Steel Industry in South Staffordshire from 1760–1950' (unpublished thesis, Birmingham University), p. 81.

[48] Samuel Lloyd, *The Lloyds of Birmingham* (Birmingham, 1907), p. 88.

[49] Evans, *op. cit.*, p. 88.

[50] Lloyd, *op. cit.*, pp. 199–201.

[51] For the early history of the Patent Shaft and Axletree Company see J. F. Ede, *A History of Wednesbury* (Wednesbury, 1962), pp. 243–5.

[52] Gale, *op. cit.*, p. 17, citing W. Matthews, *South Staffordshire Ten Yard Coal and its Mode of Working.*

[53] Evans, *op. cit.*, p. 95.

[54] *Ibid.*, p. 99.

was suitable for iron smelting. Moreover, to consider another vital raw material, there was now the need to import limestone from Llangollen and Derbyshire.

It was here, upon this very question of transport costs, that the Black Country iron industry was at its greatest disadvantage; if it was necessary to pay railway or canal charges upon the import of raw materials, of course, there was also the need to rail the finished products to the ports. But the Black Country was landlocked. Thus, despite a feeble agitation to reconstruct its canal communications in order to break free from the stranglehold of the railway freight charges,[55] the district could not hope to compete under all these adverse conditions against the coastal areas. The future of the Black Country was not to be redeemed by the adoption of the basic Bessemer process in the 1880s and 1890s at Wednesbury, Bilston and the Round Oak works near Dudley; in fact, the competition of steel was eventually to reduce the wrought iron industry—the mainstay of the district—to the role of small-scale specialist. The few puddling and Aston Byers furnaces of the present time now provide malleable iron for the limited demand of the naval chain and anchor makers. The furnaces and forges have been replaced by many vast engineering shops, the Black Country has become less black and the district's economy is more securely based upon a great variety of metal-using, light and heavy industries.

3 THE YORKSHIRE (WEST RIDING), DERBYSHIRE AND NOTTINGHAMSHIRE IRON INDUSTRY

This area has come to have its centre in Sheffield with a fringe of associated works in Lincolnshire, Derbyshire and Nottinghamshire. Today there are blast furnaces at Park Gate to the north-east; this is the only iron producing plant in the Sheffield area. Further out there is Appleby-Frodingham, in Lincolnshire. To the south and south-west there are blast furnaces at Stanton, Chesterfield, Renishaw and Clay Cross. The steel works are clustered between Sheffield and Rotherham upon the River Don.

The early ironmaking in the Sheffield area has been admirably described by Dr. Mott.[56] He makes clear that, apart from the

[55] Evans quotes from a report to the British Iron Trade Association of 1896: 'A reduction of ¼d. per ton per mile on mineral traffic would reduce the cost of pig iron in the South Staffordshire district by about 7s. 6d. per ton, a reduction which would revolutionise the trade and would, from the increase in traffic, speedily more than recoup the railway companies.'

[56] R. A. Mott, 'Early ironmaking in the Sheffield area', *Trans. Newcomen Society*, Vol. XXVII, 1956, pp. 225–36.

favourable geological disposition of the outcrops of iron ore between Leeds and Nottingham, the important factor leading to the metallurgical importance of the district in the seventeenth century was in the first place its woodland of the Sheffield Park and of Hallamshire (amounting to nearly 4,000 acres), and secondly, the availability of water power. The first charcoal furnaces were built in the Sheffield district between 1575 and 1587. Further furnaces were subsequently erected at Barlow in North Derbyshire and Chapeltown.

However, we are concerned here with the coke-smelting industry begun with the blowing-in of the Walkers' Rotherham furnace in 1765; the Chesterfield furnace in 1777; Chapeltown in 1778; and Renishaw in 1782.[57] By 1806 there were 11 blast furnaces in Derbyshire with an output of almost 9000 tons.[58]

The largest at this time was the Butterley ironworks listed as belonging to Outram and Company. The history of this firm is published under the title *Through Five Generations*.[59] It gives details of the growth of the partnership. This was formed in 1792 'to establish an ironworks and to exploit the ancient coal workings on the Butterley Hall estate'.[60] The four partners were firstly, Francis Beresford, who 'contributed £6000 to the partnership of 1792' and was the father-in-law of John Wright, the second partner. Wright came from a Nonconformist family of bankers in Nottingham, who were apparently spare-time iron merchants, and he 'was the solid capitalist of the concern'. Both the other two ironmasters were technical men—engineers and surveyors. The venture was originally promoted by Benjamin Outram, who was concerned with developing transport communications, as was William Jessop.[61] Outram planned and built several tramways in this part of England; Jessop, who came from the West Country, also made the important innovation in tramway engineering of transferring the flange from the rail to the wheel, but in addition he was responsible for constructing the Cromford canal. This was to play a vital part in linking Codnor Park (where the company was later to have another ironworks) and Butterley. But,

[57] Mott, *loc. cit.* Also G. C. Hopkinson, 'The charcoal iron industry in the Sheffield region 1585–1775', *Trans. Hunter Arch. Society*, Vol. VIII, Part III, 1958. In F. Nixon's article 'Notes on the engineering history of Derbyshire', *Edgar Allen News*, Vol. 38, No. 439, January 1959, it is stated that the first coke iron furnace in Derbyshire was erected at Morley Park, near Belper, by Francis Hurst of Alderwasley Hall. This furnace and another built in 1818 were in blast until 1874.

[58] *Boulton and Watt MSS* (Assay Office), List of Furnaces, 1806.

[59] R. H. Mottram and Colin Coote, *op. cit.* (London, 1950).

[60] Mottram and Coote, *op. cit.*, p. 45.

[61] Jessop's father had co-operated with Smeaton in the construction of the Eddystone lighthouse and he himself was a pupil of the great civil engineer.

as Mottram and Coote point out, Jessop's works of engineering covered Britain from Ayrshire to London.[62]

Thus the firm from the very first was very active in supplying cast iron rails for these tramways. Butterley, too, supplied the ironwork for several bridges, including the Vauxhall bridge, London, and for buildings and dockyards.[63] During the Napoleonic wars, however, the foundries had turned out cannon and shot and shell: after Waterloo, as with many other ironworks in this district (including Walkers of Rotherham and Newton Chambers at Thorncliffe), the manufacture of water pipes was a peace-time substitute. Nevertheless, according to Glover, 'a vast number of steam engines, sugar mills and machinery [was] sent by the Butterley Company to the West Indies, the Mauritius and other Colonial states'.[64] By 1811 the firm had extended itself to Codnor. Thus by 1831, when Glover published his local history, Butterley was a large integrated concern employing nearly 1500 men among its four furnaces, foundry, bar iron works and collieries, iron mines and limestone quarries. By 1813 the Outram interest had been bought out and two years later a new £30,000 partnership had been drawn up between John Wright and the younger Jessop.[65]

We cannot, of course, follow the history of this company in detail. In 1844 the large establishment was described in the *Penny Magazine* and extracts from this are to be found in the history of the company.[66] These describe the patriarchal arrangements of the Butterley ironmasters, including a model village with the unfortunate name of 'Ironville', as well as the technical equipment of the works and its production, which now included heavy forgings of paddle-shafts for steam-packet ships.

The Wright family interest in the concern has survived until the present day; but gradually the descendants of Jessop sold out their investments. In 1858, for instance, a two-fifteenths share was bought by Francis Wright for £58,000; thus the total capital, at this rate, must have been £435,000.[67] Four years later the company was producing 700,000–800,000 tons of coal from fourteen pits; it owned seven blast furnaces and it was noted for its girders for armour-plated steamships. Butterley, however, still engaged in great works of

[62] See Charles E. Lee, 'Early railways in Surrey', *Trans. Newcomen Society*, Vol. XXI, and B. Baxter, 'Early railways in Derbyshire', *ibid.*, Vol. XXVI.

[63] S. Glover, *History and Gazetteer of the County of Derby* (Derby, 1831), p. 200.

[64] *Ibid.*, p. 201.

[65] Wright owned two-thirds and Jessop one-third.

[66] Mottram and Coote, *op. cit.*, pp. 60–3.

[67] Mottram and Coote, *op. cit.*, p. 77.

constructional engineering and among its contracts was the St. Pancras railway station built in 1869, in wrought iron. Steel had not yet superseded iron here.[68]

For the development of these districts in general, we are fortunate in having two accounts of the iron industry in Derbyshire and Yorkshire in 1825 and 1828. During the Napoleonic wars these ironworks had participated to a great extent in the demand for shot and shell. Cannon manufacture was a speciality at Low Moor, near Leeds, and the firm continued making this product after the war, for the East India Company was a good customer, and in the Crimean War it is said the output of warlike stores amounted to several hundred tons a week.[69] The inevitable post-war decline in this manufacture inevitably hit some firms. We have seen how Newton Chambers, who had in fact been manufacturing not guns but pipes during the war, felt the depression—what is more, Newton reported attempts at the burning of haystacks by the aggrieved colliers.

We have in Thomas Butler's diary detailed accounts of the progress of the industry in the 1820s when it was finally emerging from the post-war slump.[70] To take the Yorkshire works of Low Moor, Bierley, Shelf and Bowling first. All the works were thriving in 1825; their 10 blast furnaces were making about 400 tons of pig iron a week. Of this, the Shelf ironworks were 'melting 250–270 tons into shot and shell'. The Low Moor Company was especially busy in the steam engine department (although held up by 'a rebellion amongst their Engineers and fitters up of Engines', as was the Bowling Works). The Bierley works, eventually acquired by Low Moor in 1854, were at this time confining themselves to the production of pig iron. Butler thought that these works were 'realising more profit than any of the other works according to the capital employed'. They had two furnaces in blast making 75–80 tons a week of pig iron. The South Yorkshire district, with 13 furnaces in blast, producing just over 300 tons a week, was one of small-scale works, many old and inefficient. Some furnaces were making 25 tons a week. Newton

[68] The firm continued as a family concern; in 1873, Francis Wright died (he had through his railway connections built up the wealth of the company through the lucrative rail business). In 1883 the last of the Jessops sold out—capital then was valued at £500,000, and in 1887 the concern became a company limited by guarantee. Shares and debentures were issued to the extent of nearly £3/4 million.

[69] *Souvenir of Low Moor, 1791–1906*. For an account of the establishment of this famous ironworks see Ashton, *op. cit.*, pp. 102 and 222.

[70] *Kirkstall MSS*. (i) Tour amongst the Yorkshire and Derbyshire Iron Works, April 1825. (ii) Journey amongst the Ironworks (Bradford district, Wakefield, Barnsley, Sheffield and Rotherham), September 1828.

Chambers at Thorncliffe was apparently the most prosperous concern, being engaged in the London gas-pipe trade. Trade had certainly improved for this firm, although Butler evidently was not willing to believe all the proprietors said. 'Chambers', he wrote, 'has sold Gas pipes in London at £9. He can get now £16, he says, or *perhaps lower*. He is [a] swaggering dog, says he has lately refused orders for 7000 tons of castings. I dare say this company has done very well here. If we dare believe half of what Chambers says they are doing wonders. . . .'[71]

In contrast, Butler had several sorry tales of ruin to tell of several ironmasters and ironworks. Chapeltown had been formerly the property of Richard Swallow, but 'they did him no good',[72] now they were in the hands of John Darwin. whose convivial ways have been described by Ashton in his book upon Peter Stubs.[73] Darwin, also the owner of the Elsecar furnaces, was now in a tight financial position. Hence, as Butler said, he was always willing to make a bargain 'for ready Rhino'. Twenty years earlier Darwin had been boasting that he was sitting on velvet, but 'Alas, poor John, since then has been sitting on Thomas', remarked Butler. The Holmes furnaces of Samuel Walker and Co. at Rotherham were 'very much inconvenienced by want of materials'. Their furnaces were making only 21 tons of iron each week. But apart from preparing to again make guns—the old manufacture of this works—the rest of the concern was 'at a stand'. Butler reported that 'it is talked that all the works are to be pulled down and then a large spinning or weaving factory is about to be erected'.[74] One source of loss had been the construction of the Southwark Bridge at an estimated cost of £287,000;[75] however, it seems that Walkers had to be content with an offer of 'taking 10s. in the pound upon [the] debt'.[76]

In Derbyshire, apart from Butterley which has already been discussed, there was another integrated concern. The firm of Smith and Co., owning five furnaces (at the Adelphi, Calow and Chesterfield

[71] 1825 Tour, 2 May—Thorncliffe.

[72] Then Swallow built three furnaces at 'Swall Hill' and they nearly ruined him.

[73] T. S. Ashton, *An Eighteenth-Century Industrialist* (Manchester, 1939), p. 42.

[74] For the earlier history of the Walkers' concern see A. H. John's introduction to the edition of the minutes of the original partnership. According to Guest in his *Historic Notices of Rotherham* (1879) the social pretensions of the new proprietors had led to the decline of the once flourishing iron and steel works. The steel business was given up by the partnership in 1829, although it seems that one member of the family was probably going to take it over (John, *op. cit.*, p. 30).

[75] Guest, *op. cit.*, p. 491.

[76] John, *op. cit.*, p. 28 (Minutes of 1821).

works), although producing the goods common to this area—pipes, shot and castings—was in a dilapidated state and almost in ruin.[77]

These Derbyshire works, having fifteen furnaces in blast, produced only 350 tons of pig iron a week. According to Butler, Renishaw ironworks of Appleby and Co. was the best regulated work in Derbyshire. At least one work, Butler thought, should be 'razed to the ground and rebuilt upon a modern principal (*sic*)'.

1825, then, was a period of prosperity when the smallest and, apparently, the most inefficient works could manage to make a living for its owner. On his second visit, in 1828, Butler found circumstances very different. In Yorkshire, the four works had nearly 11,000 tons of iron on hand, at a time when the ironmasters had been forced by competition to sell their iron at prices even lower than those charged in Staffordshire.[78] In South Yorkshire, Darwin by this time had gone bankrupt and Lord Fitzwilliam, the landlord of the Milton and Elsecar furnaces, had either bought up or foreclosed upon these works, and they were in the hands of other tenants. According to Butler, several ironworks were nearly abandoned, including the Vickers works at Sheffield, Attercliffe, Brightside and several others.[79] The Holmes furnaces, silent except one, provoked Butler to remark: 'Oh dear, what desolate places are these works—hardly the least hum of business noises where formerly many hundreds of people were engaged!' Yet among this desolation there was one example of new investment and lavish expenditure which, as we shall see, survived the depression.

One way of development common to many industries is for a firm established in the finishing trades to acquire at some later stage the sources of its own raw or partly worked materials—the backward integration of processes. This method of industrial expansion was not, indeed, typical of the iron industry, as although the cupola had liberated the foundry from the furnace, on the other hand, with the technical innovations of the late eighteenth century, the tendency had been for the hitherto independent forge to be drawn to the blast furnace. Moreover, geographical and geological conditions often imposed a priority upon the construction of a blast furnace. However, this form of development—the acquisition of mines and blast furnaces by the foundry—became more possible as the industry expanded in all its branches and was exemplified in South Yorkshire by the history of the Park Gate Iron Company.

[77] For this firm, see Philip Robinson, *The Smiths of Chesterfield . . . 1775–1833*, 1957.

[78] Journey amongst the Ironworks etc., 1828.

[79] Journey amongst the Ironworks, etc., 1828.

The firm was established in 1826 at Park Gate, near Rotherham, by Charles Sanderson and John Watson, both of Sheffield, described as ironmasters.[80] The works were described by Thomas Butler as being 'erected at great expense and with great taste'.[81] This was in 1828 when the land on which the works were situated, previously leased, was bought and a new deed of partnership drawn up. The firm, confining itself to the manufacture of wrought iron, castings and tinplates, managed to survive the slump which followed the first railway boom of 1825. However, Butler, with the prosperous experience of Kirkstall Forge behind him, observed: 'Whatever could induce these people to lay out such large sums of money in forge and rolling mills after such proof they have before their eyes of some of the best works and the oldest establishments in the neighbourhood have been discontinued or nearly so—I cannot tell.'[82]

In 1831 the works were the subject of a 'poem'—'Greasborough Ings or an Excursion to New Park Gate'[83]—in which, after the poet rehearses a catalogue of the usual plant of a tinplate works—the puddling furnaces, the hammer, the forge, rolls, etc.—the processes of sheet rolling and tinning are described. Despite this attention of the Muse, however, Butler's fears were confirmed, for the concern ran into financial difficulties in the early 1830s and, the partnership having been dissolved in 1831 and Sanderson, the subsequent sole owner, having become bankrupt in January 1832, the works came into the possession of two Midlands mortgagees, Joseph and William Bennett of Dudley. Two years later the works were conveyed to a Midlands partnership of Birmingham and Smethwick businessmen who had formed the Birmingham Tinplate Company in 1833.[84] The firm now appears to have enjoyed a period of prosperity, and the plant now included a blast furnace, which first came into operation during 1837 or 1838. The neighbouring Holmes furnaces were taken on lease in March 1841. However, this development was again disturbed by reorganisation of the company's proprietorship. The Birmingham partnership was dissolved in 1840 in favour of the sole ownership of William Scholefield, but he, it is said, was forced by

[80] Guest, *op. cit.*, p. 688. I am also very much indebted to Mr. T. W. Edwards for information taken from the company's legal documents.

[81] Journey amongst the Ironworks, etc., 1828.

[82] Journey amongst the Ironworks, etc., 1828.

[83] By P. Rodgers, quoted by G. P. Royston, *A History of the Park Gate Iron Company 1823–1923* (privately published), p. 27.

[84] The partnership was extended in 1836, to admit further partners, but Thomas Pemberton and Joseph Gibbons, mentioned by E. Edwards's *Personal Recollections of Birmingham Men* (1877), p. 126, do not appear among the partnership records.

the financial strain of the 1842 slump in the iron industry—when iron fell from £10 to £5 per ton—to admit the banker-ironmaster Charles Geach and Samuel Beale into co-partnership.

Henceforth, coming under the influence of Geach with his many connections in the iron and railway industry, the prosperity of the firm, now making iron rails instead of the original nail rods, was ensured. Geach, who began as a banker before he branched out into the iron industry, during the railway mania of 1845 'took three orders for 30,000 tons of railroad iron, at £12, which did not cost over £6 per ton'.[85] By 1860 the Park Gate forge and rolling mills had 60 puddling furnaces.[86] One other new product of the forge at this date was armour-plate.

The Geach family connection was maintained in the firm, which was now styled Samuel Beale and Co., until 1864, when the company was converted into a limited liability company, under the name of the Park Gate Company Limited. The authorised capital was then £300,000 although only a matter of £90,000 was paid up.[87] Park Gate was a typical prosperous business of this period.

To turn again to Derbyshire, the history of the Clay Cross Company is illustrative of two important factors in the development of the wrought iron industry—the association of iron with coal and with the railways.[88] The North Midland Railway, in a new line from Derby to Leeds, which was being constructed in 1837, caused this company's establishment, both in creating the opportunity for the enterprise and also in providing the personalities who shaped its fortunes. The occasion was the revealing of the mineral wealth of coal and ironstone to the astute railway engineer George Stephenson, during the cutting of a tunnel at Clay Cross. He, realising the possibilities of this discovery, formed a company in 1837, with himself as chairman. Other members were his son, Robert Stephenson; G. C. Glyn (later Lord Wolverton), a director of the North Midland Railway; George Hudson, the 'Railway King'; William Jackson, M. P. S. Morton, M.P., and E. L. Betts, all partners in the firm of Peto, Brassey, Betts and Jackson, constructors of railways throughout the world; Joshua Walmsley, M.P., and Joseph Sandars. Thus railway interests were predominant from the beginning.

Coal, however, not iron, was the first marketable produce of the company, and indeed it is significant to note that the North Midland Railway itself had great expectations of the traffic from the Clay

[85] W. E. Edwards, *op. cit.*, p. 127.

[86] Royston, *op. cit.*, p. 27.

[87] Information supplied by Mr. T. W. Edwards.

[88] *A Hundred Years of Enterprise 1837–1937* (privately published).

Cross collieries and coke works,[89] though coal for the London market could only travel to Rugby by rail, the remainder of the journey being by canal.[90]

It was not until 1846 that it was resolved to raise £24,000 to erect two furnaces, though in 1838 the company had started to work limestone quarries. In this connection, another development of the Derbyshire iron industry is illustrated by the history of this firm; though local ironstone was at first smelted, nevertheless when the extension of the railway through Northamptonshire revealed richer seams there, the Jurassic ore was imported to the Derbyshire furnaces. The Clay Cross Company later owned mines in Northampton (Brixworth) and Oxfordshire (Bloxham).

Finally, the occasion of the severance of the Stephensons' connection with this enterprise might be remarked upon. Ironically enough, although there was this intimate relationship between the company and the railway, through the directorships, it was the concern of Robert Stephenson for the interests of the railway that led to his departure. He refused to ratify an agreement between the railway and the company granting preferential rates for the carriage of 60,000 tons of coal to London. He would not, it is said, sacrifice the interests of the railway company to which he was consulting engineer, to his own private interest as a colliery proprietor and ironmaster.

The Kirkstall Forge and foundry[91] was one of several around Leeds which formed an industrial metal-working district, second only to the Black Country. It is probable that the continued prosperity of the district as an iron smelting centre owed much to this local market; and it is noticeable that, where the iron industry had developed without attracting into existence a body of associated subsidiary trades engaged in manufacturing the metal, such districts succumbed more easily to unfavourable economic forces. Only where there was opportunity for securing high prices for finished goods of quality and precision could a primary industry, with limited and costly raw materials, survive. Thus, because of the value which accrued to an iron-made product from a particular engineering skill, it was observed that, in 1867, the cost of a finished article at Leeds was about double the cost of similar articles made elsewhere.[92] The numerous foundries, such as Kitsons, producing spinning and weaving machinery,

[89] Report to the shareholders, 1841.

[90] The first direct rail transport of coal to London was in 1844.

[91] R. F. Butler, *The History of Kirkstall Forge* (2nd edn., York, 1954).

[92] W. Williams, 'A Sketch of the British Iron Trade' in *Handicraftsmen and Capitalists* (Birmingham, 1867), p. 100.

stationary engines and brewing apparatus, made Leeds a great engineering centre, which also supplied the ironmasters with their equipment—steam hammers, blowing engines and rolling mills.[93] In 1858 an estimate was made that in this district about 75,000 tons of pig iron were annually consumed in the forges and foundries, which employed over 2000 hands.[94] At Kirkstall production was very much concerned with supplying the needs of the links of communication. Not only were bridges, railway locomotive wheels, axles and waggons made, but also the new means of communication which removed the need for the physical movement of people or even of the penned word—the telegraph wire.

As for the primary production of pig and wrought iron, the district maintained a reputation for the quality of its bar iron, especially that of Low Moor. Because of this, the district held its own in the face of the severe competition of its Cleveland neighbour with its low-cost production.[95]

4 SOUTH WALES

In no other district is the changing importance of the factors of location better to be seen than in South Wales. In the Elizabethan era, ironworks had been set up by Sussex ironmasters in consequence of the local difficulties, among them the costliness of fuel in the old centre of the industry, the Weald. The wealth of the South Wales forests was such that the charcoal-fired forges survived until well into the nineteenth century. Charcoal iron for tin-plate remained a local speciality until the Siemens-Martin mild steel put them out of production.[96]

The new point of departure for the Welsh iron industry was the establishment of the coke-smelting furnaces along the north-eastern fringe of the bituminous coalfield following the example of the Merthyr Tydfil ironworks. These were Tredegar in 1764, Sirhowy in 1778, Beaufort in 1779, Abercarne in 1783 and Ebbw Vale in the years 1786–9. Nevertheless, as late as 1788 there were in Monmouthshire, Glamorgan and Carmarthenshire seven charcoal furnaces

[93] See E. K. Kitson, *Kitsons of Leeds* (privately published).

[94] James Kitson, 'The iron trade of Leeds', *Report of British Association*, 1858.

[95] However, in this century the Low Moor, which had been converted into a limited company, was taken over and amalgamated with the North Staffordshire firm of ironmasters of Robert Heath and Company. But following the failure of that joint concern, the Low Moor works were bought by a Sheffield firm of steelmakers, Thos. W. Ward Ltd.

[96] W. E. Minchinton, *The British Tinplate Industry*, p. 37. The last charcoal iron for tinplate was made at Lydney in 1886.

producing 4300 tons of pig iron and six coke furnaces in Glamorgan turning out 6600 tons.

The industry was built upon the perfection of Cort's methods by Crawshay, so that the manufacture of wrought iron made in this way was known as the 'Welsh method'. The South Wales industry was, in fact, geared to the production of bar iron which became the staple product, both for immediate sale to the local tin-plate makers and to the London and provincial markets and, from the 1820s onwards, the rolling of the rails for the network of iron communications of the world. In 1831, in a survey of the Welsh industry, it was noted that 'the Welsh ironmasters confine themselves more to this regular trade (i.e. common bar making) than those of Staffordshire.'[97] It was true that some works possessed foundries (e.g. the Neath Abbey works which made steam engines), but it was reported that, 'a great deal of Welsh foundry iron has actually been sent to Staffordshire for the purpose of being run into saucepans, nails, etc.'[98] Because of the extra profit, which the author saw the Staffordshire ironmasters deriving from these secondary trades, he thought that this finishing side of the industry would 'probably be extensively entered into in Wales before very long'.[99] He was wrong in his surmise, South Wales did not develop an extensive foundry trade as the Black Country had done.

It should perhaps be emphasised at this point that geology had divided the coalfield into two main divisions; in the west, the anthracite measures, in the east, the bituminous coal. The Vale of Neath was an approximate dividing line, with the semi-anthracite and steam coals forming an area where the divisions merge. The iron industry was distributed all over the coalfield, but the eastern district, concentrating upon Merther Tydfil and district, as we have seen, saw the greatest development. The other, the anthracite area, had a delayed and unsuccessful spell of exploitation, which depended upon the application of Neilson's hot blast to the hitherto unusable fuel.

In 1815 the Merthyr district of the coalfield, with its mammoth works of Dowlais, Cyfarthfa, Penydarren and Plymouth making about 50,000 tons a year,[100] was the major producing area in South Wales. Future development, in fact, largely amounted to the consolidation of established firms who, increasing the numbers of

[97] Anon., 'On the manufacture of iron' (1831), p. 25, *Institution of Civil Engineers' Tracts*, Vol. 56.

[98] *Ibid.*, p. 26. [99] *Ibid.*, p. 25.

[100] T. E. Clarke, *Guide to Merthyr* (1894). Cyfarthfa 18,200 tons, Dowlais 15,600 tons, Penydarren 7800 tons, Plymouth 7800 tons.

furnaces with their greater and greater yields, accounted for the major part of the expansion of the Welsh output. In 1820 the four Merthyr works were sending down the Glamorgan canal to Cardiff 46,756 tons of iron (chiefly wrought iron) out of a total of 150,000 tons of pig iron output for the whole of South Wales. Twenty years later the comparable figures were 109,777 tons and 505,000 tons respectively.[101]

The periods of railway construction, creating their enormous demand for wrought iron rails, chairs and other items of track and rolling stock construction, were undoubtedly the greatest stimulant to the expansion of the industry in the decades around the middle of the nineteenth century. From the profits of this lucrative trade, the additional capital resources of the huge establishments of Dowlais and Cyfarthfa were obtained for the purchase of larger and more efficient blast furnaces and rolling mills. In 1844 it was reported in *The Times* that: 'In anticipation of an increased demand for railway iron, Mr. William Crawshay has decided upon the erection of an extensive steam rail mill, capable if necessary of the completion of 2500 tons of rails per month, at an estimated cost of £25,000.' Dowlais in 1836, according to Guest, a witness in an important legal case brought by the Stanhope and Tyne Railroad Company against another firm of Welsh ironmasters, had made 20,000 tons of rails. The defendants, Harfords of Ebbw Vale and Sirhowey, with an establishment of 3000–4000 employees, were themselves making from 7000 to 10,000 tons a year of railroad iron. Fortunately for the Welsh iron trade, the defendants' use of cinder in their iron, which kept the cost of manufacture so low, was upheld by the court.[102] More will be said later concerning the effect of this demand upon the expansion of the industry, but here we may note how crucial this outlet for the production of iron was to the great works like Dowlais. For example, in 1843, during a depression in the trade at home, Evans, Guest's representative abroad, 'got an order for 12,000 tons of rails at £6 on their way through Poland'.[103] When Lady Guest was writing in 1852, that 'at present we have hardly an order to go on with', a contract of 60,000–70,000 tons of rails for Russia was in the balance; eventually, however, this was secured and the works, which was in danger of being closed down, went into full production again.[104]

[101] Figures of the Merthyr consignments compiled from the returns of shipments in Lloyd, *op. cit.*

[102] Snell, *Report of Case*, Barnard *v.* Harford (London, 1838).

[103] *Lady Charlotte Guest's Diary*, 16 December 1843, p. 157.

[104] *Ibid.*, pp. 286–90 *passim*; Elsas, *op. cit.*, pp. XVIII–XIX, gives a list of railways supplied with iron by Dowlais from 1821 to 1857.

The history of the development of the western district of the coalfield is very different from that of the unceasing exploitation of the bituminous coal area of the north-east. For a detailed account of this area the reader is referred to L. T. Williams's *The Economic Development of Swansea and of the Swansea District.*[105] Williams divides this area into three sub-regions: the first, the anthracite zone of the north-west and north, of the Upper Tawe valley, in which the chief ironworks were the Abercrave, Ynyscedwyn and Ystalyfera establishments. The second sub-region is the Maesteg–Tondu–Bridgend area of the south-east—an area of bituminous, steam and house coals, the mining of which was the chief industry. Iron working was a modest affair connected with the Maesteg ironworks formed in 1826; according to Williams, 'it was not until 1839 that iron smelting on a large scale really commenced'.[106] Other ironworks were the Llynvi, which were amalgamated with Maesteg *c.* 1860–5, and the Tondu works, which are interesting as affording an example of Manchester capital coming into the area. The Brogden family in 1853–4 bought up the derelict works and greatly developed the neighbourhood.[107] Finally, there is the coastal area. This was scattered with forges in the eighteenth century, but in this author's words: '[It] cannot be described as a thriving and extensive iron producing district until after 1840.'[108] At Briton Ferry in the 1840s blast furnaces were built, but the most important branch of the industry was the forges and rolling mills of Llanelly.

This area was, of course, to assume a new importance after the successful commercial trial of the Siemens open-hearth process of steel manufacture at Landore in the years 1868–75. However, before looking at the invasion of Wales by steel, something should be said about the effects of hot blast upon this district. For the most thriving concerns in this western district were Ynyscedwyn and Ystalyfera of Crane and Budd, the pioneers in exploiting the iron and coal resources of this district. Budd, 'the Guest of the Western Coalfield', possessed, in 1870 at Ystalyfera, an establishment of 11 furnaces and 42 puddling furnaces; although this expansion was not altogether based upon the exploitation of anthracite. In fact, bituminous coal was brought to Ystalyfera by canal.

To review briefly the attempts to smelt iron ore with this pure yet very hard coal, it may be said that as early as 1804 a patent had been taken out to use anthracite; but since this was with cold blast it must certainly have been a failure. In the years 1824–6, the British Iron Company at the Abercrave ironworks, with the technical

[105] *Op. cit.* (Cardiff, 1940), pp. 94–126.
[106] Williams, *op. cit.*, p. 107. [107] *Ibid.*, p. 184. [108] *Ibid.*, p. 110.

assistance of Mushet, was equally unsuccessful. The furnaces were blown out after it was found they could not cover their costs which exceeded the selling price of the iron by £6–£8 a ton.[109] The only use that was made of anthracite in this area was as a mixture, in small quantities, with bituminous coal at Landore in 1826–9. It was not until 1837 that 'George Crane at Ynyscedwyn solved the problem which had for half a century engaged the attention of all parties interested in the stone coal district; and by the application of hot blast rendered the mineral at once a fuel equal to any hitherto used for the manufacture of iron'.[110] Crane, incidentally, was able to make use of Neilson's patent, since this did not specify any particular kind of fuel, and he believed that anthracite would give results similar to those obtainable from smelting with charcoal fuel.[111] But this was not so; outputs were small and the furnaces would not work properly. J. P. Budd, it is true, made some improvements in the 1840s, using a higher pressure blast, but Welsh methods never achieved the success which the American smelters had with this fuel. In 1853 there were 36 furnaces in the district but only 19 were in blast. In contrast, the U.S.A. in the early part of 1840 had 50 furnaces at work.[112] The Welsh experiment was not a success, as Williams observes: 'After 1860 iron smelting in this (anthracite) belt decayed rapidly and ceased entirely before the close of the century.'[113]

Nevertheless, there was to be a change in the technical character of the South Wales industry. First, after 1843, when Neilson's patent expired, several ironworks introduced the hot blast system, including Dowlais itself. In 1869, the Tredegar, Blaenavon, Aberdare and Abernant works were using hot blast.[114] Secondly, the native ores, which had long been supplemented from Cumberland, Lancashire, the Forest of Dean, Somerset and Devon,[115] had reached the point of uneconomic working. Thus the pneumatic conversion process of steelmaking of Bessemer, which promised economy in fuel with a greater value product, was a great attraction to the Welsh iron-masters. We shall see the intimate connections between Bessemer and Dowlais, and the rival efforts of Parry at Ebbw Vale concerning the early experiments with the invention later. However, when the

[109] *Crane v. Price and Others, Report of Trial*, 1840, p. 31.

[110] *Children's Employment Commission*, Vol. II (1842), p. 668.

[111] Crane tried to interest the Dowlais Company in the invention, but without success (*Dowlais MSS*, George Crane—T. R. Guest, 14 April 1837).

[112] H. Scrivenor, *History of the Iron Trade* (London, 1841), p. 304.

[113] Williams, *op. cit.*, p. 98.

[114] *Mining Journal*, October–November 1869.

[115] C. S. Orwin, *The Reclamation of Exmoor Forest* (London, 1929), gives details of the prospecting of G. T. Clark of Dowlais in the 1850s.

process had been improved, several of the South Wales works, from 1866 onwards, adopted the manufacture of steel. By 1870 the Dowlais works had six converters and the Ebbw Vale works seven, all of 5 tons capacity. Ten years later Rhymney and Blaenavon had adopted the Bessemer process; but by this time the converters had been made larger to take 8 to 10 tons of pig iron.[116] Nevertheless, it was not Bessemer steel but the 'mild steel' of the open-hearth furnace which became important in the South Wales tin-plate industry.

Another major change in the Welsh industry resulted from the demands, which followed, for the phosphorus-free haematite ores. In the words of Colonel Mitchell, the historian of the Monmouthshire iron and steel trade: '[These demands] soon began to tell upon the supply [available locally,] while its price made its use almost prohibitory, so that, in order to keep up the make and cost within bounds, fresh sources of supply had to be sought.'[117] Thus the first importation of ore from Spain began in 1868–9,[118] and with that shift in the balance of costs the way was being prepared for a move of the South Wales iron industry to the coast.

In 1890 Lady Charlotte Schreiber paid her last visit to Dowlais, and Lord Wimborne, her eldest son, was at the head of the firm. The move of the furnaces to a site at East Moors, near Cardiff, had already been decided upon. In the following year, in 1891, the blast furnaces were set up at Cardiff; nevertheless the steel rail rolling, which was the chief activity of the firm, was still retained at Dowlais, even when steel was made at the coastal site.[119] Thus the move to the sea was never complete; a regard for the dependent community made Wimborne oppose a decision made purely upon economic considerations. Even today, Guest Keen and Baldwin has a large foundry at Dowlais.[120]

5 SCOTLAND, LANARKSHIRE AND AYRSHIRE

There were in the nineteenth century, according to Kohn, three schools of iron smelting, according to the natural conditions of the supply of the raw materials. First, there were those depending on rich ores and dear fuel, such as the ironmasters at Barrow-in-Furness and West Cumberland. Second, in the Clevelands, there was the combination of poor ores and cheap fuel, which needed the skill and

[116] Williams, *op. cit.*, p. 99.
[117] Mitchell, *The Monmouthshire Iron and Steel Trade* (Newport, 1904), p. 17.
[118] *Ibid.* [119] D. L. Burn, *op. cit.*, p. 236.
[120] The last merger taking in the Guest enterprise was made in 1929.

science of the engineers and blast furnace managers. Finally, there was the best-favoured district of all, Scotland, where there was to be found together rich ores and cheap fuel.

Not that there was, by any means, a great expansion of the iron industry in Scotland in the first two decades of the century. It is true that the Carron Company (which imported the Cumberland ores and Welsh iron) steadily extended its production.[121] Carron, indeed, kept up her manufacture of armaments—the contract with the East India Company came to an end in 1822, but the works still supplied the British Ordnance Department until 1848.[122] In 1829, when *The Morning Herald* described its activities, there were four furnaces in blast and four cupolas with an output of about 300 tons of finished goods a week. 'Field pieces' were being made at the rate of twenty a week.[123] However, from 1806 to 1830, the industry made very slow progress, merely increasing its output of pig iron from 22,840 tons to 39,500. The chief difficulty was the high cost of production as a result of the unsuitable quality of the Scottish coals. The consumption of fuel, on the average, seems to have been three to four times as great as in England and Wales. Professor Hamilton, indeed, asserts in his *Industrial Revolution in Scotland* that it was only the natural protection of transport costs (on the carriage of iron from South Wales) which enabled the Scottish producers to sell their product in competition with the cheap Welsh iron.[124]

Nevertheless, several ironworks came into existence during this period, including the important Calder and Gartsherrie ironworks.[125] But it was the revolutionary effects of the application of the hot blast process from 1830 onwards which set off a great burst of

[121] Another early concern was that of the Wilson brothers (one of whom was a director of the Swedish East India Company), who exploited the mineral resources of the banks of Mousewater (Lanarkshire) and established a community which was named Wilsontown. Despite launching into the profitable munitions business in 1781, as a result of litigation between the brothers, the firm was financially embarrassed in 1808 and was taken over by 54 gentlemen, 'possessed of goodwill and public spirit', of Scotland and England. In 1812 the whole works, mines and quarries were shut down. Not until 1821 were they reopened by William Dixon of Govan Colliery and the Calder Iron Works. The ironworks was finally broken up in 1845–6 (See P. M. Ritchie, 'The Romance of Wilsontown', *Journal of the West of Scotland Iron and Steel Institute*, Vol. XLVI, 1938–9.) For Dixon's coal interests, see P. L. Payne, 'The Govan Collieries, 1804–5', *Business History* Vol. III, 1961.

[122] Cadell, *op. cit*., p. 183. [123] *Ibid*., p. 188. [124] *Op. cit*., p. 173.

[125] J. Cleland, 'The account of the former and present state of Glasgow', *Trans. Glasgow and Clydesdale Statistical Society*, 1836. For a detailed account of the establishment and development of these and other ironworks see the important book of Professor Hamilton previously referred to, and Andrew Miller's *The Rise and Progress of Coatbridge* (Glasgow, 1864).

activity in exploiting the mineral resources of the northern kingdom, which were now found to be of great wealth.[126] For perhaps more important than the direct benefits resulting from Neilson's invention itself were its complementary consequences. It was possible to exploit the rich mineral resources of the blackband ironstone. This ore, which was to be found between the Firth of Forth and the Clyde and southwards, was a very rich kind of iron ore, containing 50–70 per cent of iron and, in the best specimens, a certain percentage of coal. The area of greatest concentration was at Airdrie, where, early in the century, Mushet had discovered this ore, only to have it ignored for about twenty years.

In parenthesis, it may be noted that Mushet had been a partner for a short time in the Calder ironworks, originally erected in the 1790s by an association of Glasgow stocking weavers. In 1802 this works came into the hands of William Dixon, who had migrated from Northumberland at the age of seventeen to seek his fortune.[127]

The first works to use the blackband ironstone was the Monkland Company in 1825,[128] but in the next decade many furnaces were being erected in this locality. In 1836 alone there were twenty-two more furnaces being erected;[129] by 1840 there were thirty furnaces all using blackband and hot blast.[130]

The expansion of the Scottish output, most of which came from Lanarkshire—with Ayrshire producing about one-quarter of the whole, the second largest area—has already been outlined. Here we should try to note the causes of this phenomenal expansion. One of the chief factors determining the character of the iron industry in Scotland was the influence of the very low costs in encouraging large outputs. This cheapness of product—derived, as Clapham points out, from the threefold 'geological, technical and commercial' advantages—was both the cause and consequence of Scotland's gigantic production. The Scottish ironmasters, until the establishment of its wrought iron side of the industry in the 1840s, relied for the most part on a small return on a large turnover (prices of Scottish pig iron slumped to as low as 32s. 6d. per ton). Hence it was here that speculation, which often created a fictitious demand and then disastrous periods of below-cost prices, had the ironmasters upon the rack of falling prices and an enforced expanding output in order to cover standing charges.

[126] R. H. Campbell's article on 'Investment in the Scottish pig iron trades', *Scottish Journal of Political Economy*, Vol. I (October 1954) looks at this development more closely than has been possible here.

[127] Ritchie, *loc. cit.*, and Miller, *op. cit.*, pp. 104–7.

[128] Miller, *op. cit.*, pp. 107–8. [129] Cleland, *loc. cit.*, p. 50.

[130] Hamilton, *op. cit.*, p. 182.

Moreover, the abundance of natural resources led to prodigality in their use and hence very little progress in technique. Thus when the tide of competition, which had been behind Scotland's price leadership for thirty years, turned against her, it was found that the Scottish ironmasters had much to learn from the Cleveland district, where the cost of fuel had more than encouraged economy of this sort. Kohn observed: 'There can be no doubt that the growing competition of the Cleveland district will before long bring the Scottish ironmasters down from their present aristocratic habits to a rational economy of fuel.'[131]

Associated with an industry geared to high outputs were the large-scale concerns. Of these the Bairds' concern at Gartsherrie and elsewhere was the largest in this period. McGeorge's book *The Bairds of Gartsherrie*[132] contains the memoirs of James Baird, a son of the founder of this great iron and coal concern, and these throw a great deal of light upon the building up of the ironworks to the stage when, in 1867, there were sixteen blast furnaces at Gartsherrie. In all, at this time, the firm owned more than fifty furnaces at the different associated establishments.

Alexander Baird, a farmer, had branched out into coal mining and in this way had leased several collieries along the Monkland canal. In 1827, Baird owned six pits in the Cartgill coalfield. The following year a lease was signed for the exploitation of the iron mines near Gartsherrie, and, in the words of James Baird, who was a partner in the family enterprise of the father and the three sons: 'In 1828, we commenced the construction of the Gartsherrie ironworks. Early in the spring of that year the foundations were excavated, and in the summer the building was commenced. Stones having been found on the ground, the work progressed rapidly.'[133] The first furnace was put in blast upon 4 May 1830—the Bairds having had to build it without the assistance of experienced men. Indeed, if we can believe these memoirs, the Bairds were forced to keep an eye upon the furnace for twenty hours out of the twenty-four hours at first, until they could employ 'a working mechanic'.[134] As we have seen, the life of the early ironmaster was extremely arduous. The furnace in the first year of its production, incidentally, had an output of 3100 tons.

Expansion was rapid, more furnaces were built—by 1839 there were eight furnaces in blast. During these years, too, the hot blast

[131] Kohn, *Iron and Steel Manufacture*, p. 9.

[132] A. McGeorge, *The Bairds of Gartsherrie: Some Notices of their Origin and History* (privately published, 1875).

[133] McGeorge, *op. cit.*, pp. 55–7.

[134] *Ibid.*, p. 60.

had been adopted, but, for reasons which we have seen, Baird wrote down the benefits to be derived from Neilson's patented system.[135]

William Baird was, apparently, the driving force behind this expansion and, according to James, his brother, he secured the foundations of the business by insisting upon cash payments for the mounting sales of pig iron. (One wonders if this unaccommodating principle was always adhered to.) At all events the profits were able to finance a further large-scale extension to the plant. Another eight furnaces were planned alongside the Monkland Canal, and a year's reserve supply of ironstone having been laid by, in November 1840 four furnaces were 'put in blast at the same instant . . . a thing unprecedented in the annals of iron making'.[136] When the sixteen furnaces were complete their capacity exceeded 100,000 tons of iron a year.

Naturally with such a rate of exploitation the mineral resources of the Monkland field of blackband ironstone were nearing exhaustion. The high fixed rental of the Airdrie House estate leased by the Bairds was, of course, an inducement for them to remove its riches as quickly as possible.[137] In consequence, the Bairds moved south to Ayrshire in 1844 and, having persuaded a noble lord that a lucrative income was more enjoyable than his estate and that the aristocratic name 'Eglinton' would look very well on the iron,[138] they set up the ironworks of that name and began a policy of expansion comparable to that pursued at Gartsherrie. Later, in the 1850s, the Bairds acquired the Blair ironworks, the Muirkirk and the Lugar works. All these had been unsuccessful, but were made profitable by the energy and organisation of the Bairds who, when they had completed their purchases, commanded a capacity of 300,000 tons. It is small wonder this family, whose wealth came also from banking and railways, was able to acquire landed properties in eight counties.[139]

The history of the establishment of the Bairds' concern is one which conforms to what one might term the classical mode of entry into the industry—from the land, to coal, to the blast furnace. In

[135] Baird points out the difficulty in keeping up the heat of the blast 'for more than 2 or 3 shifts' and the frequent breakages of pipes; he claims that they invented and used a new stove on a different principle. As a result of their invention, he claimed they had been able to increase the output of a furnace, working with a Neilson hot-blast stove, from 60 tons a week to 250 tons. They also took out a patent for using the waste heat in 1849, but the plan was abandoned.

[136] McGeorge, *op. cit.*, p. 75.

[137] Baird reveals that there were in the 1840s 60 blast furnaces on the coalfield.

[138] The remark of Baird's representatives upon this subject is said to have caused some embarrassment: 'It will do very well for the pigs, my lord.'

[139] A. Miller, *The Rise and Progress of Coatbridge* (Glasgow, 1864), pp. 109–11.

the case of the Coltness ironworks, on the north-east fringe of Lanarkshire, which in the 1860s had twelve furnaces in operation, however, its establishment was the result of the transfer of enterprise from the cotton industry to that of iron. The ironworks had begun as a small workshop for the repair of machinery at a Glasgow mill and had gradually extended its activities until it assumed an independent existence as a foundry. It was at this stage that Houldsworth in 1837 decided to leave the cotton trade to his relatives in Manchester and become an ironmaster.[140]

Apart from the early attempts to make bar iron in Scotland with the puddling process, which had apparently not been too successful, the malleable iron industry was still in its infancy in the 1830s. The chief reason for this failure to develop this side of the industry seems to have been that the proper quality of pig iron was not produced at that time. However, there had been one small ironworks—the Muirhill ironworks with a capacity of 20 tons weekly—and from 1839 onwards the manufacture was taken up, possibly as a result of using the recently invented 'wet' puddling methods of Hall of Staffordshire. In that year it was reported that three of the largest makers in Scotland were directing their attention to the manufacture of bar iron: the Monkland works, the Dundyvan works of Dunlop and Wilson, and William Dixon's Govan ironworks.[141] However, once under way the trade expanded rapidly. In 1842 the output was 35,000 tons; in 1848, it had reached the figure of 90,000 tons.[142] By 1867 the production of bar iron was 142,000 tons, from about 400 puddling furnaces.[143] As Professor Hamilton says: 'The malleable iron trade was to gain for itself a position comparable to that which casting had occupied for the best half of a century.'[144] This expansion of the iron industry was accompanied by an increase in the engineering industries in the district which again sprang largely from the growth of iron shipbuilding upon the Clyde. A great part of local demand, in fact, was for heavy forgings, such as propeller shafts, and the framework of the iron boats, besides the iron plate which had superseded the copper-bottomed wooden planks.

There was, of course, to be a major change in the shipbuilding industry as a result of the inventions of Bessemer and Siemens-Martin. In 1871 the Steel Company of Scotland, which had previously experimented with Bessemer steel, introduced the open-hearth

[140] J. L. Carvel, *The Coltness Iron Company* (privately printed, 1949).

[141] J. Johnson, 'The Iron Trade of Great Britain', *Mechanics Magazine*, April–September, 1839, p. 346.

[142] J. Barclay, *Statistics of the Scotch Iron Trade* (Glasgow, 1850).

[143] D. Bremner, *The Industries of Scotland* (Edinburgh, 1869).

[144] Hamilton, *op. cit.*, p. 187.

process to Scotland. Between 1873 and 1876 the company erected twelve open-hearth furnaces.[145] This steelworks, with which James Riley, the associate of Bessemer at Dowlais, was connected, was one of the leading works in the country in the 1880s.[146] The firm had no blast furnaces of its own and James Riley, who was a go-ahead innovator, built up its reputation upon the basis of making mild steel for ship's plate. In 1880 the company bought the Blochairn Malleable Iron Works to convert it to the production of steel.[147]

The Siemens process was, incidentally, according to a past president of the Iron and Steel Institute, closely associated with malleable ironworks in order to utilise the existing rolling mills.[148] But from this time other new steel works had their beginning, among them David Colville and Sons at Motherwell, who were later to dominate the Scottish steel industry. Finally, it may be noted that it was the disaster of the wrecking of the wrought iron Tay Bridge in Scotland, by a storm, which forced engineers to reconsider their views about steel. With this new demand for constructional purposes, production expanded rapidly. Already by 1885 the output of steel ingots amounted to 241,000 tons.[149]

[145] C. Erickson, *British Industrialists*, pp. 159–60.

[146] D. L. Burn, *op. cit.*, p. 220. C. Erickson, *op. cit.*, p. 169, for details of the career of Riley. He was 'exceptional in being the only manual worker to gain a position of leadership in the steel industry during its revolutionary period'. See also the article by E. F. Gibson, in the *Scottish Journal of Political Economy*, cited below.

[147] Hamilton, *op. cit.*, p. 189.

[148] Presidential Address of J. Craig, *Journal of the Iron and Steel Institute*, 1940. It is to be noted that the Siemens process had been adopted at Crewe because it was possible to convert iron rails into steel ones with only a very small waste.

[149] Hamilton, *op. cit.*, p. 189. A very thorough assessment of the Scottish steel industry, which examines the reasons for the responses of the blast furnace owners and of the malleable ironworks to the new technology, is given in E. F. Gibson, 'The establishment of the Scottish steel industry', *Scottish Journal of Political Economy*, Vol. 5, 1958. 'By 1885, the Scottish steel industry consisted of ten firms, with 73 open hearth furnaces and an annual output of over 240,000 tons, equivalent to some 42 per cent of the total British make of Siemens steel.'

CHAPTER IX

The Advance of Technology in the Wrought Iron Industry

'Industrial progress . . . depended upon two indispensable conditions: first, the possibility of increasing the output of the products . . . without a proportionate increase of industrial costs; and second, the finding of a market without which the larger output would have been not worth attaining.'[1]

To this may be added another quotation: 'Economic motives are present but do not become effective until many technical resistances have been overcome. The actual achievement is thus dominated by the timing of the technological advance.'[2]

In the hundred years covered by this book—from Cort to Gilchrist Thomas—the British iron and steel industry underwent at least three major revolutions. The first we have already discussed, the second was the application of the heated blast to the furnace; the third was the inventions of Bessemer, Mushet, Siemens, Martin and of Gilchrist and Thomas. These last-named developments transformed the industry and, in so far as steel is a different metal from iron, the age-old product of first the bloomery and then the puddling furnace was superseded by a metal made by science.

With such a heavy product as iron or steel it is obvious that the lay-out of plant and the mechanisation of the machinery which handles it through its many processes is a major factor limiting the productivity of the melting and rolling shops. The Americans were first in the metallurgical field to realise this and apply these methods to steel-making, and such imitation and adoption of up-to-date American and continental methods as there was in the 1870s and 1880s, by such firms as Bolckow and Vaughan and the Frodingham

[1] W. H. B. Court, 'Economic progress in the eighteenth-century Midlands', *Trans. R.H.S.*, Vol. XXVIII, (4th series), p. 89.

[2] A. P. Usher, 'The industrialisation of modern Britain', *Technology and Culture*, Spring 1960, p. 109.

Company, has been described in a masterly way by Burn in his book.[3] But in the first half of the century there was little theoretical consideration of the organisation of industry, hence we know little about this aspect of production. Nevertheless, it must be admitted, it is apparent that the original sites were ill chosen for expansion, for they quickly became cluttered with puddling furnaces around the blast furnaces.

THE BLAST FURNACE IN THE NINETEENTH CENTURY

We are already aware of the phenomenal increase in output over the period; what is even more remarkable is that the expansion of production from 1806 to 1880, from a quarter of a million to eight million tons, was achieved by only 400 more furnaces by the latter date. It is apparent that there was an outstanding development of blast furnace technique during this period. The major inventions of the eighteenth century had provided the motive forces which were to determine the pattern of future development. Although water-power persisted in some works (the water-wheel still driving the bellows for the blast and the forge-hammers), steam-power and coke were to be the chief weapons of the nineteenth-century ironmaster. The technical innovations of the later period, it is true to say, did not introduce any new elemental force to the furnace until the introduction of the electric-arc furnace by Siemens in the 1870s; the industry was still dependent upon the inventions of Abraham Darby and James Watt. Indeed the outstanding invention of the first half of the century, the hot blast of James Beaumont Neilson (1828), was but an extension of the steam-blast coke-fired furnace; nevertheless, its consequences were revolutionary.

A modern furnace is a complex piece of engineering giving the technicians control over the processes in accordance with the metallurgical science and the inherited practical skill of the smelter or keeper. At the beginning of the nineteenth century, however, the blast furnace had not evolved far beyond its original Tudor prototype.[4] The oldest type of furnace in use during the Napoleonic wars was a crude structure, a square stone tower, slightly tapering, with a cylinder on top. These were often built into the hillside so as to provide a natural platform for feeding the furnace from the top. In

[3] D. L. Burn, *The Economic History of Steelmaking* (Cambridge, 1940), chap. IV.

[4] See H. R. Schubert, 'The first English blast furnace', *Journal of the Iron and Steel Institute*, February 1952. For the first known description of an English blast furnace (Barden, Sussex, 1646), see *Publications of the Scottish History Society*, 3rd ser., Vol. L, 1958, pp. 146–52.

1815, it is true, the height of the furnace had been increased to between 35 and 50 feet.[5] Outputs varied between 25 and 60 tons a week. In the country backwaters of the industry, low outputs continued to be usual, but, as we have seen, in South Wales, which concentrated on the manufacture of bar iron, the ironmasters were anxious to expand the output of their furnaces by increasing the burden. Later when hot blast had given a great impetus to the exploitation of the West Scottish coalfield, it was there that the 'gigantic crucibles' (as they were called) were built to give the largest outputs. They in their turn were eclipsed by the Cleveland furnaces.

The first and the most obvious source of a larger output was by increasing the size, and particularly the height, of the blast furnace. Where the nature of the raw materials would allow it (depending upon the strength of the coke, to support the increased burden), the ironmaster, when rebuilding his furnaces or erecting new ones, added several feet to their height. Often, improvements were made in the shape of the interior to secure the speediest and most efficient firing of the charge to avoid 'scaffolding' in the furnace. As a result, there was an extreme diversity in the size and shape of blast furnaces.

In 1827, when Dufrenoy and Beaumont reported on the manufacture of iron in this country,[6] the average size of furnaces was 45 feet to 50 feet in height, though at Dowlais and the Plymouth works, in South Wales, the furnaces were 60 feet high and producing 80 tons of pig iron a week. In South Staffordshire, where there was more concentration on quality than quantity, this growth in the furnace capacity was less marked, but even here, among the large owners, there was this same drive towards large-scale production. A furnace erected for J. Gibbons of Pensnet, in 1832, 60 feet high, with many improvements in contour and with a round hearth, was capable of producing the unprecedented output of 151 tons a week.[7] By 1848, some of the largest Scottish furnaces were yielding 150 and even 200 tons a week, according to a paper by J. P. Budd, the South Wales ironmaster.[8]

Nevertheless, there was a limit to this means of expansion. Kohn reported in his *Iron and Steel Manufacture* that there had been a trial at Gartsherrie of furnaces which were 60 feet high, but that the furnace charge failed to support itself above the tuyères.[9]

[5] Anon., *On the Manufacture of Iron* (1831).

[6] 'La fabrication de la fonte et du fer en Angleterre', *Annales des Mines*, 1827, 2me Série, Vol. II.

[7] J. Gibbons, *Practical Remarks on Staffordshire Blast Furnaces* (1844).

[8] *Report of the British Association for 1848*, p. 75.

[9] Kohn, *op. cit.*, p. 9.

It was in the Cleveland district, however, where considerations of fuel economy and the relatively low yield of the ore compelled the ironmasters to look to their costs, that the construction of the blast furnace was most advanced. From 1864 onwards furnaces were being erected 100 feet high,[10] though, at the same time, opinion was not unanimous about the economies resulting from increased heights. Certainly by 1871 it was thought that the maximum manageable height had been reached though by then there were other considerations to be taken into account, such as the closed-top. At this time the most efficient furnace made 350 tons a week. The output of a large modern furnace is about 4600 tons.[11]

This method of increasing outputs was implicit in the competitive nature of the industry; taller furnaces were built by ironmasters, when firms expanded, to outstrip their neighbours, and this development was not in accordance with any principle or patent. With the Neilson process, however, it was another matter. This undoubtedly was the most important single innovation in the industry in the age of iron.

James Beaumont Neilson (1792–1865),[12] the son of a colliery engine-wright, after working a few years at Govan colliery, had been apprenticed in his brother's foundry at Glasgow. His first job as colliery engineer was of short duration and in 1817, without qualifications, he was appointed superintendent and foreman of the Glasgow gasworks. It is interesting to note that, on being appointed, he attended evening classes at the University to acquire the necessary knowledge of the sciences for his job. Thus, with this typically Scottish regard for education, he made a success of his job; and, in fact, being promoted to the post of manager and engineer, he stayed with the company until his retirement in 1847. Although, therefore, he had received his training in an iron foundry, it was only indirectly that Neilson became connected with the iron industry.

Having gained experience of purifying coal gas, he was approached in 1824 by an ironmaster, who asked him if a similar process could be used to free from sulphur the air which was to enter the blast furnace. As Neilson explained later: 'I perceived that he imagined the presence of sulphur in the air to be the cause of blast furnaces working irregularly and making bad iron in the summer months.'[13] In

[10] E. Williams, 'Cleveland Ironworks', *Proc. of South Wales Institute of Engineers*, Vol. VI.

[11] P. Dunsheath, ed., *A Century of Technology*, p. 13.

[12] This account of his life is based on 'The life of J. B. Neilson, F.R.S., the inventor of the hot blast' compiled by T. B. Mackenzie (*West of Scotland Iron and Steel Institute Trans.*, 1928).

[13] *Trans. Inst. Civil Engineers*, Vol I (2nd Edn., 1842), p. 81. David Murray,

1825 he read a paper on the problem before the Glasgow Philosophical Society, in which he came to the conclusion that this defect was due to a lack of oxygen (the air being rarefied by the heat of summer) and, equally important, the absorption of moisture by the air in its passage through the pressure-equalising water vaults. However, his first remedy, to use calcined lime as a drying agent, was not put to trial, for just then another problem arose which was to set him upon the direct road to his great invention. This was an enquiry from the Muirkirk Ironworks about the weakness in the blast arising from the distance of the blast-engine from the furnace. It then occurred to Neilson that, simply by heating the air to increase its volume, the force of the blast would be improved. Experiments with coal gas at the Glasgow works, and with a blacksmith's fire, confirmed this idea. This was the principle of his hot blast system.

Naturally, Neilson wished to apply his theory to the blast furnace, but here he was frustrated for some time as a result of the intense conservatism of the ironmasters, who feared to meddle with the furnaces. Eventually, however, he was allowed to conduct further experiments at the Clyde Ironworks of Colin Dunlop, who, with Charles MacIntosh and others, owned shares in the patent (No. 5701) which Neilson took out in September 1828[14] for 'Application

---

Steel and Iron (London, 1959), pp. 32–3, gives another version of the Neilson innovation:

'Just then it so happened that a water chamber which regulated the air pressure for some furnaces on the edge of the moor at Wilsontown in Lanarkshire cracked and leaked.

Neilson was consulted about repairing the tank and about putting in a pump to lift cold water from coal workings in the vicinity. Really cold water was thought to be absolutely essential, to avoid taking the chill off the blast air. As it happened, Neilson was a little delayed in getting to the remote ironworks. In the meantime, to tide over pending his arrival, lukewarm water from the condensers of a steam engine which drove the air blowers was run into the leaking tank. This warmed up rather than cooled down the blast, and, according to the best blast furnace doctrine at the time, the result should have been less, and worse, iron. In fact, as the workmen observed, the slight increase in blast temperature made for better iron, and a little more of it. Pressed for an explanation Neilson kept his own counsel. The most he ventured was that he would look into it, and that something might come of it. That something was the hot blast which established his name in the books, and, after a good deal of legal wrangling, his ample fortune. It may have been that Neilson, as is now revealed perhaps for the first time, quietly pinched the idea. . . . On the other hand, it is to his lasting credit that, in the spirit of the times, he drew a fundamental working principle out of a more or less casual observation.'

[14] Dunlop, who owned three-tenths of the patent right, retired and eventually Neilson owned six-tenths of the patent, MacIntosh three-tenths and Wilson of Dundyvan one-tenth.

of air to produce heat in fires, forges, or furnaces where bellows or other blowing apparatus are required'. These experiments were made in a founder's cupola, when the temperature was only raised by 50°F, but in the next year, when the heated blast was applied to the furnace for the first time by means of wrought iron chambers (with a fire below) at each of the tuyères, the temperature was raised to about 200°F. The complete specification was patented on 28 February 1829, though it was not until 1832 that a really satisfactory form of blast stove was evolved and the difficulties about leakages and tuyères melting under the great heat were overcome. This last improvement came, not from Neilson, but from James Condie of the Blair Ironworks, who, in 1834, adopted an old idea, a spiral tuyère with a cast-iron water jacket to enable cooling by a continuous flow of water.[15]

The immediate consequences of the adoption of the invention at the Clyde and Calder Ironworks showed its great importance to the Scottish industry. The consumption of coal, hitherto very high, was reduced by about two-thirds. This was with the blast heated to a temperature of 600°F, which became the heat usually adopted; when the temperature was only 200°F, as at the Clyde works, the saving was only 2½ tons of coal per ton of iron.[16] Moreover, furnaces gave greater yields because they could be tapped more frequently than the previous once a day. Nor were Scottish claims in this respect unfounded, for they received confirmation on many points in the report of M. Dufrenoy, a representative of the French Department of Mines, whose country's iron industry was then so far behind that of Britain. Dufrenoy's evidence from the trials at the Calder furnace showed that the daily production with cold air and coke, in 1828, had been 5 tons 12 cwt.; in 1831, with heated air (200°F) and coke, the output was 6 tons 12 cwt.; and in 1833, using coal and a temperature of 612°F, the daily output had increased by 50 per cent from 6 tons to 9 tons.[17] Further, he reported that there had been a reduction in the consumption of flux, and a saving in labour and in steam power for the blast; also, the method produced iron of better quality —a point on which there was to be much subsequent discussion. This report was translated in all probability by Lady Charlotte

[15] William Jessop, of the Butterley Ironworks, at the trial *J. B. Neilson* v. *W. Baird* (1843), stated that the water-cooled tuyère was known long before 1800.

[16] *Trans. Royal Society of Edinburgh*, Vol. XIII (1836), p. 373. Paper by Dr. Thomas Clark. The process had also been tried out, in 1830, at the Low Moor Ironworks in Yorkshire.

[17] M. Dufrenoy, 'Rapport à M. le Directeur général des ponts et chaussées et des mines sur l'emploi de l'air chaud dans les mines à fer de l'Écosse et de l'Angleterre', *Annales des Mines*, Vol. IV, 3me Série (1833), pp. 446 and 451.

Guest, the wife of the prominent South Wales ironmaster, and its contents were widely publicised; possibly because of this, the process was widely adopted from the first.

As early as 1833, Dufrenoy listed 67 blast furnaces working on hot blast; though he regretted that the new process had not only to overcome the conservatism of ironmasters, but also a widespread prejudice that it was not practicable to use raw coal (viz. uncoked coal) in the furnace because of its sulphur content.[18] Dr. Thomas Clark, previously mentioned, forecast that the use of raw coal (which Neilson's hot blast had made possible for the first time) would be given up, because trials at Wilsontown had shown that the furnace became 'gobbed' and produced iron of inferior quality.[19] This proved to be wrong, however, and the practice of using coal soon spread. Mushet's discovery of blackband, the doubly rich ore, had been neglected for thirty years. When its exploitation was once begun then the resources of the Monkland coalfield in Lanarkshire were quickly consumed in order to expand Scotland's output of iron from a mere 37,700 tons in 1828 to 241,000 tons in 1840.

In Scotland, the chief among the exponents of the new process was the ironmaster, William Baird, who entered into an agreement to pay the patentees 1s. per ton royalty in 1832. However, he soon came to regard this as an exaction, and combined with the ironmasters in an attempt to invalidate the patent, meeting the cost by a levy on output; then he entered into a lawsuit to have the patent annulled.[20] This case revealed many significant details about the effects of the invention. In the course of the proceedings, it was shown that Baird's output had increased in five years from approximately 1000 tons a month to nearly sixteen times as much, and that, in the period 1833–40, the firm had made profits amounting to £269,655.

Soon after the announcement of the discovery, it was over-optimistically forecast: '. . . The new process by the hot blast will add immensely to the quantity of iron produced in Great Britain, and in a short time I am confident the price of iron must fall very much: indeed many are of opinion the price of Bar Iron will be at £3 before the end of two years.' However, the information about the spread of hot blast, from ironmaster witnesses from any part of the country, shows that its reception in England was not wholly favourable. Hot blast had been adopted in Derbyshire between 1832 and 1835, but in Shropshire it was viewed with disfavour; in Staffordshire, too, there was a strong prejudice against it, though the first works to adopt it, Lloyd. Foster and Company of Wednesbury, had done so as

[18] Dufrenoy, *loc. cit.*, p. 435.
[19] Clark, *loc. cit.*, p. 373.
[20] *Neilson* v. *Baird*, 1843.

early as 1834. One ironmaster in this district had adopted the hot blast in 1835 and then reverted to cold blast in 1839.[21]

Staffordshire, indeed, eventually specialised in cold blast iron. South Wales, too, at this time, was another district where the hot blast was not taken up at first. The different qualities of coal available in each district, of course, determined the local attractiveness of Neilson's invention. In South Wales, so low was the consumption of fuel that the very small saving of coal through the use of hot blast (11 cwt. to a ton of pig iron) was almost all absorbed by payment for the royalty. However, Harfords of Ebbw Vale had attempted, in 1841, to overthrow the patent rights, claiming that Thomas Botfield, the Shropshire ironmaster, had preceded Neilson with the same idea. In fact, the Shropshire patent was not the same; moreover, Botfield's process was never exploited on a large scale. In passing, the tradition may be noted that John Wilkinson had invented the hot blast; but the evidence brought forward in the Baird case, when the defence first tried to disprove Neilson's originality, is inconclusive on that point.[22]

Despite the imminent and catastrophic reduction in the price of bar iron anticipated by the Scottish iron merchants, which if it had been realised would probably have put the Midlands forge-masters out of business, the Staffordshire ironmasters were, on the whole, tenacious in their belief in the quality of cold blast iron. Their conservatism was to some extent confirmed by the superior price which the established product could command; although there was a complaint that 'hot blast has seriously injured us in foreign markets'.[23] Some ironmasters used a mixture of hot and cold blast, since it was thought that this compromise would give the quantity and quality of both systems. In 1843, 42 furnaces operated on hot air and 38 on cold blast, thus it is evident that the process was strongly established by that time. Indeed, an estimate for the whole country computed that two years earlier about half a million tons of hot blast iron were produced, as against 800,000 tons of cold blast iron.[24] It says much for the productivity of Neilson's system that only 162 furnaces, out of a total of approximately 400, operated on hot blast. After this date, incidentally, the patent rights expired and

[21] All these facts come from the report of the trial, which involved 163 witnesses. Their evidence has been described as 'a compendium of the principles of ironmaking, as practised throughout Scotland and England between the years 1792 and 1843'!

[22] Dr. Raistrick in his book upon Coalbrookdale suggests that Abraham Darby I had been experimenting with a pre-heated blast (*op. cit.*, p. 105).

[23] *Midlands Mining Commission*, 1843, p. 66, evidence of James Boydell.

[24] *Neilson* v. *Baird*, 1843, evidence of David Mushet.

naturally ironmasters were then less unwilling at least to try out the Neilson system; thus hot blast was taken up in districts, like Wales, where once it had been considered not worth while.

On the technical merits of the process, a sharp division of opinion lasted for a long time. In 1842, a Yorkshire ironmaster, Henry Hartop, was able to deny its advantages.[25] Yet as late as 1855 Truran, of Dowlais, was demonstrating that 'the saving in fuel, and the increase in make due to it, is not in general one fourth of that which has been asserted to have been effected'.[26] It was true, as he said, that economies arising from improvements in the preparation of the fuel and ironstone, as well as in the furnaces and blowing engines, had to be taken into account. In the Cleveland district, when the waste gases of the furnace were utilised in the blast stoves of Edward Cowper and Thomas Whitwell, temperatures double that finally advocated by Neilson were being used in the 1860s. Here, when a temperature of 1300°F produced 1 ton of pig iron from 22 cwt. of coke, it was claimed that there was a saving of 1 cwt. for every 100°F increase in the heat of the blast.

The field of fuel economy in the iron and steel industry, perhaps the most important aspect of practical metallurgy today, has of course developed greatly since the foregoing inventions. As W. A. Bone remarked in his book *Coal: Its Constitution and Uses:* 'As a result of a century's effort, a blast furnace has now produced as much iron in a day as one did in a week in 1829, and at about one-fifth the cost in fuel.'[27] This was a German furnace, which achieved the remarkable output of 7520 tons in one week in 1928–31. This was perhaps the zenith of that expansion of output which had its origin in Neilson's invention.

In parenthesis it might be added that, as with other British innovations, their greatest value revealed itself in foreign countries. Although hot blast was of some consequence in making the anthracite coals of South Wales usable for iron smelting for a short period, it was in the United States that hot blast enabled the iron industry to be built up so quickly on the basis of the same kind of coalfields.[28]

The further exploitation of the Neilson hot blast principle depended upon improvements in the blast stoves by Cowper and

[25] H. Hartop, 'A paper on the relative properties of iron made by the use of cold and hot air blast in the smelting furnace' (*W. Riding Geological and Polytechnic Society*, July 1842).

[26] W. Truran, *The Iron Manufacture of Great Britain* (London, 1862), p. 142.

[27] *Op. cit.* (London, 1930), p. 471.

[28] The process was taken over by a Welshman. For details of the hot blast anthracite iron industry in South Wales see pp. 169-70.

Whitwell. The most important invention in this field was the Siemens regenerative furnace. Since its significance, however, is most properly associated with the development of the steel industry and the openhearth process of the Martin brothers, discussion of this invention is reserved for a later chapter.

Nevertheless, in order to obtain the greatest efficiency from his fuel the ironmaster had to have a proper regard for the great loss of heat at the top of the furnace. In this respect Britain in the first half of the century was behind the most advanced practice of the time. This extravagance was almost certainly brought about by the abundance of cheap coal, an asset which the Continental countries did not possess, so that the furnaces were open-topped and so much of the flame and heat only went to illuminate the otherwise murky skies.

Although the idea of using the waste heat had been discussed in this country and one engineer-ironmaster had tried to place boilers on top of his puddling furnaces, it was in France where the greatest advances had been made. For instance, a patent taken out in this country by a Forest of Dean ironmaster, Teague, for a reverberatory furnace on the top of the blast furnace had been anticipated in France as early as 1810.[29]

The first step of practical significance towards the economical use of the waste gases seems to be that taken by J. P. Budd of Ystalyfera, near Swansea. The account of his experiments is contained in the *Report of the British Association Meeting* for 1848, but three years earlier he had taken out a patent (No. 10475) for 'the application of the heat, flame and gases of the waste heat to the raising the temperature of the blast in stoves'. According to Budd, there had been some earlier attempts to apply some part of the waste heat to the raising the temperature required for the hot blast, but these had failed either because the arrangements for collecting the gases at times necessitated the blowing out of the furnaces; or with another scheme, the saving had been cancelled out by the need for a blast engine to provide the necessary temperature for the gas furnace. It was the difficulties and expenses of smelting with anthracite which led Budd to adopt a means for drawing off some of the exhaust gases through a blast heating stove. However, he did not attempt any combustion of the gases, as their heat on leaving the furnace was more than sufficient to heat the blast to 600°F, but he did use some of the heat to raise steam in the boilers. His claims about the savings which resulted, however, are surely exaggerated; for he made the assertion 'it would

[29] H. W. Dickinson, 'Utilization of waste heat from industrial operations', *Trans. Newcomen Society*, Vol XXIV, pp. 1–11.

appear to be more profitable to employ a blast furnace if as a gas generator only, even if you melted nothing in it, and carried off its heated vapours by lines to your boilers and stoves'. Budd, in spite of his limited application of the waste heat, nevertheless recognised the full potentiality of these economies, and envisaged a time when the many stages of manufacture, from the ore to the finished product, malleable iron, would be carried out by the full use of the calorific content of the coal which was part of the charge in the blast furnace.

As with Neilson's patent, the innovation was attacked by the conservative ironmasters, who thought like Truran that, if the charging of the furnace was done correctly, 'the advantages of superior economy would be attained without having recourse to the large outlay of capital incidental to all plans for economising superfluous products'.[30] There was doubtless some justification for this remark. Yet, despite examples of unsuccessful experiments at Dowlais, Ebbw Vale and other works in South Wales, events again proved Truran wrong in his opinion that it was 'not profitable nor expedient to control the escape of the gases'.[31]

It was in the Cleveland district again, as with superheated blast, that the most progress was made with the closed top of which the bell and hopper was the most favoured kind; although some ironmasters preferred to keep the tops open and take off the gases through holes in the side of the furnace, from fear of explosions and to improve the quality of iron. In South Staffordshire, where the process had not reached the same stage of perfection, it was thought that the escape of some gas was inevitable, since, with furnaces in regular work, there was an overabundance.[32] Here, the utilisation of waste heat had not been extended to the puddling furnace. But an improved method of two local ironmasters, Addenbrooke and Millward, had been adopted in 1864.[33] However, Dr. Percy put his finger upon a vital factor when, in reply to a question about fuel economy in South Staffordshire, he explained: 'When fuel is becoming exhausted and the iron ore too, it is very difficult to induce men to lay out money in accordance with new ideas.' Thus it is true, as Burn makes it evident, that in general the ironmasters were not as concerned about this problem as their competitors on the Continent.[34] But, where there

[30] W. Truran, *op. cit.*, p. v. [31] *Ibid.*, p. 155.

[32] *Proc. Inst. Mechanical Engineers*, 1863, p. 235.

[33] F. W. Hackwood, *Wednesbury Workshops* (Wednesbury, 1889), p. 119.

[34] D. L. Burn, *op.cit.* p. 7, citing evidence given to *S.C. on Scientific Instruction* (1867–8). Further evidence which is to be found in the 1871 *Royal Commission on Coal* (B.P.P. 1871, XVIII) is summarised by Burn (pp. 5–10), who treats this problem very fully.

were conditions of low yield ore (as in the Cleveland district) or of expensive coke (as at Barrow-in-Furness) and, especially, new investment and modern management, the blast furnaces were made as efficient as scientific knowledge knew how to make them at that time.

THE FORGE AND ROLLING MILL

The blast furnaces were, and are still, indispensable in iron and steel manufacture. However, in terms of value, the production of the forge, rolling mill and the melting shop is much higher than the value of the primary product. Pig iron may have slumped in price to 30 shillings, but the value of wrought iron was always more than 100 per cent higher than that; with steel, of course, the increase in value, as a result of successive processes of conversion, was even greater.

Thus the wrought iron industry probably accounted for the greater part of the value of the total product of the iron industry in the first half of the century, especially when we take into our reckoning the millions of pounds' worth of rails which were produced in that period. Unfortunately we have no statistics, apart from export figures, of these values of production.[34a]

Nor do we have any spectacular invention to which we can point as revolutionising the wrought iron industry. Indeed the technique of puddling remained much the same at the end of the period as it did in Crawshay's day. The maximum capacity of the puddling furnace was something less than 5 cwt., and 5 charges were possible in a shift of 8 hours, thus with a furnace working 24 hours a day, 7 days a week, the weekly output would only amount to roughly 20 tons. Undoubtedly, the increase in outputs was accounted for by a multiplication in the numbers of puddling furnaces. It must be remembered that puddling was essentially a manual process and an exceeding arduous one at that.

The principal objection, however, to the technique of puddling as it stood was the loss of iron in the course of the process. As much as 25 per cent was wasted in the working of the iron in the puddling furnaces, and although the situation was improved to some extent by the substitution of iron bottoms in the furnaces in place of sand by Samuel Rogers of Nantyglo[35] in 1816–18, the problem continued

[34a] B. R. Mitchell ('The coming of the railway and U.K. economic growth', *Journal of Econ. Hist.*, Vol. XXIV, no. 3, Sept. 1964, p. 325) gives estimates of the proportion of *pig* iron to total output consumed in railway construction.

[35] S. Rogers, *An Elementary Treatise on Metallurgy* (1858).

to occupy the attention of improvers for many years. Rogers claimed also to have increased the output per furnace from 8 to 20 tons per week and sometimes to 24. However, Griffiths in his *Guide to the Iron Trade*[36] indicates that the method was only coming into general use by 1825. In 1822 an improvement on Rogers' method was patented by R. S. Harford (No. 4624) 'to protect the cast iron bottoms or floors of puddling furnaces' by means of charcoal.

One important means of cost reduction in the refining and puddling processes was the abandonment of the old method of allowing the pig iron to cool and then remelting it. The process of the molten charge —a practice which is a feature of fuel economy today in modern furnace management—was introduced. At first, it was thought sufficient to put the metal into the puddling furnace in a heated state,[37] but ten years later, in 1832, the Midlands ironmaster, James Foster of Stourbridge, took up the idea and was bold enough to try feeding the puddling furnace direct from the blast furnace. This improvement was remarked upon by the Midlands Mining Commissioner in 1843: 'Mr. Barker (the managing partner of the Chillington works) showed me a great improvement in puddling iron immediately from the smelting furnace, for which a patent had been taken out some years ago, but the opposition of the men to its introduction was such that it had been suffered to lie dormant until the severe distress in the iron trade had driven both the employers and workmen to the necessity, the one of adopting, the other of substituting the cheapest means of producing the article manufactured.'[38]

The following year a patent embodying the same idea to be applied to the refinery furnace was taken out by J. J. Guest (No. 6379). However, there were many difficulties and disadvantages attached to these methods. As Truran pointed out, in existing works the necessity of reconstructing the forge and bringing it close to the blast furnace was a great objection to its use. Moreover, the great heat as a result of the proximity of the forge to the furnace was a great disadvantage. Then, too, there were technical difficulties with the separation of the cinder, when the puddling furnace was fed from the refinery.[39] Nevertheless, the method was still being used in Staffordshire in 1846, fourteen years later; although then other difficulties had appeared. It was reported to Guest's manager, 'They do not think so highly as they did of the puddling furnaces that take the

[36] *Op. cit.* (1873), p. 168.

[37] Patent No. 4713 of W. Jones, 1822.

[38] Abridgment of Specifications, Iron and Steel Patent No. 6300. *Midlands Mining Commission*, 1843, p. 633.

[39] W. Truran, *The Iron Manufacture of Gt. Britain* (London, 1862), pp. 210–12.

iron direct from the blast furnaces as the men of late get drunk and neglect their work very much and by so doing they waste a larger quantity of iron than they do at the common puddling furnaces.'[40] The puddlers, who were considered to be the aristocrats of the iron-workers, were also the most troublesome to the employers; strictly independent and resenting interference with their art, and also given to stealing, if paid a bonus by output. They were not the least of the ironmasters' incentives to get more economical methods of working the iron. It is not surprising that a great number of the improvements patented in the period relate to puddling and refining.

The most important of these was surely the method of 'wet puddling' or pig boiling perfected between the years 1825 and 1832. This was to give a great boost to the Staffordshire iron district which concentrated on the production of finished iron to meet the requirements of the highly developed metal-using trades in the district.

As with so many improvements in the industry, the idea had been mooted before and even tried with success. It is a futile occupation to discuss the degree of originality of inventions, the crucial test is when the process becomes commercially realisable. However, David Mushet had fettled puddling furnaces using oxides of iron and such haematite iron ores to improve the yield and quality since 1815, and Anthony Hill, who used this method with success, had himself, in conjunction with his patent for using finery cinders in the blast furnace,[41] taken out a patent for adding lime in the puddling furnace. The inventor of the 'wet puddling' process, Joseph Hall recognised the value of lining the furnace bottom with cinder instead of sand, a practice, he said, which had come about through the adoption of the iron bottom.[42] Hall had been experimenting at Bloomfield works, Tipton, since 1826, although his patent for 'tap cinder' obtained from puddling crude iron and used as protection for the bottoms and sides of the furnaces was not taken out until 1839. His object was to use a furnace lining which would be a reagent in the puddling process, in order to remove the necessity for a preliminary refining with its attendant waste. He achieved this by using the prepare 'tap-cinder' or 'bull-dog', as the calcined cinder was called, as a fettling or lining for his furnaces. This served the double purpose of protecting the iron bottoms and sides of the furnace, and of more consequence, the process of reduction by oxidisation was materially speeded up. 'With

[40] *Dowlais MSS*, Wm. Evans—E. I. Hutchins, 26 October 1846.

[41] This method, which became general, became a very important means of increasing the output. However, it led to a decline in quality.

[42] J. Hall, 'An Old Shoe Heel' (1856), reprinted in *Edgar Allen News*, June–September 1948, p. 75.

this method the iron became melted at one stage, and this gave rise to the term "wet-puddling" as distinct from Cort's process where the iron only became pasty. The other term "pig-boiling" arose also from the fact that at a certain stage in the process, when the oxygen in the furnace bottom or fettling combines with the carbon in the iron, the molten metal virtually boils with bubbles of CO gas.'[43] According to Gale, only a few firms in the Black Country continued to use the refinery. Hall's own company confined itself to the finishing process,having no blast furnaces of its own. In the 1860s the firm at its three works had over one hundred puddling furnaces in operation.[44]

As late as 1870, however, it was observed by Bessemer: 'Among the many important problems still awaiting for solution in the malleable iron manufacture, is the process of mechanical puddling.'[45] The urgency of the need to improve a technique which had remained unchanged for over fifty years had been heightened by recent increased demands for wrought iron for armour plating, which advances in contemporary artillery required. There were many attempts to produce machinery which would assist the manual stirring of the iron—a hollow puddling bar with an air pump attached, invented by an ironmaster named Richardson, was said to speed up the process by 18 minutes a charge;[46] but the main difficulty with any mechanical contrivance was to secure that degree of adjustment in the working which only the skilled puddler could achieve. Again, the introduction of steam and air into the puddling furnace was tried by several inventors, including Nasmyth, but here, success was not to be achieved until Bessemer took up the idea and used it in the converter to make steel. These were attempts to produce a rotating puddling furnace with a scraper to stir the metal inside it. In 1864 several of these were being tried, as a result of a recent invention, by H. Bennett of the Wombridge Iron Works,[47] in Shropshire and the North of England, with a little success. Nevertheless, in 1869 a committee of the leading members of the iron industry was set up to investigate the various methods of mechanical puddling including the American invention, Danks' rotary furnace. By the time that any mechanical method would have been realisable, however, the compe-

[43] W. K. V. Gale, 'Notes on the Black Country Iron Trade', *Trans. Newcomen Society*, Vol. XXIV.

[44] Gale, *loc. cit.*, p. 18.

[45] Presidential address of Iron and Steel Institute, *Journal of the Iron and Steel Instititute*, 1869–70, p. 366.

[46] D. Bremner, *The Industries of Scotland* (Edinburgh), 1869, p. 52.

[47] H. Bennett, 'On puddling by machinery', *Proc. Inst. Mechanical Engineers*, 1864, pp. 298–305.

tition of steel was stifling incentive to develop this branch of the industry further. Thus mechanical puddling was never perfected and adopted on a large scale.[48]

Of the mechanisation of the industry in general, it is impossible here to treat the subject in any detail. There are no statistics of the use of steam power applied to furnaces and forges, comparable to those for the cotton industry.[49] Many instances of the amount of steam power employed at individual works could be cited,[50] and in 1855 Truran estimated that there were 2120 steam engines employed in the industry with a total power of 242,000 h.p., but it is difficult to apply these figures to give a general picture of the industry. Certainly steam power was essential for producing the blast, and particularly in the finishing side of the industry, where only steam could give sufficient speed for rolling, to keep up with the increased pace of production.

Precision, too, was attained by machinery in rolling; the guide rolls were introduced in the late 1820s. This invention made what was a costly skilled process, to secure the accuracy which was becoming more and more important, the result of one machine process. Yet in many ways, apparently, the iron industry was slow to apply that necessary refinement of precision to its own machinery. The practice had long been for the works to make their own machine castings and lay down mills without calling in mechanical engineers. It was not surprising that the machinery tended to be massive and clumsy and heavily geared, which could not help but increase the costs of production. It was not until 1869 that the *Mining Journal* could remark: 'At no previous time in the history of that craft was there displayed so much attention as now to the adaptation of improved machinery.'[51] Improved means of feeding the furnace by hydraulic and steam hoists

[48] J. L. Carr and W. Taplin, *History of the British Steel Industry* (Oxford, 1962), p. 57, note that mechanical puddling eliminated phosphorus to a much larger extent than hand puddling did; therefore it was hoped that it might enable some of the commoner qualities of pig iron to be used for steel making. After 1880 interest in mechanical puddling died. J. S. Jeans's epitaph sums up the history of the puddling furnace: 'It will probably not be denied that the skill, experience and inventiveness of practical men have seldom been so barren of useful results. The puddling furnace of a generation ago is in all its essential features the puddling furnace of today. Both have been condemned as crude, barbarous and wasteful, and yet both continue to enjoy a measure of vitality and appreciation to which their merits have certainly not entitled them!' (*Jnl. of Iron and Steel Inst.* 1882, quoted in *op. cit.*, p. 57.)

[49] Except for the *Factory and Workshops Returns* of 1870.

[50] E.g. in 1828 Crawshays, according to Dufrenoy and Beaumont, employed 350 h.p. to provide blast for 12 furnaces.

[51] *Mining Journal*, 16 January 1869, p. 46.

and many other devices were then being introduced. There were exceptions of course, to this rule, as for instance in the development of the hammer—the essential tool in the charcoal forge long before Cort's rolls—and much later, squeezers which came to dispense with its services. As the heavy engineering projects of the Industrial Revolution increased the demand for large and massive forgings for bridges, crankshafts for steam paddle-boats, etc., so the hammer came into its own. The most common form of hammer, even after Nasmyth and others had brought out direct acting steam engines, was the belly helve. 'For plain ordinary work it is not surpassed in efficiency, even by the direct-acting steam hammer', it was said.[52] It was used for large forgings, 3 to 4 feet in diameter. In 1839 Nasmyth invented his steam hammer and in 1842 took out the patent for it in this country. At first he was unsuccessful in getting ironmasters to adopt his hammer—the Low Moor Company turned it down in 1843—but a few years later this company had adopted and used it for the large forgings of the steamship *The Great Eastern*, and in 1845 it had been introduced into the South Wales industry.[53] It was introduced into the Midlands in 1850, but there a prejudice against the steam hammer prevented its more rapid application. It was alleged 'with a steam hammer the impurities contained in the iron could be wrapped up in the ball by gentle blows, instead of all being driven out; but by the heavy blows of the old helve all noxious matter was at once driven out and improperly puddled iron detected'.[54]

Yet, by 1867, the use of the steam hammer was a part of general up-to-date practice. In fact, in steel making it was being superseded. To quote Burn once more: 'Some of the most fundamental innovations of these years in the mechanical processes immediately following the casting of the ingots had their origin in England'[55] and this was chiefly in the rolling of steel rails. Modern methods here were, first, the substitution of roller 'cogging' instead of consolidating the ingot by hammering;[56] and secondly, the introduction of reversing mills. (Apparently, Wilkinsons's early reversing rolling mill had never been taken up by other ironmasters.) The pioneer of this innovation was Ramsbottom of the railway steelworks at Crewe in 1866, and it was quickly taken up at Dowlais. As a result of this invention

[52] W. Clay, *Wrought Iron in Large Masses* (London, 1867), p. 314.

[53] E. J. Jones, *Contributions towards the Economic History of Wales* (Cardiff, 1928), p. 56.

[54] F. Smith, 'The Helve *v.* the steam-hammer', *Proc. Inst. Mechanical Engineers*, 1866.

[55] Burn, *op. cit.*, p. 56.

[56] By 1867, Crewe and Dowlais had adopted this method. (*Ibid.*)

'the mill capacity was more than doubled'.[57] although, because of the load, more powerful steam engines were required.

Thus, by the 1870s, the technical resources of the British iron and steel industry were such that a large-scale transformation was imminent. Bessemer steel had by now passed out of the experimental stage; and the process, in the hands of competitors who were not handicapped by a conservative concern for quality or by the amount of capital vested in wrought iron manufacture, was being developed for large-scale production. Technical initiative in the steel industry had passed to the United States. British firms, realising the challenge, were forced to reorganise upon modern lines, if capital reserves would allow it.

At the same time, however, the British iron industry had been built up over a long period not for the production of steel, but for wrought iron. This manufacture, unfortunately, had put a premium on manual skill and the introduction of machinery seemed to be of little value.[58] In these circumstances the iron and steel branches of the industry were parting company wrought iron manufacture could only make a stand by confining itself to a limited market for a product of high quality; steel, on the other hand, was committed to expansion. Just as, earlier, steam power and the empirical enterprise of Britain's ironmasters had established the modern large-scale iron industry, so now, mechanisation and organised metallurgical science were to create the age of steel.

[57] *Ibid.*, p. 57; W. K. V. Gale, 'Rolling of iron', *Edgar Allen News*, October–November 1965.

[58] Except to free the forgemasters from the troublesome demands of the puddlers; but, of course, the mechanical furnaces still required operators to work them. Because of the labour factor, experimentation, improvements and capital expenditure were directed to giving greater production, higher efficiency and greater regularity from the blast furnace.

CHAPTER X

The Growth of Large-scale Industry: Capital and Enterprise in the Iron and Steel Industry in the Nineteenth Century

The preceding accounts of the establishment and early history of the ironworks, given in Part I, Chapter V, have already indicated the problems facing the pioneering ironmasters in providing capital for their blast furnaces and forges. Before discussing this provision of capital, however, we should comment upon another important factor which often enabled ironmasters with very little capital to make a venture of a concern. This was the mineral lease, which, as John has pointed out, was 'the basis of industrial activity . . . the concession upon which nearly all industrial concerns of any size were founded'.[1] It is impossible to decide now how far the expansion of the industry was brought about by the initiative of landlords in offering land for exploitation and to what extent the landowners were persuaded by prospecting individuals, or other groups with capital to invest, to lease their land for industrial purposes. In some cases, as at Haigh, for instance, the landowner was forced by economic circumstances to develop his estates by actively taking part in an industrial enterprise.

The lease was the economic price for the key to a supply of minerals, the actual extent of which was not always very well known. Not until the second half of the nineteenth century were geologists employed to estimate the reserves of coal and ore: hence so long as as the cost of the mineral rights was cheaply bought by the ironmaster, who took a gamble upon the abundance of cheaply worked mineral, the limits to the expansion of the industry at first were largely physical ones.[2]

[1] John, *op. cit.*, p. 164.

[2] Restriction of output to maintain prices was not really effective, as we have seen.

One important factor, indeed, in the great expansion of some of the vast concerns in South Wales from 1760 onwards, was the very low rents which the ironmasters had to pay. The case of Dowlais, which was able to sub-let an area of its land to the neighbouring Penydarren works at a great profit, and which was threatened with ransom terms by the Marquis of Bute in 1848 on the occasion of the renewal of the lease, is well known. Nevertheless, in Scotland, the Bairds had to pay a fairly high standing rent, therefore they ransacked the coalfield as quickly as possible in order to exploit the lease to the full. As time passed, of course, landlords became aware of the potential value of their minerals. For instance, the Plymouth works was leased in 1799 at a basic rent of £2477 for an output of 2000 tons of iron a year and an additional 5s. for every ton produced in excesss of that amount.[3] This, undoubtedly, was the best sort of lease for a landlord to make. Conversely, as we have seen, the shrewd ironmaster realised the burden of paying mineral rents, and Crawshay, for example, bought out Anthony Bacon's interests in the land and mines upon which Cyfarthfa was situated.

Finally, in respect of the influence of the landowners in facilitating or promoting the establishment of the iron industry the following comment of a contemporary upon the growth of the Midlands iron industry is illuminating. South Staffordshire had many small iron and coal masters and their origin is explained by Gilpin: 'In Staffs. landed property is very much divided and naturally all the proprietors [are] desirous of turning their coal and iron mines to immediate account. Hence,there is a colliery in almost every field. As there is no sale for such an immense quantity of coal and limestone, several of these little proprietors unite together and build furnaces; clerks from the neighbouring manufacturers are taken in as partners to direct the concerns; the tradesmen of the town in the vicinity who can raise a 100 or 200 pounds form part of the firm, and it is in this way that the ironworks have been multiplied in that county. The proprietors embark all their property and all that they can borrow in these establishments.'[4] Gilpin, in this quotation, made it clear how capital was scraped together to establish an ironworks at this time.

What was the size of investments in ironworks at this time? How much capital was it necessary to raise? The most abundant evidence is for South Wales. Apart from Cyfarthfa and Dowlais the concerns

[3] J. Lloyd, *Old South Wales Ironworks*, p. 75. The landlords, if they had made a bad bargain, presumably did not need any lessons in business matters from the ironmasters, but if they did, they could learn from their tenants' example. At Aberdare, in 1799, the rent was £40 a year; in the following year, the lease was re-conveyed, this time at £1000. (*Ibid.*, pp. 115–16)

[4] *Dowlais MSS*, G. Gilpin—J. J. Guest, 3 October 1819.

cannot be described as large. The amount of capital involved appears to have been very modest. Clydach, which had a capacity of 2000 tons of iron a year, together with another plant, the Beaufort works, was run by a £16,000 partnership in 1795.[5] Sirhowey in 1797 had a capital of £12,000 and that of the Plymouth works (one of the four Merthyr ironworks) in 1803 was £20,000.[6] The evidence is not so precise for Staffordshire, but Attwood's statement, in 1812, that 'at least 10 establishments . . . have cost at least £50,000 . . . '[7] implies that the average was well below this. Carron and Coalbrookdale companies were, on the other hand, giants in the iron industry, both in fame and in size, but there were others too. For instance, Walkers of Rotherham, the large steel and armaments partnership, in 1796 had a capital of £214,393.[8]

Nevertheless, these few figures of capital accounts of the ironworks of the Industrial Revolution do not take us very far towards an understanding of the investment in industry at that period. What is more interesting is the problem of where the capital came from to launch and finance these concerns, for this throws a great deal of light on how the money required for the large-scale investment necessary for this revolution was 'injected' into the industrial system.

In the first place, even with the relatively small capital needed for the establishment or lease of a blast furnace and its equipment, the burden of investment was too great for individual ownership to be common. In fact, it is difficult to call to mind one example of a blast furnace being owned by one owner. Even John Wilkinson could not collect enough personal capital for the running of his many concerns. Despite the difficulties of partnership this was by far the most common form of enterprise in the iron industry, although the Carron Company was unique in the industry in obtaining a charter to authorise its joint-stock character. Even if concerns established by individual enterprise in the early years of the Industrial Revolution were retained by family interests, inheritance and trusteeship eventually subdivided the responsibility for and liability of investment.

Of the capital for initial investment and for expansion four sources were available. First the owner of the mineral resources, that is the landowner, might launch into iron production himself. Secondly, further capital might 'be spun out of the bowels' of the concern itself,

[5] Lloyd, *op. cit.*, p. 180.

[6] Lloyd, *op. cit.*, pp. 80 and 148. John gives details for other furnaces: Penydarren in 1786, £14,000; Abernant in 1802, £40,000; Tredegar in 1800, £30,000; Aberdare in 1801, £15,000; and the Forest of Dean works of Redbrook and Lydbrook, £20,000.

[7] *Minutes of Evidence against Orders in Council*, 1812.

[8] Ashton, *op. cit.*, pp. 43–48.

by retrenchment and the retention of profits. Thirdly, there was the flow into the industry of merchant capital; and lastly, loan capital might be procured from people with surplus wealth or, more usually, bankers who made available otherwise idle capital. These categories, of course, are not exclusive of one another and all were resorted to in turn or in combination. These methods of financing industrial development persisted until the coming of the registered limited liability joint stock company of the second half of the nineteenth century.

There is no need to add further comments to those made earlier in connection with the Haigh ironworks of Lord Balcarres concerning aristocratic ironmasters, although Lord Dudley in South Staffordshire and Earl Granville (a descendant of Earl Gower) in North Staffordshire and Shropshire were prominent ironmasters in the middle of the century.

The ploughing back of profits, it seems (for we have no overwhelming body of evidence upon this point), was the most common way of financing the further growth of an industrial concern. We have noticed this policy being carried out at Dowlais and Cyfarthfa, and Professor Ashton has investigated the course of capital-building by this method at the Thorncliffe ironworks of Newton Chambers in South Yorkshire for the period 1795–1831.[9] He has shown how the partners built up their capital from small beginnings by the limitation of payments to them to 5 per cent interest on capital. Working partners, of course, received modest salaries; Thomas Chambers and George Newton paid themselves £80 a year for the first decade of the firm's existence. Profits were only enjoyed indirectly by way of interest payments on the increased capital.

The merits of such a policy were made obvious by the ability of the firm to weather the losses arising from the depression of trade, and later, in 1866 and 1869, the paralysing colliery strikes in the district. For instance, in 1817, the partnership made a loss of £3856 on a capital of £51,090, when desperate efforts to secure a reduction in the local rates and in the ironstone royalties had to be resorted to. Of course, such a policy of abstinence was only possible if the partners were able and willing to let the profits remain in the business. The decision could, on occasion, give rise to plain speaking between the members of the partnership. Robert Scott was reminded by Newton, in March 1813: 'We think we have good grounds for expecting that for the privilege of being a sleeping partner (which in these times

[9] Ashton, *op. cit.*, pp. 156–61. See also R. L. Downes, 'The Stour Partnership 1726–36', *Econ. Hist. Rev.*, 2nd Series, Vol. III, No. 1 (1950), for an earlier example of this accumulation of profits.

is certainly a great privilege!) you ought to let your capital be made up to the strict letter of the Articles of Partnership, especially as we do not require you to make any real advance, but permit it to be made up with interest which is spun out of the bowels of the concern itself.' Difficulties arose when partners wished to withdraw capital, as did Scott in 1828. Scott threatened to dissolve the partnership because he thought he should have a larger return on his capital. Eventually, the dispute was settled by an independent arbitrator who wrote urging Newton Chambers to agree to this settlement with all urgency: '... If this business go to court, it will not come out in all probability without the ruin of the whole concern.'[10]

Later, the competing attractions of other investment, and the demands of capital which had become free of trust, worried the managing partners. Even where capital could not be withdrawn, it was used as a negotiable security against loans, and thus passed into the hands of outsiders, who desired dividends commensurate with the prosperity of the firm. One such person, Samuel Plimsoll, M.P., who had secured an interest in this way, made a great issue of this to reverse the policy of ploughing back profits. In the course of his campaign conducted in 1878[11] he revealed the astonishing resilience and earning capacity of the company as a result of this long-established policy. He showed that, despite losses of 130 per cent during 1869–71, during the next four years an average profit of 44 per cent a year was maintained, which was retained by the company. The capital increased by 169 per cent from returned profits during the period 1862–75.

Finally, this situation, complicated as it was by difficulties of this kind and by problems of trust moneys, was cleared up by the formation of a limited liability company in 1881, though this solution had been delayed for ten years by legal problems.

Moreover, such a prudent policy was not without its dangers. This especially during an age when a body of efficient and trustworthy managers, with a strict code of behaviour, such as now animates the highly-paid executive of the present-day impersonal large-scale business concern, had not been developed. The obvious danger of malversation of profits was shown to a startling degree by the conspiracy of the Stainton family in the Carron Company.[12]

[10] Quoted in the Report on the Newton Chambers archives in *Bulletin* of the National Register of Archives, 1963, p. 16.

[11] S. Plimsoll, *Newton Chambers and Company's Affairs Considered*, 1878, quoted in 'The Story of Thorncliffe' (a history of this firm, as yet in typescript). I am very much indebted to Mr. Elliott, the Public Relations Officer of the company, for the use of this valuable document.

[12] *The Carron Company. Extract from some professional recollections by a*

The joint stock company, which is the modern method of tapping industrial and financial capital and applying it to industrial development, rapidly assumed predominance in the second half of the nineteenth century. Nevertheless, in the first flush of the railway boom of 1825, there was a mushroom growth of mining companies; Hunt enumerates 74 mining companies with a total capital of £370,000 projected in 1825–6.[13] But the only existing companies in the iron industry were the British Iron Company with £650,000 paid up on a nominal capital of £2m. (later to be reduced to £1m.) and the Welsh Iron and Coal Company with a fully paid up capital of £250,000; a smaller company, the Cheshire Iron and Coal Company, projected with a capital of £50,000; and another, to work in Sweden, which was abandoned.[14] In common with the rest of the promotion of this time, most of this in the iron industry was mainly speculative; for instance, in 1827, English notes that the value of the British Iron Company's stock had fallen to £120,000 and that of the Welsh company to £80,000.[15]

In the 1830s several companies were promoted in South Wales. These were apparently attracted by the expectations of the huge profits awaiting the exploiters of the rich mineral resources of South Wales. According to Joseph Parkes, before the 1843 Select Committee on Joint Stock Companies,[16] upwards of four millions sterling had been invested in Monmouthshire and Glamorgan in this way. But if, as *The Times* announced, 'each member of every company so formed (expected) to be at least a Bailey, a Guest or a

---

former member of the Council of Incorporated Law (1883). See also Campbell, *op. cit.*, pp. 161–94. As Campbell says: 'The criticism which may be levelled against Henry Stainton and Joseph Dawson is not for their accumulative policy, but, for the way in which it was carried out and for the use of the funds acquired. Instead of being employed to finance further capital development at Carron, they were simply held as liquid reserves, much greater than ever could have been justified. . . . The most damaging criticism is that these activities may have prevented Carron Company from expanding even more than it did in the nineteenth century. From 1850 so much effort was occupied in uncovering the mysteries of these machinations, that for fifteen years little else could be achieved.'

[13] B. C. Hunt, *The Development of the Business Corporation in England 1800–67* (Cambridge, Mass., 1936), p. 45.

[14] H. English, *Complete View of Joint Stock Companies in 1824* and *1825* (London, 1827).

[15] There was to have been a parliamentary enquiry into these bubble concerns, but the only one to be investigated was the Arigna Mining Company, an Irish company (capital £300,000), which attracted some English capital, including that of James Brogden, the Manchester financier who had interests in South Wales and who, later, controlled the Tondu ironworks.

[16] *Loc. cit.* (B.P.P. 1844 (119)), evidence of Joseph Parkes, Q. 2370.

Crawshay', then disappointments were many. Some facts from the histories of the Monmouthshire Iron and Coal Company and the Victoria Iron Works throw much light upon the speculation of this period.

The first had an authorised capital of £300,000 and a directorate of three railway directors and a few Somerset gentlemen issued its prospectus in October 1836. In this they pointed out the favourable situation at Abercarne for smelting Devon and Cornwall ore which could be obtained in exchange for coal. Moreover, the expected demand from their fellow railway promoters led them to anticipate a profit of 32 per cent.[17] In a second report it was announced with similar optimism that 'being about to erect a town of considerable magnitude in connexion with the Lower Ebbw Vale Iron Works, the Directors deemed it a proper compliment to her present Majesty to call the Town by her name' and that 770 employees were on their books.

However, there the appearance of another future Dowlais or Cyfarthfa ends. In the next report, there appeared the ominous foreboding of the engineer's report which spoke of 'the protracted accomplishment of our expectations'; there had been slowness in collecting calls of capital and contracts had not been fulfilled. To meet this contingency the directors were borrowing £18,000. The last report reveals that £54,485 capital had been paid up, but the slump of 1839 had overtaken the tardy appearance of the company's production and there was no call for its large stocks of iron.[18]

From the account given of its successor, which itself enjoyed no better fate, we learn that the affair ended in disaster. With their capital exhausted and the price of iron having fallen by a half, the directors, it is reported, left the sinking ship, finding refuge in death or exile abroad.[19] Perhaps such fates sound too melodramatic; however, the Monmouthshire and Glamorgan Banking Company had the ironworks assigned over to it in repayment of a debt of £12,000. This banking firm, whose chairman, according to the *Mining Journal* of 1845, had dabbled heavily in iron and mining concerns, endeavoured to make a success of the venture. Robert Mushet, the industrial con-

[17] *Prospectus of the Monmouthshire Iron and Coal Company 1836.*

[18] *Reports of the Directors of the Monmouthshire Iron and Coal Company 1836–9.* The shareholders of this company, or more possibly in its successor in 1840, included such people as brewers, a fruiterer, soapboiler, bone-crusher, coalman, timekeeper and butler, as well as Yorkshire cloth manufacturers (*Memorial of the Mons. Iron and Coal Company*, 17 October 1840, P.R.O. 54/12332).

[19] Anon. *Plain Facts: History of the Victoria Iron Works from its commencement being No. 1 of the Bubbles of South Wales* (London, 1846).

sultant of his day, was called in. His report was not very optimistic; with the existing equipment, in which £160,873 had been sunk, a bare profit of one farthing a ton could be obtained on an output of 3994 tons, with the prices prevailing in that year, 1841. Nevertheless, the present depression could not last, 'the continued expansion of the Railway system at home, on the Continent and in America, gives good reason to hope . . . that the concern, despite its past history and present prospects, could yield £15,000 per annum'.[20] However, this was not to be; labouring under the initial obstacle of poor mineral resources and the threat of ejection by its landlord, Sir Benjamin Hall, the bank was no more successful than its predecessor and eventually the works were sold to another company for £55,000.[21]

A similar story of unfulfilled expectations could be told of the Blaenavon Iron and Coal Company. After fifteen years of existence, in 1854, its balance sheet showed a paid-up capital of £402,385. Yet there was an inexplicable deficiency of £233,483 and only £40,000 had been received in dividends in the last ten years.[22]

Nevertheless, the character of enterprise among the Welsh ironmasters was soon to change in the direction of joint stock ownership, for it was not a far step from the partnership to the joint stock company. In 1837, a new company with the Lord Mayor of London on its board and with a capital of £500,000 absorbed the Rhymney Iron Company,[23] which had beloned to Forman and Co., and its neighbour, the Bute ironworks. In 1839, the Plymouth works were to be reorganised into a company and a prospectus was issued by the Plymouth, Pentrebach and Duffryn Iron Company, which was to have a capital of £400,000. The reason given for the reorganisation was the old age of the previous partners, Richard and Anthony Hill, who had no heirs to succeed them in the business.[24] At the Ebbw Vale works, the occasion for the formation of a company was the bankruptcy in 1842 of the Bristol house of Hartford and Co., ironmasters and iron merchants.[25] Finally, it may be noted that as part of the negotiations between the Marquis of Bute and Sir John Guest, over the renewal of the Dowlais lease, a joint stock company with a capital of £950,000

[20] R. Mushet, *Report on the Victoria Iron Works* (Bath, 1841), pp. 11–12, 17.

[21] *Statement by John Jones concerning his loss of interest in the Victoria Iron Works* (1854). Jones had been a speculative builder on the site and taken up 42 shares in the company.

[22] *Report on the Blaenavon Iron and Coal Company* (1855).

[23] Thomas Jones, *Rhymney Memories* (Newton, 1938), pp. 13–14.

[24] *Prospectus of Plymouth, Pentrebach and Duffryn Iron Company 1839.* The works at this time consisted of 8 blast furnaces, refineries, forges and mills, etc. It is noteworthy that water-wheels were still being used for blowing the furnaces.

[25] J. Lloyd, *op. cit.*, p. 156.

was proposed.[26] However, it is said a panic in the city of London upset this scheme to combine the Dowlais and Penydarren ironworks.

According to Parkes,[27] the following joint stock companies had actually come into being by 1844:

		Capital £
British Iron Company		1,600,000
Rhymney	,,	650,000
Blaina	,,	320,000
Victoria	,,	170,000
Pentwyn	,,	220,000
Blaenavon	,,	500,000
Cambrian	,,	250,000
Maesteg	,,	100,000
Abergar	,,	100,000
Ystalyfera	,,	100,000

As the experience of the Victoria and Blaenavon Companies shows, there was much substance in Parkes's observation, 'the majority are notoriously losing concerns'. By far the most speculative of these was the British Iron Company. This company had been formed in 1824 and had been involved in heavy losses between 1825 and 1827 in attempting to exploit the anthracite deposits of South Wales at Abercrave. Besides this establishment, it is said that the company had nine other concerns under its control.[28] Of these, the chief appear to have been the Corngreaves Ironworks in South Staffordshire and the Ruabon works in North Wales; on these it was alleged to have spent £550,000 and £134, 952 respectively. Although its managing committee included David Mushet[29] and a member of the Ricardo family, all the company's concerns appear to have made losses or at best very small profits. For instance, at Corngreaves between 1825–30, the average profit was said to be only £5750 a year.[30] At Abersychan the profits were, according to Cort, only 4 per cent on a capital of £267,616 (the report of the directors said 12 per cent). With shares in arrears—only £180,000 was paid up in 1826[31] of the original £2,000,000 capital—it is not surprising that the market price of shares fell to as low as 20s. per £30 share. An extensive study of this company's finances, if it were possible, would throw much

[26] *The Diaries of Lady Charlotte Guest*, pp. 168, 172, 175.

[27] S.C. on Joint Stock Companies *1843–4* (B.P.P. 1844 (119)), evidence of J. Parkes, Q. 2370.

[28] R. Cort, *A letter to the Shareholders of the British Iron Company*, etc. (London, 1826), p. liv.

[29] *Ibid.*, pp. 13 and 18.

[30] *Case in part relative to the Frauds practised on the Shareholders of the British Iron Company by some of their agents* (London, 1841).

[31] Cort, *op. cit.* p. 24.

light on the development of business enterprise at this period. Eventually, in the 1840s, the company was reorganised on more modest lines. In 1840 the *Circular to Bankers*[32] belatedly warned its readers about the British Iron Company: 'We have frequently cautioned our subscribers against being decoyed by high names to invest their money in enterprises, without examining and ascertaining what substantial claims to their patronage any project might have, and we could point to no other undertaking than the British Iron Company which would more emphatically confirm the necessity of caution. . . .' 'The two causes of failure in the establishment of profitable iron-works', said Wilkie, 'were firstly, the injudicious selection of sites, as affecting carriage, quality of minerals, royalty etc., and secondly, want of judgement in carrying out works';[33] the British Iron Company appears to have blundered in all these directions.

Nevertheless, despite the notorious character of some of these companies, each period of boom connected with the railways threw up fresh enterprises of this sort. In 1844 there were fifteen companies in the iron industry, and the return for 1857–8 of companies registered under the Act[34] of 1856 with limited liability[35] reveals a steadily increasing number, including old-established companies newly converted to the joint-stock principle and new companies to work the new sources of ore in Northamptonshire and Wiltshire. In the following years, between 1856 and 1865, there were, according to Shannon, ninety-nine companies devoted to iron manufacture, of which twenty-one were abortive and small,[36] although the return of 1864[37] includes only forty-two and not all these were in operation.

In contrast with the earlier companies' capitals, these were quite modest, the maximum, £500,000 belonged to the Weardale Iron Company; the majority had capitals under £100,000, and some (e.g. the Weedon Iron Company, Northamptonshire) possessed capitals of £20,000–25,000. In the next two years, however, the Ebbw Vale Company declared a nominal capital of £4,000,000—its capital actually received, though, amounted to £798,355 from 486 shareholders. *The Times* commented at this period:

> 'The extensive iron works of the country were up to within the last few years entirely in the hands of private capitalists, but since the passing

[32] *Op. cit.*, 14 February 1840, p. 276. It mentioned that members of the Stock Exchange had invested large sums in the company.

[33] G. Wilkie, *On the Manufacture of iron in Great Britain* (Edinburgh, 1857).

[34] B.P.P. *Accounts and Papers* (XLVII).

[35] B.P.P. *Accounts and Papers* 1857–8 (LIII).

[36] H. A. Shannon, 'The Coming of Limited Liability', *Economic History*, 1932, p. 422.

[37] B.P.P. *Accounts and Papers* 1864 (LVIII), 452.

> of the Limited Liability Act, the joint stock principle has been gradually adopted, not only in opening new works, but many of the proprietors of old iron-making establishments have disposed of their interests to Limited Liabilities Companies. . . . About twenty years ago the iron works of the Kingdom fell into the hands of joint stock companies, and a disastrous failure was the result in nearly every instance and some predict a like result again, while others maintain that . . . discreet and honest managements are, the only requisites to insure success for a large proportion of the iron making companies.[38]

The rise of the joint stock company has been studied in some detail by one investigator, who portrays this trend in the financial organisation of industry as the victory of the investing classes over the industrialists.[39] There were four reasons, according to Jefferys, for the conversion of established concerns and it is true that not all of them arose from the needs of technical expansion. The first, however, explains the changeover to limited liability as the outcome of the very great resources needed to finance the new steelmaking processes of Bessemer and Siemens-Martin.[40] Also, the increased cost of ore and the provision of facilities for importing ore (including the ownership of mining rights abroad) necessitated an increase in the working capital of a concern. Secondly, and linked with this 'legitimate' reason for conversion, was the desire upon the part of some concerns to extend the possible sources of capital. A good example of this is Bolckow Vaughan's, which we shall discuss later; in this way they were able to draw upon Manchester capital. The third motive behind conversion was the desire of partners to retire and the inability of concerns to find replacements to take on the active direction. Both the will and the ability of the new cadets of industry to carry on the large-scale concerns of this period was much rarer than it had been during the Industrial Revolution. Jefferys mentions that the Scottish partnership of Merry and Cunningham of Lanark in their prospectus of 1872 for a new company with £1 million capital pleaded the retirement of the former owners[41] as the reason for selling out to a new company. Finally, there was the most disreputable reason—to avoid bankruptcy. This, apparently was a potent reason in the minds of the Black Country ironmasters, who were all too conscious of the impending ruin of the trade and for whom 'selling out' was a wise move.

[38] *The Times*, 10 June 1865.

[39] J. B. Jefferys, *Trends in Business Organisation in Great Britain since 1856* (unpublished Ph.D thesis of the University of London, 1938), p. 53.

[40] See below, p. 358.

[41] *Ibid.*, p. 80. We have already seen the same reason for conversion given by the Hill brothers in 1839.

But, undoubtedly, another factor which put pressure upon the ironmasters, who, after all, as we shall see, were not yet deprived of loan capital by the bankers, was the activity of professional company promoters. As D. L. Burn has pointed out, prominent in the field of iron and steel industrial finance was the Manchester firm of accountants, Chadwick and Adamson.[42] David Chadwick, who had been involved in the fraudulent flotation of the Blochairn Iron Company in 1873, and who revealed to the 1877 Committee on the Companies Acts of 1862 and 1867 that his firm acted as auditors for 62 joint stock companies and private firms, 'representing upwards of 30 millions of paid up capital',[43] was responsible for the Ebbw Vale fiasco, as well as the other conversions.[44]

Among these was the former partnership of Bolckow and Vaughan, converted to limited liability in 1864, with a nominal capital of £2,500,000 in 25,000 shares of £100 each. The paid-up capital in March 1867 was £813,737 and an analysis of the company's returns to the Registrar of Joint Stock Companies for that year shows the predominance of Manchester capital. Benjamin Whitworth of Prestwich, a cotton merchant-converter, was one of the chief shareholders with 5000 shares, and other cotton-spinners and merchants were included in the directorate. Of 90 shareholders possessing 50 or more shares, 56 came from the Manchester district. From the evidence of this company's finances, it certainly looks as if the prospects of the iron industry were very attractive to Lancashire capitalists.

The original purchase of the Middlesbrough iron works by the company had, incidentally, been partly paid for by the issue of 8000 vendors' shares with £50 paid up on each share.[45] Moreover, a dividend of 10 per cent had been guaranteed to the other members of the company, by Bolckow and Vaughan agreeing not to receive dividends until 10 per cent had been paid on the other shares for five years.[46] Thus it seems likely that the Middlesbrough partners had no fraudulent intentions in converting their concern into a limited liability

[42] Burn, *op. cit.*, pp. 254–5.

[43] *S.C. on Companies Acts of 1862 and 1867* (B.P.P. 1877, VIII, 365), Q. 1947.

[44] Nevertheless, not all the financiers of this period of close-to-the-wind company promotion, which prompted the Select Committee just referred to, were so fortunate in extricating themselves. For instance W. S. L. Price, an accountant and liquidator, told the Committee, 'I am at this moment carrying on some ironworks in Staffordshire and have been doing so for a considerable time with my own money' (Q. 1055).

[45] Plus another £200,000 approximately to be paid in instalments.

[46] *Report and accounts of Bolckow, Vaughan and Company Ltd., and proceedings of the ordinary Annual General Meeting* held at Manchester, 27 March 1868 (Manchester, 1868), p. 35.

company. Nevertheless, as with Ebbw Vale, the amount of nominal capital was considerably over-estimated, and already by 1867 it had been decided to reduce the capital figure to £1½ million.[47]

However, despite the unfortunate circumstances surrounding company promotion, which had done little to dismiss the previous suspicion attached to them after the failure of the 1840s, the pressure of economic circumstances, especially upon the old-established family concerns, by the 1880s made the joint stock company an integral part of the industry's financial structure. According to Jefferys, who had made a calculation based upon the *Mineral Returns* of 1882, 41 per cent of the concerns working blast furnaces were limited liability companies.[48] The greatest concentrations of the new impersonal organisations were in the North-east and Sheffield. The Cleveland companies' shares were quoted on the Manchester Stock Exchange from 1875 onwards and half of the firms controlling a majority of the furnaces were limited companies in 1883.[49] In Sheffield, several of the largest firms were converted in the 1860s, very early after the 1862 Act, and according to *The Accountant* (which Jefferys quotes) in 1886 there were 44 companies with a paid-up capital of £12,065,757.[50] Finally, a computation for the whole country in 1885 worked out that of 102 iron, steel and coal companies each had an average of £300,000 to £400,000 paid-up and loan capital.[51] The implications of this radical change in the character of the ownership and direction of the iron and steel works of the 1870s and 1880s will be discussed later.

There were available to industrialists, however, sources of local capital for expansion and consolidation. We have already seen how Dowlais, at first, depended upon hand-to-mouth loans to keep itself in existence and even Coalbrookdale, with the wealth of the Darby family behind it, had to resort to loans from Richard Reynolds. The loan had a vital part to play in the building up of industrial concerns, and where partners were not always to be found on account of the

[47] The 1867 Act enabled companies to reduce their capital and the 1877 Select Committee lists the companies who availed themselves of the opportunity. In the iron and steel industry only two did so: (1) The Mersey Steel and Iron Company, formed in 1864, which had its capital reduced twice from £800,000 to £240,000 of which £200,233 was paid up; (2) the Titanic Steel and Iron Company which was eventually wound up after its capital had been reduced to £172,000 from £360,000.

[48] Jefferys, *op. cit.*, p. 75.

[49] *Ibid.*, pp. 82 and 75 (citing G. T. Jones, *op. cit.*, p. 283).

[50] Jefferys, *op. cit.*, p. 75. Henry Pochin, an industrial chemist (and Mayor of Salford), who was on the Board of Bolckow's, was also prominent upon other boards of directors, including Browns, Staveley and Ebbw Vale.

[51] *Burdetts' Intelligence* (1885) quoted in Jefferys, *op. cit.*, p. 75.

great risk of unlimited liability to their fortunes, it was certainly easier to find persons who would lend money on security.

The normal recourse of the industrialist today if he is in need of temporary accommodation for financing some short-term exigency is his joint stock bank. However, this is by no means a modern development in the financing of industry, and in the eighteenth century many ironmasters sought the help of their bankers (not a few ironmasters, of course, were themselves bankers and were able to attract loan capital in this way).[52]

The Bank of England itself made loans to ironmasters after the Napoleonic wars and later, in the 1820s, when they opened branches in the provinces, they financed the iron industry particularly in South Wales.[53] There are many examples of the importance of loan capital obtained from banks. Often they illustrate the failure of the ironmasters to justify the bankers' confidence in them. Lloyd in his *Old South Wales Ironworks* mentions several instances, concerning the Hirwaun, Plymouth and Clydach ironworks.[54] In the Midlands, when the post-war slump brought down several ironmasters, again the dependence of the industry upon the banks was revealed. One concern owed £16,000[55] and in another instance it was reported that '(the) Government has granted an extent to the Dudley Old Bank against Gibbons & Stokes for £8000'.[56]

Later in the century when, as we have seen, an attempt was made to raise capital by means of the machinery of joint stock companies, a machine which was not yet 'run in' as it were, the problem of proiding working capital became even more acute. As Wilkie observed: The main difficulties in such undertakings may be considered as mostly arising from erecting works of such magnitude as to swallow up at once so large a proportion of the whole capital of the undertaking as to leave little or nothing for conducting the commercial business of the enterprise.'[57] It was here that banks often came to the rescue and John, in his study of the South Wales companies, has suggested that it was those having directors who could draw upon the reserves of the London banks which survived.[58] He instances several

[52] See L. S. Pressnell, *Country Banking in the Industrial Revolution* (Oxford, 1956).

[53] R. O. Roberts of the University of Wales, Swansea, has kindly given me this information, and he tells me that the Bank of England's records contain a great deal of material illustrating the industrial role of the central bank.

[54] J. Lloyd, *op. cit.*, pp. 16, 83 and 192.

[55] *Dowlais MSS*, T. Blakemore—J. J. Guest, 13 June 1816.

[56] *Ibid.*, G. Wood, Jnr.—Guest, 20 April 1816.

[57] Wilkie, *op. cit.*, p. 134.

[58] John, *op. cit.*, p. 56. Window-dressing by securing a City Alderman or

companies supported by London bankers—the Llynvi Company, the Pembrokeshire Iron and Coal Company, and the Blaenavon Company.

Incidentally, it is significant that at this time bankers were not unwilling to sink their assets in such intractable investments with apparently small regard for liquidity.[59] The joint stock banks themselves went out for business in new districts and in order to promote the exploitation of new processes. Crick and Wadsworth, the historians of the Midland Bank, report the resolution of the Cumberland Union Bank in 1865 to open a branch at Holborn Hill as there were 'large iron mines in operation and several blast furnaces in the course of operation'.[60] During the next few years, when, as they say, 'the minute books are eloquent in the number of loans to haematite companies, collieries and rolling mills' to exploit 'the red gold of the West', there was a Scottish invasion by the Clydesdale Bank in 1874. This bank opened three branches at Carlisle, Whitehaven and Workington.[61] There is evidence, too, that the North-eastern banks helped to finance the introduction of the Bessemer converters into the steel industry.[62] Thus, as Burn points out for the succeeding period when large sums of money were needed for technical modernisation: 'Many members of the industry did not feel that their sources of their capital supply were restricted'[63] and, indeed, there was no reason why they should. The financiers of the British industrial system had their own provincial centres, as well as London, and although British

M.P.'s for the directorate was known to promoters well before Chadwick's more infamous abuse of the practice.

[59] A striking example of the extent to which bankers sometimes allowed their ironmaster customers to run up huge overdrafts is the case of the Derwent Iron Company, later to become the Consett Company. A northern banker, Jonathan Richardson, was the promoter of the first company which was built upon a lavish scale. Soon they had 18 blast furnaces under their control, but when the Northumberland and District Bank crashed in 1857 it was discovered that this company had overdrawn its accounts by nearly one million pounds. As a result of this failure in its supply of loan capital the company had to be reorganised. See H. W. Richardson and J. M. Bass, 'The profitability of the Consett Iron Company before 1914' (*Business History*, Vol. VII, 1965).

[60] W. F. Crick and J. E. Wadsworth, *A Hundred Years of Joint Stock Banking* (London, 1936), p. 132.

[61] *S.C. on Banks of Issue*, 1875, evidence Mr. Readman, Q. 3675. He revealed the Bank had strong connections with the iron trade of the West of Scotland and he said many large Scottish ironmasters had establishments in Cumberland. See also W. H. Marwick's *Economic Development of Victorian Scotland* for the business and commercial interests of the Scottish ironmasters.

[62] Crick and Wadsworth, *op. cit.*, p. 226.

[63] Burn, *op. cit.*, p. 252.

bankers did not provide long-term loans as did the German and American banks, the supply of working capital from the local banks and branches of the joint stock banks was more plentiful than has hitherto been realised.

So far, it has been assumed that it is evident that there was a trend towards the growth of large-scale enterprise in the iron and steel industry. It is true, of course, that the increases in the capital flotations of the companies previously mentioned and the examples of integration make this tendency clear. Indeed, much of this capital expansion is to be explained by the increased integration of the large-scale concerns. Even in the eighteenth century, coal and iron had so interpenetrated one another that it is not at all easy to separate economically the two distinct industries.[64] Ironmasters were seldom foolish enough to venture into an enterprise without possession or a good lease of the mines; and possessing coal and iron mines, what was more natural than that they should become coalmasters by building up an extensive land-sale of coal? Thus the Botfields in Shropshire, the Dowlais Company and many others had large coal interests.[65] Moreover, when in the 1850s the South Wales coalfield had begun to show diminishing profits in the production of iron and new opportunities appeared in the steam-coal trade, then the ironmasters like the Hills of the Plymouth works and the Bailey brothers started exploiting the new riches.

This intimacy of the two industries arose naturally out of geological circumstances.

But obviously the most usual way of extending one's industrial enterprise was to begin manufacturing the finished products either of the foundry or of the engineeering shop. Throughout the history

[64] The types of concern in the iron industry which are the subject of this discussion have been confined to those engaged primarily in producing the raw metal from the ore by the basic processes, in furnaces and forges. In the second half of the century the large scale production of steel in the converters or open-hearth furnaces increasingly became the basic production of the industry. For a discussion of the problem of definition of iron manufacture in 1841 see H. D. Fong's *Triumph of Factory System in England* (Tientsin, 1930), p. 127.

[65] See J. H. Morris and L. J. Williams, *The South Wales Coal Industry*, 1841–75 (Cardiff, 1958), pp. 82 ff., for an enlightening discussion of the entry of the South Wales ironmasters into the South Wales coal trade, from the 1840s, and the continuation of the coal trade to decline fortunes of the iron industry. In 1859, William Menelaus of Dowlais, had come to the conclusion that a capacity of 2000 tons of finished iron a week should be the limit to risk for one family in the iron industry. In 1866, William Crawshay II had decided: 'I really think it would be wise to give up the iron trade entirely and sell the coal.' Dowlais followed Menelaus's line of policy: in 1868 the profits from its coal pits constituted one-third of the total net profit of the concern.

of the modern iron and steel industry and in almost every district this extension in the range of the ironmasters' output took place. At first, as we have seen in the case of Coalbrookdale, it was the manufacture of the steam engine and the cast iron bridge which provided the market on the doorstep for the blast furnace and forge; later and elsewhere, it was the rolling of rails and the forging of locomotive parts; and finally, in the age of steel, although the demand for engineering products for construction and manufacture continued, the forging of steel guns and armour plate—the manufacture of armaments—assumed a great importance in the industry.

CHAPTER XI

The Iron Trade

1 THE USES OF IRON

> 'I consider that a part of the timber now imported for Dock Yards and public buildings may be advantageously dispensed with and iron substituted. That (the) Govt. may use advantageously large quantities for floor plates in public buildings, that public rail roads may be formed; that probably with encouragement foreign markets may be obtained.'[1]

Lord Liverpool, Britain's Prime Minister, has been credited with this far-sighted plan for relieving the post-war distress of the British iron industry in 1817. As we know, in pointing to the potentialities of substituting iron for wood in architecture and engineering construction; in advocating the promotion of railways—even if he meant tramways—and in looking to the expansion of British exports of iron overseas, this spokesman of the iron industry was setting out a blue-print for the future of the industry.

Of course, this practical piece of advice was not based on purely speculative crystal-gazing. We have looked at the gradual and cumulative process of substitution of wooden structures by iron, symbolised by the famous iron bridge. As Dupin, a French observer of England during the Napoleonic wars, commented: 'This continuous substitution of wood by iron is not at all the result of a craze or passing fancy, it stems from a comparison of the low price of this metal with the dearness of timber, which is excessive throughout England.' He was seeing evidence at places like Liverpool of the use of iron in shipbuilding and dock construction, for capstans, bridgeways, pulleys. 'All these things,' he said, 'traditionally made of wood were, in the most modern pieces of building, being constructed completely of iron.'[2] However, as architectural historians

[1] *Add. MSS* 38367 f.121 (*Liverpool MSS*), 1817; J. P. C. Dupin, *Voyages dans la Grande-Bretagne* (1817), Book VI, p. 247, has a description of the Glamorgan canal where there were small iron boats '4½ metres long, 2½ wide'. This traveller commented: 'The solidity of these boats means that they rarely need repairs and their long life makes them economic.'

[2] J. P. C. Dupin, *Voyages dans la Grande-Bretagne* (1817), pp. 26–7.

are now beginning to demonstrate,[3] the story of building with iron and glass—an innovation used to revolutionary effect in the railway stations of the Railway Age as well as the spectacular Crystal Palace in 1851—is a remarkable and fascinating demonstration of the exploitation of new technological possibilities. As Hitchcock says: 'The Industrial Revolution produced a parallel but gradual revolution in building methods; even today, after some two hundred years, the results of that revolution have not been fully realised.'[4] The first use of iron for constructional purposes pre-dated its most spectacular and most evident use as members of cast iron bridge structures: according to Hitchcock, the 'first cast iron members of truly architectural scale introduced into a building of any consequence were probably the pillars used to support the galleries of St. Anne's (Church), Liverpool, in 1770–2'.[5] The next stage taken by architects, with a view to reducing the fire-risk of timber-framed and floored buildings, was to build in a hollow 'flower-pot' cellular construction, floors supported by iron stanchions. This was adopted in the famous cotton mills of William Strutt at Derby and Belper in 1792.[6] The logical development of this principle—to form the complete interior skeleton of iron beams and stanchions—was soon carried through, first, on a small-scale at the Ditherington Flax Spinning Mill, near Shrewsbury in 1796–7,[7] and then three years later at Lee's Cotton Twist Mill built at Salford in 1799–1801. This was a seven-storeyed structure employing beams with a 12 ft. 10 in. span. We need not dwell here upon other examples of the use of iron columns—sometimes it was used 'frankly and freely' and functionally, even incorporating systems of central heating; sometimes the columns

[3] See in particular H.-R. Hitchcock, *Architecture, Nineteenth and Twentieth Centuries* (London, 1958), Ch. 7; *Early Victorian Architecture in Britain* (London, 1954), and other detailed studies by him. Hitchcock traces very clearly the Continental influences which had such an important bearing upon this development. See also H. R. Johnson and A. W. Skempton, 'William Strutt's cotton mills, 1793–1812', *T.N.S.*, Vol. XXX), and A. W. Skempton and H. R. Johnson, 'The first iron frames', *Architectural Review*, March 1962.

[4] *Architecture*, p. 115.

[5] Hitchcock, *Ibid.*, p. 116.

[6] *Ibid.*, p. 117. However, small lock-keepers' houses had been built of iron slabs in 1790. Here, of course, the slabs provided the integral strength to support the roof.

[7] See Skempton and Johnson, 'The first iron frames', for a detailed analysis of the building and of the context in which it came to be built. As they say: 'No other region in England, or in the world, could show so much work in structural iron. . . . ' There were churches with iron columns at Wellington and Shrewsbury; in 1795, under Telford's direction, Buildwas Bridge across the Severn and a canal viaduct at Longdon, in which William Reynolds co-operated, were erected.

were cast in the forms of trees such as palm trees with naturalistic leaves. It is clear that engineers were exploiting the new material with marked effect; moreover, as Skempton points out, the cast iron beam had attained 'a rational form and was subject to theoretical analysis'.[8] So we see the desire of iron-minded ironmasters—Wilkinson was not alone in this—to build complete structures such as churches of iron. However, architectural conservatism decrying the frank exposure of the metal on the outside of buildings—they should be proper and present a stone or brick façade to the world—delayed the further development of iron buildings. Meanwhile, corrugated iron sheeting had been invented in the early 1830s and doubtless it was exported for colonial construction as a cheap roofing material; possibly, too, it formed part of the 'elegant sea coast cottages of iron, costing £250' mentioned in *The Times* of March 1839.[9] The most functional employment of this sheeting, as panels between the structural members of the building, had to wait until the 1850s when the designer of the Boat Store, Sheerness, created a startlingly modern building.[10]

Often, of course, there was an incompatibility if not a conflict between the implications of this new method of building and the traditional and popular efflorescence of artistic adornment with which the simple forms were overlaid. These elements were nowhere more evident than in the famous Great Exhibition building of 1851. This has often been described as a masterpiece of prefabrication by Joseph Paxton, the Chatsworth gardener. But, as we have implied above, there was a move towards this form of standardisation and simplification of small structures such as cottages and houses in the 1830s and 1840s. According to Hitchcock, the Victorian prefabrication of iron structures begins with a palace for a king, supplied to an African chieftain, one Eyambo, to provide accommodation for his 320 wives and the royal personage.[11] Certainly, by the 1840s lighthouses were frequently erected in ironmasters' yards and then dismantled for shipment to the West Indies. Sir George Grove, of musical fame, figures at this period as an exporter of these corrugated iron lighthouses. Liverpool and Manchester were centres of this

[8] Johnson and Skempton, *loc. cit.*, p. 194.

[9] 22 March 1839. However, when this type of structure had been discussed some years earlier in the *Mechanics' Magazine*, 1823–4, it was observed, with admirable precision, 'wherever such houses may be erected they must be covered with ivy, or some other evergreen creeper, to moderate the effect of changes in the interior temperature'.

[10] A. W. Skempton, 'The Boat Store, Sheerness (1858–60) and its place in structural history', *T.N.S.*, Vol. XXXII, 1959–60.

[11] Hitchcock, *Early Victorian Architecture*, p. 515.

prefabricated construction. In 1844, it was said that 20,000 tons of iron was shipped to the port for this purpose, and the *Annual Register*, with how much precision we do not know, described St. George's Church there as 'nearly the first iron church erected in Great Britain. The whole of the framework of the windows, doors, groins, roofs, pulpit, ornamental enrichment are of cast iron.'[12] At Manchester, E. T. Bellhouse, a leading iron founder and exponent of the fireproof qualities of iron construction—in 1848 he brought out a fireproof hoist—successfully manufactured houses for emigrants to Australia. It is not surprising that we read that the Prince Consort ordered a prefabricated iron ballroom for Balmoral Castle to be modelled on these exports.[13]

It was in such 'palatial' rooms, naturally, that ornamentation came into its own. At last the use of iron was accepted and revealed as a technical necessity, and it could be rendered beautiful by the intricate castings and traceries now being mass-produced by the famous foundries of Coalbrookdale. And it is quite likely that the modest and unoriginal innovation of Paxton, using iron-framed glass windows for the Exhibition building, was as easily overlooked as an outstanding example of 'mechanical elegance', on account of its own inappropriate decorations as from the luxuriance of all that it contained.

That being so, one should remember the technical achievement of the railway engineers, such as Brunel working in collaboration with architects—for example, M. D. Wyatt—who, in their new palaces of transport, the London railway terminus stations, surely had a greater impact on the public imagination. For, after all, the novelty of travelling by a thrilling and exciting new form of transport had not yet died away. In this context, then, of breakneck speed, the architects of iron and glass—ferrovitreous construction—achieved an exhilarating feeling of lightness and space.

These, doubtless, are architectural not technological qualities. However, the rapid rise to popularity of the cast and wrought iron[14]

[12] *Annual Register* of 1841. It is interesting that twenty years earlier, when the iron trade was still in a depressed state, a petition for authority to build churches partly of iron, was presented to Parliament. (*Journals of the House of Commons*, 1821, p. 169)

[13] Hitchcock, *Architecture*, p. 122. The demand for prefabricated dwellings, halls and churches, of course, swelled as a result of the Californian and Australian gold rushes. In Sydney, the present Legislative Council Chamber is a relic of this boom.

[14] Wrought iron began to supersede cast iron for building structures after the disastrous collapse of cast iron bridges (e.g. the Dee Bridge) in the 1840s and the thorough analytical investigations of William Fairbairn into the strength of

building reached its peak in the 1850s, and, ironically, it was as the result of entrusting the design and construction of the South Kensington Museum of Science and Art to a firm of contractors employing cast-iron prefabrication methods, that there was a wave of disapproval against this example of industrial ugliness, christened the 'Brompton Boilers'. To quote Hitchcock's summary of this phase: 'After this time British and continental interest in iron construction waned rapidly. . . . Structural steel began to be used here and there from the early sixties, but the serious beginnings of the Age of Steel, an age which still continues, lay a full generation ahead.'[15] Again, as Hitchcock says, 'The line of technical development which runs from the cast-iron-framed textile mills of the 1790s in England to the steel-framed skyscrapers of the 1890s in America seems to posterity a simple and obvious one. But, in fact, various lags and cul-de-sacs make the story long and complex. The most significant technical advances in iron construction of the first half of the century were not in the commercial field. . . .'[16] This architectural story cannot be pursued here. We must now attempt to describe the far-wider ramifications of the use of iron in engineering, railway and shipbuilding construction.

The use of iron and steel beams, cross-members, stanchions, etc., depended upon the ironmaster providing suitable architectural sections where strength was nicely balanced against reduction of weight and cost. The finishing side of the iron and steel industry which was much developed in the United States and which led directly to the possibility of erecting the skyscraper, was again the result of the introduction of power-driven and specially designed rolls and shears into the ironworks. The special needs of the railways for wrought iron rails of the famous T section profile had occasioned the invention of rolls to simplify and reduce the number of operations for the rolling process. As we have seen, the power-driven Nasmyth steam hammer had resulted in certain economies.

The iron and steel industry itself then, in having its efficiency transformed by these innovations capable of dealing with their greater working loads at far greater speeds than human muscle could move, was a major consumer of its own products. Not the least significant development of the Industrial Revolution was the advancement of the mechanical engineering industry.[17] Indeed, it has been suggested

wrought iron when used in constructional work. The famous Britannia Bridge of Robert Stephenson, across the Menai Straits, built in 1845–50, was the most striking demonstration of the superiority of wrought iron.

[15] Hitchcock, *Architecture*, p. 129. [16] *Op. cit.*, p. 233.

[17] J. B. Jefferys, *The Story of the Engineers* (London, 1946); W. H. G. Armytage, *A Social History of Mechanical Engineering* (London, 1961), Chapter 12.

that the three main products of the iron industry, in 1850, were textile machinery, locomotives and steam engines. Later, marine engineering, armaments and, to a lesser extent, agricultural and mining machinery became more important. Some ironmasters had always owned workshops making machinery, and in at least one case, that of the Coltness ironworks in Scotland, the engineer had through integration taken in the primary smelting processes. In the case of Bessemer steel, too, the armour plate and railway spring and buffer manufactures of Sheffield were to take the lead in developing the new method of manufacturing steel. Fairbairns and Galloways of Manchester, Armstrongs of Elswick, engineers and makers of locomotives, millwork machinery, dock installations and the armour plate for the new naval ships, themselves now constructed of iron and steel, all became smelters as well as 'manufacturers' (i.e. finishers) of steel. One might say that it was the tendency for the iron districts, where smelting became unprofitable, to concentrate eventually on engineering industries, manufacturing iron and steel into many different products. The fortunes of such areas as the Black Country, Shropshire, Yorkshire, pointed to this need for the ironmasters there to make their iron products more capital-and labour-intensive.[18]

Nevertheless it was of course the railways, running like a spider's web almost directly from the furnace across the face of all the developing continents of the world in the nineteenth century, which demonstrated the vital interdependence of the new industrial system. Again Hitchcock aptly sums up this development: 'The railways, as they expanded, demanded even more and more iron for their rails, their locomotives and their sheds, thus continually stimulating iron production. And it was the railways that brought together the raw materials for that production, moved the semi-finished products from plant to plant, and finally distributed iron members of all sorts—including building components—all over Britain.'[19]

The construction of the British railway system and of the great American and Continental lines had a two-fold significance on the development of the British iron industry; first, in the actual physical demand for the products of the industry—rails, chairs, rolling-stock, bridges and stations; secondly, as an extremely potent psychological element in boosting business expectations and industrial activity during the decennial periods of boom which agitated the course of industrial progress. Some indication can be given of the vast extent of the ambitious enterprises conjured up by the railway promoters, when it is realised that in the first of these major booms in 1835–7,

[18] South Wales appears to be the one serious exception to this generalisation.
[19] Hitchcock, *Early Victorian Architecture*, p. 495.

a capital expenditure of £36 million was authorised to lay 1,200 miles of track with a consumption of about 500,000 tons of iron. The direct impact of this vision of prosperity for the iron industry is to be seen in the movement of prices of bar iron. From £6 10s. in 1834, by 1836 the price had jumped up to £11 a ton.[20] The ironmasters had long used iron tram roads, and indeed the first use of iron rails had been the idea of the Shropshire ironmaster Richard Reynolds in 1767, and one of the partners of the Butterley ironworks, Jessop, had been the pioneer in the development of the edge railways, which replaced the tramways. The Butterley ironworks in the early nineteenth century had laid down several railroads, not only serving the collieries in Nottinghamshire and Derbyshire, but also public railways (including the Derby, Ashby, Blisworth, Peak Forest, Croydon railways and the principal railways in Wales, viz. the Sirhowy Railway, the Merthyr Railway).[21]

In 1817, railways were seen to have possibilities for relieving the post-war depression. In 1821 William Crawshay had suggested 'a general line of railroad along the main roads of England by Parliament'.[22] But, as John says, the large and more conservative concerns would not venture into enterprises of this sort, during the first spell of railway activity in 1824–5, when the London to Edinburgh and the Liverpool to Birmingham railways were being projected. Guest, for instance, refused, despite Crawshay's interest, to venture with a railway in the north-west to link up with the projected Liverpool railways, to exploit the iron ores of the Furness area.[23] The importance of the demand for iron for these railways was, however, appreciated elsewhere: the *Mechanics' Magazine* at that time, in fact, estimated that about 200,000 tons of iron would be required. These expectations were of course premature; nevertheless in 1829–30 the Ebbw Vale works and the Penydarren works were supplying the Stockton and Manchester railways respectively.[24] It was not until 1835, however, that railways orders began to dominate the corre-

[20] G. R. Porter, *On the progress, present amount and probable future condition of the iron manufacture in Great Britain* (Report of 16th meeting of British Association 1846).

[21] *Dowlais MSS*, W. Jessop—J. Cairns, 9 June 1821.

[22] *Crawshay MSS*, W.C.I—W.C. II, 22 November 1821. 'Deputies will do no good here about the French Duty, no more than was done with American. England has no power even if she had any right to dictate to these powers. The best relief to the iron trade would be a general line of rail road along the main roads of England by Parliament. Such an improvement in land carriage would be a consumption for all the spare produce of iron.'

[23] *Dowlais MSS*, Harrison Ainslie and Co.—Guest Lewis and Co., 10 November 1824.

[24] John, *op. cit.*, p. 103.

spondence of the great ironmasters. The following year, the Dowlais works had an output of 20,000 tons of rails alone.[25] Not only was the home programme of £36 million of railway construction being met, but the American railways were being supplied through the agency of the Baring Bros. and others. This branch of the export trade at once assumed a great importance. An American agent, who was concerned with placing contracts for some of the projected railroads—the Pennsylvania; Delaware, New Jersey and the State of New York Railways—claimed to have bought upwards of 40,000 tons of rails at this time.[26] With truth he remarked that 'all countries throughout the world must get their railway iron in England where it is manufactured with economy, such rapidity and so perfectly that it is useless to pretend to compete with this branch of industry'. As an illustration of the British ironmaster's eager response to this demand, he recited how one order of 1800 tons of rails had been manufactured and shipped across the Atlantic within four months of the order leaving America.

It is not surprising, in view of this anxiety of the ironmasters to make the most of their monopoly in the supply of rails, that, in 1836, contracts for 400,000 tons of rail road were said to have been signed with American houses in England.[27] Not only to America did the British ironmasters supply rails; many houses shared in the supply of the Continental lines undertaken by British contractors, such as Brassey and Peto. Dowlais rails, in fact, were laid on the Breslau and Freiburg, Saxon-Silesian, Chemnitz, the Turin and Genoa and other Italian railways. The diplomacy of securing rail contracts was one of the chief concerns of the magnate ironmasters at this time.[28] Investment in railways, moreover, besides being one way of extending sales on credit, was also an attractive outlet for the surplus capital of the ironmasters. The first motive was far from unimportant; Crawshay in 1853 held £249,099 of American railway mortgage bonds,[29] and the House of Baring played an important role in this form of investment abroad. They not only provided short-term credit to enable American railroads to make their purchases, but also induced ironmasters to accept bonds in part payment.[30] In 1849, the Welsh house of Thompson and Forman

[25] *Ibid.*, p. 104.

[26] *Mechanics' Magazine*, 1835–6, p. 244.

[27] *Mechanics' Magazine* quoting *American Railroad Journal*, April–September 1836, p. 176.

[28] Addis, *op. cit.*, p. 157, App. II, gives details of William Crawshay II's holdings of American railway bonds.

[29] John, *op. cit.*, p. 105.

[30] Addis (p. 157) cites 'the supposedly mythical anecdote' of the colloquy

accepted bonds in the Baltimore and Ohio railway in half payment of a contract for 22,000 tons of rails.[31] The Barings themselves secretly controlled the Weardale Iron Company which supplied the American market with bar iron. Hidy points out, too, that these bankers exploited the financial weakness of some of the British producers—expanded capacities and inadequate capital reserves—when prices were low during the slump of 1847, by encouraging American purchases at this period at low rates of commission.[32] It should be pointed out, in explanation, that at this time the British ironmasters seldom dealt direct with the American buyers, but with banker-correspondents such as Barings, or merchant correspondents, such as Ralston of London, whose activities have been mentioned earlier. However, in the second half of the century, when competition between the merchant bankers in the financing of the American railroad iron became more acute, some of the ironmasters, e.g. Bailey Bros., dispensed with the bankers by authorising American firms to make contracts to deliver rails direct to the railroad companies. Ironmasters in the early 1850s, as Hidy points out, were shipping rails on consignment from Newport at £5 10s. a ton. The freight rate was 12s. per ton to New York and the insurance 5 per cent.[33] Before that date however, the iron trade had had some misgivings whether it could possibly supply all the demands made upon it. There were, of course, evident dangers to the industry arising from this trade. Porter, the statistician, in fact raised the question of excess capacity and asked whether the British ironmasters might not 'be exposing to ruinous depreciation establishments, for the formation of which, vast capitals have been sunk,[34] through the cessation of the demand for rails. However, the problem of surplus capacity, which from the numbers of furnaces out of blast during periods of depression should already have given the ironmasters pause for the future, was neglected and instead there was only eagerness to supply all railway ventures. In 1845, during the height of another railway boom, for instance, it was feared that the railway programme for 2000 miles of railways at home and an export

between an American railway magnate and a Welsh ironmaster discussing these negotiations. 'Said the ironmaster, speaking of rails, "They were pretty poor stuff we sent you." "Yes", said the railway magnate, "but not worse than the bonds we gave you for them." ' Cited in Harold Cox (ed.), *British Industries under Free Trade*, p. 289.

[31] Hidy *op. cit.* pp. 402 and 594, note 37. The ironmasters charged an extra £1 a ton for this concession.

[32] *Ibid.*, pp. 401–2.

[33] Hidy, *op. cit.*, p. 587.

[34] G. R. Porter, *op. cit.*, p. 99.

demand of 500,000 tons for overseas railways, would leave a deficiency of yet another 500,000 tons over the estimated make of 1845 of 1,330,000 tons. Such estimates are of course of little value, especially as yet another forecast a surplus production of 443,000 tons.[35] However, some light was thrown upon the consumption of iron in other fields of enterprise at this time: 'the cotton trade takes ordinarily one-sixth of iron for machinery' and some 480,000 tons were required for 'general consumption of iron in Great Britain in the shape of bar iron, castings, water and gas-pipes, steam engines and hardware'. As to the actual consumption of iron by the railways at this time, Beck calculates that by 1850 England must have consumed 1,853,880 tons of iron for rails alone.[36] One important factor here, as Clapham points out, was the increased weight of the rails. In 1841 the weight of iron, in rails and chairs, used for a mile of single track was 156 tons as against 53½ tons in the early days.[37] Railway demand then dominated the home market—an influence which was not altogether salutary to the development of the industry. To the very last days of the demand for wrought iron rails, in the 1870s, when steel was beginning to supersede iron, it might be said that this concentration on this product was the chief cause of depression in the industry. Nevertheless, the actual demand coming from railway construction was undoubtedly the principal cause of growth, particularly in those areas concentrating on the production of railroad material. Without the receipt of such large contracts as the construction of the railways provided—for example in 1846, South Wales works shared orders for 120,000 tons of rails (of which the Plymouth works alone took 40,000 tons at prices of more than £13 per ton)[38]—it would have scarcely been possible for ironworks to sink the huge sums they did in equipment to extend their capacity.

As an indication of the disruptive effects of the periodic bursts of railway promotion, which swept the iron industry first into feverish activity and then sudden calamitous depression, a brief survey of the trade during the period 1835–47 will show irregular development of the iron industry all too clearly. In this period two major railway booms and slumps occurred. After 1830, the threat of Scotch competition made itself felt more and more in the English and Welsh markets. This and the ever-present tendency to increase production with the continuing depression of prices,[39] helped to prolong the

[35] *The Times*, 29 March 1845 *et seq.* [36] Beck, *op. cit.*, p. 550.

[37] Clapham, *op. cit.*, p. 428, citing Tomlinson, *History of the North-Eastern Railway*, pp. 405–6.

[38] *The Times*, 27 January 1846.

[39] 1833, *S.C. Mfres., Shipping and Commerce*, Q. 9671, evidence of W. Matthews.

slump. In 1834 *The Times* reported, 'That important branch of national industry, the iron trade, is in a drooping condition.' It went on to say that Shropshire's best forge bars would only fetch £4 and 'Staffordshire bars are at any price the manufacturers can command'. But soon the news of prospects of great orders was in the air, and, when the spate of railway legislation of 1836–7 made the prospects more immediate, the ironmasters abandoned their efforts to regulate output and to bring stability to the industry. They launched into the manufacture of rails with feverish anticipation. In this atmosphere speculation could, of course, thrive, driving up the price of bar iron from £6 10s. per ton in 1834 to £11 in 1836.[40] As we have seen, this period of boom brought into existence several joint stock companies in South Wales with a lavish investment of capital; however, the demand for iron, which to a large extent was fictitious, evaporated before they could get into production. As *The Times* remarked of this development, 'If it had been a law of nature that the whole world should for a given number of years, be clothed in iron, be domiciled in iron, sleep upon iron, ride upon cast iron horses, and in cast iron chariots, no greater zeal to give effect to the law could have been displayed than that which has prompted the production of a vast mass of iron . . . under the circumstances which we have attempted to describe. The result is a collapse of the whole trade. . . .'[41] The years of depression, particularly 1839–40, were a period of bankruptcies, of a renewed exploitation of the truck system, and of strikes. In 1842 there were 190 furnaces out of blast, of a total of 429. But once more, this time after a period of restricted output, the iron industry put aside the depression and misery of those years, to pursue the illusory prosperity of the 1845 railway boom. Again a potential demand, which was estimated to be one million tons for the number of Bills passed by Parliament,[42] sent the price of iron rocketing to £12. *The Times* warned ironmasters that a headlong plunge into production at these high prices 'would totally destroy foreign trade and the manufacturer'; there was too, a symptomatic minor slump in June 1845, in the Midlands; but the industry had not developed a strongly organised body to control the price and production policies of the industry throughout the country. Restriction of output was always an unwelcome measure to individual ironmasters, who naturally only concerned themselves with their own interests, and often, because of necessity, only their immediate interests at that. Hence, such agreements were soon cast off in such circumstances as the year 1845. Strangely enough, since there was, even before the railway

[40] G. R. Porter, *The Progress etc. of the Iron Trade*, p. 103.

[41] *The Times*, 21 June 1843. [42] *Ibid.*, 19 July 1845.

boom of 1844, much evidence to the contrary, it was said, 'it is an established and well-known fact that the iron trade generally ranks among the last to feel the effect of depression'.[43] 1847 was of course a year of crisis in many trades. The severity with which 'the prostration of railway interests' hit the ironmasters had its conclusion in the sorry story of reductions in make, wage cuts, of unemployment, strikes, of panic selling and great reductions in the price of iron. In South Staffordshire 62 per cent of the blast furnaces were out of blast; in Scotland and Wales the depression was not felt so badly, with 31 per cent and 22 per cent out.[44]

Finally, in considering the effects of this phenomenal expansion of the net of communications and of transport facilities, which were a necessary concomitant of industrial progress, we should look at the ways in which that other element, water, came under the sway of iron. The first major development in the improvement of transport in Britain—the canals—did not depend to a great extent upon iron. Only the occasional steam-engine required for pumping and the rare iron aqueduct made a demand upon the blast furnace and the foundry. John Wilkinson, in the eighteenth century, had experimented with an iron barge on the River Severn, but it was not until 1810 that iron was regularly used for the construction of vessels employed in canal and river navigation.[45] However, the bridge builders had readily adopted the new material. According to Gloag and Bridgwater, the authors of the *History of Cast Iron in Architecture*: 'From 1801 the cast iron girder and jack arch system of bridge building had come into use, and over 400 bridges of this type are still in existence, most of them built before 1860'. Of these, the Menai Suspension bridge completed in 1826, by the great civil engineer Thomas Telford, is perhaps the most spectacular, with its single span of wrought iron 580 feet long.[46]

The steam engine was, however, the greatest formative influence on the iron industry. Steam power had already been applied in shipping when the first ocean-going ships were built of iron. A pioneer in this field was Manby (himself an ironmaster, at first in the Midlands and then abroad, in France), who in 1820 built the *Aaron*

[43] *The Times*, 18 January 1844 (quoting *The Hereford Times*).

[44] Clapham, *op. cit.*, Vol. 1, p. 429. The 1857 crisis also revealed the dangers of railway booms (this time in America). See C. S. Orwin, *The Reclamation of Exmoor Forest* (London, 1929), quoting a letter from G. T. Clark of Dowlais: 'Rumour says that the Baileys have lost £400,000 by American dealings, that the Ebbw Vale have been driven to get a credit for £200,000 at the bank and that Crawshay has lost—for the present at least—£150,000 in Saunderson's house.'

[45] G. R. Porter, *Progress of the Nation* (London, 1851), p. 575.

[46] P. Dunsheath (ed.), *A Century of Technology* (London, 1961), p. 17.

Manby for service on the Seine.[47] But there was a great prejudice against using iron; it was thought that the iron hull would rust into holes (that fear was not dispelled until the *Great Britain* in 1846 went aground for eleven months and then resumed its service, intact).[48] Thus iron ships did not rapidly come into service. Although it was said 'the use of iron as a material for steam ships is now as general in England as the use of wood'.[49] this was far from being the case. In 1850, only 9 per cent of ships were built of iron. It was not until 1865 that the Cunard Line launched its first iron ship, the *Persia*.[50] In fact, it was not until steel had superseded wrought iron in shipbuilding, on account of its longer life and lightness,[51] that metal-constructed ships drove the wooden ships off the seas. By 1880 the triumph of the steel ship was complete; only 4 per cent of all ships built in Britain in that year had wooden hulls.[52]

SHIPS BUILT IN BRITAIN

	Iron	Timber
1850	12,850 tons	120,000 tons
1860	64,679	147,000
1870	255,000	161,000
1880	495,367	19,938

2 COMMERCE IN IRON

Speaking of Britain's trading position in iron and steel products during the eighteenth century, Professor Ashton says this of the balance of trade, which seems to have changed in the 1740s: 'At the beginning of the century England purchased from foreigners far more iron than she sold abroad; by the end of it her exports (including manufactured goods made of iron) were five times as great as her imports.'[53] The overseas market then, even in the period preceding the Industrial Revolution, had always been important when exports

[47] W. H. Chaloner and W. O. Henderson, 'Aaron Manby, builder of the first iron steamship', *T.N.S.*, Vol. XXIX, pp. 77–91. The ship was built in parts at Tipton, Staffordshire, transported to the Thames, there assembled, and sailed over to the Seine.

[48] C. R. Fay, *Great Britain from Adam Smith to the Present Day* (1944 edition), p. 169.

[49] *The Economist*, March–December 1844, p. 1443, quoting *Le Moniteur.*

[50] Fay, *op. cit.*, p. 169.

[51] In 1866 Bessemer pointed out an iron ship weighed 12 cwt. per ton of measurement, a steel ship 7 cwt. Steel ships therefore had greater load capacities.

[52] I. L. Bell, *op. cit.*, p. 330. Figures of tonnages of iron and timber ships built 1850–80 are as follows:

[53] T. S. Ashton's introduction to E. B. Schumpeter, *English Overseas Trade Statistics 1697–1808* (Oxford, 1960), p. 12.

had consisted of wrought iron,[54] hardware and nails. Although, as we have seen, imports of wrought iron from Sweden continued to increase, the technological innovations had by the end of the Napoleonic wars freed Britain from its former dependence on these foreign sources. In the years which followed Britain achieved commercial supremacy in the world.

The growth of British exports of iron products is shown in the table below. It shows the rise of British metal manufacturers supplying the greater part of the requirements of the industrialisation throughout the world in the nineteenth century.

TABLE OF EXPORTS OF IRON, 1806–1880[55]

	Tons		Tons
1806	36,925	1845	351,978
1823	46,413	1850	783,424
1825	34,372	1860	1,503,000
1830	117,420	1870	2,826,000
1835	199,007	1880	3,793,000
1840	268,328		

Thus it will be seen, especially in the period 1850–70, when the trade almost doubled itself in each decade, that the expansion of exports of iron exceeded the rate in the expansion of output; in other words, the export trade was becoming increasingly the most important element in the demand for iron. However, as Deane and Cole point out, the fastest rate of growth occurred in the period 1830–50, when outputs rose fourfold. Up to the middle of the century this expansion was to meet the growth demand for iron within Britain for the needs of industrialisation.[56]

A systematic analysis of the development of the iron trade itself,

[54] Re-exports of bar iron had always been of some significance. In 1815, when exports stood at 21,300 tons, 13,100 of these were re-exported: according to the official figures, given by Mitchell, in the following year 8,200 out of a total import of 8,500 tons were reshipped abroad. In the succeeding decade, a steady average for the re-export trade was approximately 3,000 tons.

[55] Sources of figures:

(*a*) G. R. Porter, *Progress of the Nation* (1851 ed), p. 575.
(*b*) G. R. Porter, *On the Progress of the Iron Trade etc.*
(*c*) *Iron*, Vol. I (1873), p. 294. '30 Years of Commerce'.
(*d*) Mitchell, *Historical Statistics.*

These figures include certain kinds of manufactured iron such as angle, rod and bolt iron, castings, hoops, sheets and wire. Yearly figures are given in B. R. Mitchell, *Historical Statistics*, Table 11, pp. 146–7.

[56] Deane and Cole, *op. cit.*, pp. 224–5. Table of Growth 1805–1907 and estimated value of exports as percentage of gross profits of the industry.

and of the distribution of that trade between the various countries, cannot be attempted here. Our task must be limited to sketching in the extent of the overseas trade in iron and steel to determine its influence upon the total demand for iron and upon the policies, or, rather, the responses of the ironmasters to that demand. To indicate the first point, according to Deane and Cole's calculations this rose from 16 to 50 per cent of the total value of iron production between the years.[57]

The importance of the American market to the British iron industry has been studied by Potter for the years 1815–60.[58] The chief products of the iron industry bought by the U.S. were, of course, railroad iron and pig iron for castings. In fact, although the effective American market for British iron was kept below the potential market by tariffs in the 1820s and 1830s—imported rolled bar iron bore a specific duty equivalent to about 100 per cent of value[59]—the United States was Britain's largest customer. Even before the railway boom of the 1850s, when iron exports exceeded in value both cotton and woollen textiles sent by British manufacturers across the Atlantic equalled American domestic production, American imports had risen from a quarter to nearly a half of overseas sales.

This table, taken from Potter's article, sets out the salient features of this trade.

Annual average five year periods	Exports of bar iron (1000 tons) (including railroad iron)		Exports of pig iron (1000 tons) (including railroad iron)		Value of Total iron and steel Exports	
	Total	Percentage to U.S.	Total	Percentage to U.S.	£m	Percentage to U.S.
1816–20	40	10	4·9	16	—	
21–5	41	8	7·0	17	(1·8)	6
26–30	49	15	8·5	24	1·2	17
31–5	76	28	21·6	44	1·3	23
36–40	112	38	44·0	25	2·4	25
41–5	181	23	102·4	17	2·9	17
46–50	304	49	163·0	38	4·9	35
51–5	567	57	274·0	37	8·9	43
56–60	468*	26	360·0	18	12·5	22
1861	378*	8	388·0	8	10·3	9

* Railroad iron only.

[57] *Loc. cit.* In 1871, iron and steel production is estimated as representing 11·6 per cent of the National Income.

[58] J. Potter, 'The Atlantic Community, 1815–60' in L. S. Pressnell (ed.), *Studies in the Industrial Revolution* (London, 1960).

[59] All railroad iron, however, was admitted free.

That this single powerful market had a profound effect on the rate of growth of the British industry and upon the British ironmasters' interest in U.S. railways there can be no doubt.[60]

As we have seen this phenomenal export trade was achieved in the face of tariffs imposed to protect America's infant anthracite iron industry, which itself doubled its output in the 1840s. This question of tariff protection was one of the major problems of commercial relations in the period under review. The British industry had itself been well protected by a high tariff barrier until 1825, when Huskisson's reductions of the duty to nominal amounts left the industry without a barrier to foreign competition except the overriding factor, cheapness of cost. Thus, despite prohibitive French duties, British iron was imported and could undersell the more expensive French product. Nevertheless, tariffs imposed by the Continental countries and the U.S.A. were alleged by the ironmasters to be a potent cause of depression in the industry. The British ironmasters, it should be emphasised, tried to extend their sales abroad very quickly after the cessation of hostilities in 1815. Guest of Dowlais went abroad canvassing orders and in 1817 was supplying Paris and Brest—South Wales had, of course, a geographical advantage over the other districts for supplying the French market. In 1822 a trial shipment of iron to the U.S.A. was made, but apparently the trade was not yet firmly established. However, foreign duties hampered the British trade considerably, especially as they discriminated against British iron which was rolled. Both the U.S.A. and France allowed Russian and Swedish iron, which was hammered, to enter their countries at a cheaper rate. This disadvantage was removed to some extent by a reduction on rolled iron by Congress in 1832;[61] however a delegation of ironmasters from all over Britain presented a petition to the Board of Trade in 1845, praying for official intervention to secure the remission of this discriminating duty of $8 a ton against rolled iron.[62] In the case of France, a full account of the protection of its iron industry is given in Dunham's

[60] However, despite such remarks as those of Clark of Dowlais, in December 1857: 'The only symptom of a better state of things is that we have got a rail order by which we shall only lose something under £1 per ton . . . ' (cited Orwin, *op. cit.*, p. 140), it is probable the U.S. ironmasters' propaganda for more protection (e.g. in 1849) attributed an unrealistic attitude to the British exporters. It was alleged that the British ironmasters were 'not governed by the cost of production in selling their commodities, but by the extent and urgency of the demand'. *Documents relating to the manufacture of iron in Pennsylvania*, (Philadelphia, 1850) and *Memorial to the Senate and House of Representatives of the U.S.A.* (Philadephia, 1850).

[61] *Crawshay MSS*, M. Wilson and Co.—W. C. and Co., 6 June 1832.

[62] *Add. MSS.* No. 40559 (Peel's papers).

The Anglo-French Treaty of Commerce of 1860.[63] Here, the duties against British iron would seem to be so high as totally to prevent the import of that product into France. The Bourbon monarchy had raised the rate from 10 per cent to approximately 50 per cent on wood-smelted iron, then, discovering that the British were shipping iron smelted with coal, a special duty of 12 per cent was imposed.[64] In 1835 this duty was reduced by one-fifth, and rails were admitted at the same rate as bar iron. Even with this concession the duty amounted to about £8 13s. a ton. It was only the fact that the metallurgical industries of France were so little developed that allowed British iron to enter its market, for, as a French correspondent remarked to Guest, 'which [sc. the duty] although enormously high, will not prevent the admission of English rails, as there is but one foundry in this country—the Creuzot—where they can be manufactured and that establishment could not supply them to any great extent'.[65] Such was British superiority in the production of iron that, in 1844, a duty of 200 per cent upon the cost of production in England of iron steamers was considered 'wholly inadequate' to protect the French iron shipbuilding industry.[66] It was not until the 1850s that the discrimination against coke-smelted iron was abolished, and a reduction in duties to a level of about 55 per cent made possible for the first time importation of English rails in quantity.[67] At this time, moreover, some imports of iron at half rate duty in the case of rails, and free of duty for shipbuilding, were permitted. But even when Cobden, who regarded French iron as the keystone of the protective system, had negotiated the Treaty of Commerce of 1860, with its reductions in 1864 to an average level of 35 per cent for pig iron, 38 per cent for rails, and 29 per cent on bar iron, the French industry still retained some measure of protection in the home market. Curiously enough, under a re-export system to foster French exports, partly manufactured iron from Britain was allowed in free, provided that an equivalent amount of finished goods was exported.[68] It was, however, as a result of these relaxations in the 1860s that imports of pig and bar iron jumped from 28,941 tons and 661 tons respectively in 1860 to 199,994 and 80,887 tons in 1862.[69]

[63] A. L. Dunham: *The Anglo-French Treaty of Commerce of 1860 and the progress of the Industrial Revolution in France* (Michigan, 1930).

[64] Dunham, *op. cit.*, p. 9.

[65] *Dowlais MSS*, J. Cattermole—G. L. and Co., 12 October 1835.

[66] *The Economist*, March–December 1844, p. 1443.

[67] Dunham, *op. cit.*, p. 20.

[68] *Ibid.*, p. 174. N. Senior's *Conversations with Thiers, M. Guizot and other distinguished persons* (London, 1878).

[69] *Ibid.*, p. 175.

It was always the policy of the ironmasters, who had been made free traders by Huskisson in 1825, to advocate the extension of free trade abroad. Indeed as early as 1821 they had tried to secure reduction in the American and French duties, without success. In 1833 Hill, before the Select Committee on Manufactures, Commerce and Shipping, had suggested that efforts should be made to persuade foreign countries to admit British iron at a lower rate of duty, and, in 1845, Peel was approached to secure equality of treatment for British iron by the American Customs. However, the British ironmasters did not secure official support for their free trade campaigns for other countries until 1860, when Cobden negotiated the Anglo-French treaty.

It was, of course, through reaping the fruits of the technological improvements which made possible lower costs of production, and also, if it were required, an improvement in quality, which enabled British iron to compete in the world's markets. And, again, whilst strictly it would be necessary to make a thorough examination of the movements in the prices of the many qualities of iron, all that can be done here is to indicate, by enumerating the price of 'English Merchant Bar Iron at Liverpool', the extent of those economies made by improved production.

English Merchant Bar at Liverpool—
Prices of Iron, Great Britain, 1806–45[70]

	£		£
1806	17·5	1826	11·0
1807	16·0	7	10
1808	15·0	8	9
1809	15·5	9	7·85
1810	14·10	30	6·25
1811	15·0	1	6·25
1812	14·0	2	6·25
1813	13	3	6·25
1814	13	4	7·75
1815	13·25	5	6·5
1816	11·5	6	10·5
1817	8·75	7	10·5
1818	13·0	8	9·75
1819	12·5	9	10·5
1820	11·0	40	9·0
1821	9·0	1	8·0
1822	8·0	2	6·5
1823	8·5	3	5·25
1824	8·75	4	4·75
1825	14·00	5	6·5

Thus, it can be seen that, despite the many fluctuations reflecting

[70] Mitchell, *Historical Statistics*, p. 492.

conditions of speculation, boom and slump, the prevailing level of prices in the middle of the century was roughly half that prevailing in the market at Liverpool at the end of the Napoleonic wars.

3 ORGANISATION OF MARKETING AND DISTRIBUTION IN IRON INDUSTRY

The study of the history of the iron trade, that is the merchanting of the products of the blast furnace and forge by specialist dealers, brokers and export houses, merits a full-scale treatment. In the eighteenth century, when the importing side of the business was perhaps the most important, there were certain Swedish merchant houses at Gothenburg and elsewhere who handled the trade and who had their agents in the two chief English ports handling the wrought iron, Hull and London.[71] But early in the eighteenth century the West Coast centres, Liverpool and Bristol, had established trading links with the American and West Indies markets. We have seen how, early in the history of Coalbrookdale company, exports of pots and other ironware were shipped *via* Richard Champion of Bristol; in 1745, through the Reynolds connection, they had a branch office in Liverpool. Crowley's commercial organisation embraced offices and warehouses in London, of course, and in the Midlands at Stourbridge, Walsall and Wolverhampton. Carron, established in 1760, was also induced to supply southern markets *via* depots in London.

There were, as we can see, two main sides to the iron trade, the home trade—the supply of iron to the ironmonger, blacksmith or railsmith; and the export trade—hardware of all kinds, and in the nineteenth century, and increasingly, railway iron for foreign lines. This division was blurred by the coastwise shipping of iron from Scotland in the North to Liverpool, and from South Wales to Liverpool and Cardiff.

The varied yet exhausting experiences of the ironmasters or their representatives in the 'field', contacting possible customers, during slow progress, have been mentioned several times already in this book.[72] And, of course, the lively and frank diary of Thomas Butler gives us a vivid insight into the pleasures and problems of the ironmaster away from his counting house or furnace. However, all the significant features of this unsophisticated way of trading, of having

[71] See K. G. Hildebrand, 'Foreign Markets for Swedish Iron in the eighteenth century', *Scand. Econ. Hist. Review*, Vol. 6, 1958.

[72] A. Raistrick, *Dynasty of Ironfounders*, p. 10.

a man 'on the road', can best be seen from these extracts from the Dowlais MSS. Here we see the types of potential customers approached by Dowlais, the terms of business they were accustomed to, the nature of competition meeting the Dowlais traveller in the country towns, and finally how the South Wales ironworks were reacting to this market situation.[73]

William Wood to Josiah John Guest:[74]

Portsmouth 13th July, 1823.

'Since I wrote to you last I came back from Plymouth to Exeter and proceeded from thence to Bridport. I went through Honiton, but the Coach did not stop long enough for me to see Mr. Munk to whom therefore I sent a card from Bridport, 8th July. At Bridport I called on Mr Selwood, the ironmonger, whose behaviour was somewhat rude, he wanted no iron, he sold no iron, Bristol men sold iron to the Blacksmiths. I told him I was no Bristol man, but the Traveller of Messrs. Guest Lewis & Company of Dowlais Ironworks in Wales, from whom I believed he had had a Circular—very likely, sir—very likely, sir—I have 20 or 30 tons coming and want no more now, at least said I, permit me to leave a card, he took a card in a superficial way and I left him.

Dorchester, called on *Mr. Bennett Harvey* and saw his Son. The young man was at Cardiff. He is an Ironmonger and his Father (who was not at home) is the Bar Iron Seller . . . N.B. I rather think Mr. B. H. does not sell much Iron, for I saw at Pool half a Ton weighing for him at Heaths into a Cart he had sent for it.

Weymouth, called on *Mr. W. Heath*, who Mr. Wise calls a Mercer & Draper, but he is a Slop seller, and deals in Iron. He deals with Crawshay of London and shewed me an Invoice from him of 5 tons of Swedish Iron at £24, and in the letter inviting him to buy Welsh Iron, acknowledged to be of the most superior quality at £7. 10. 0. but Heath says it is of no use for him to buy Welsh Iron, which he cannot sell. . . . the Bristol men sell to the blacksmiths at 7/6 per cwt. They all take me for a Bristol man till my Card convinces them I am not so, and till I have assured them that Messrs. Guest Lewis & Company are far too high in the Iron Trade to do business in so degrading a way . . . Mr. Harvey, an Ironmonger whose shop as well as himself wears a respectable appearance, says he sells very little Iron but what he wants he buys in Bristol . . . (This is the case with numbers of Ironmongers in different places, twenty of whose Orders would not make a good one, only fit for Bristol men who, by selling to them at low prices, put it out of the power of the real Iron Merchant to serve them) . . .

Southampton, *Charles Baker* has a large and well stocked Iron Warehouse. . . . He says the Dowlais Bar Iron is good but the Pig Iron execrably bad, so bad that 30 tons he had some time ago is mostly unsold at this

[73] *The Diary of Thomas Butler*, 1796–99 *passim* (London, 1906).

[74] M. Elsas (ed.), *Iron in the Making* (Cardiff, 1962), pp. 112–15.

time . . Mr. Baker says his wish is to deal with the Dowlais Company, but his duty to himself obliges him to give the preference to such as will sell him cheaper.

Wood's conclusions about the possibility of this kind of trading in this part of the country are interesting:

'Thus, Sir, you see that the three letters I have written you are all of the same stamp, contain the same kind of information, and all prove that no Orders can be obtained on this Coast till the influence the Bristol men have in these markets is done away with. . . .

I hope that, unsuccessful as I have been in *obtaining Orders*, my Journey has been productive of one thing necessary the exposing the state of the Trade in the West and South of England, which will enable you to judge of the means of rectifying it to your future advantage.'

The last extract, a reply from a firm of Bristol merchants to a letter from Dowlais, evidently complaining of the Bristol system of selling to 'little dealers', and which provoked a strong assertion of independence, suggests that the ironmasters were powerless, on the whole, to put pressure on their customers.

' . . . It seems an odd system to sell to little dealers, who club together to make *a small cargo*, on the same terms as those who take so large a quantity of the Works, and who run risks and divide your *Orders* from their troublesome Contingencies, whether you are disposed to sell to one and all at the same price (whether for 20 or 2,000 tons) is not for us to question, but we doubt very much that such Houses will get a preference in orders *in dull periods* and we think that such a plan is not feasible nor cannot last.

For ourselves we can only say that we shall buy as well as we can, of whom we like, without being dictated to by any body of manufacturers, *how we are to buy and sell.*

In fact, again, the only answer to meet competition in the home trade was to establish depots of their own and to eliminate the middleman. Even then competition could only be curbed, temporarily, by market agreements at the quarterly meetings.

In the case of Cyfarthfa, where the merchant's interests from the first were at least as compelling as those of the ironmaster, the position was more complicated and has been surveyed by Addis in *The Crawshay Dynasty.*[75] The resulting conflict of interests between

[75] *Op. cit.*, pp. 98–106.

the works and George Yard, when, at times, the London House sought a complete monopoly of the output at a 1 per cent commission, do not concern us here. Apart from these considerations, however, were the facts that the South Wales office, possessing a specialist knowledge of the South Wales and Bristol markets, could more efficiently serve their area, and it appears that this was the case until 1834, when William Crawshay I died.[76] The London House, centred in the headquarters of the dealers and brokers busy in promoting the export business and railway orders, acted as an important and necessary intermediary. Moreover, until 1850, when the railway system had effectively linked up South Wales with the financial capital, it was necessary to keep large stocks at London to meet these orders.[77]

London still continued to be the financial clearing house where the negotiations for the shipment of rails and taking up of railway bonds took place, yet in the nineteenth century, Liverpool rose to pre-eminence as the headquarters for the export of British iron all over the world. By the 1850s it was claimed that the north-western sea-port and commercial centre was 'now the greatest shipping port for iron of any in the world'. As Braithwaite Poole, in making that claim, said: 'We receive supplies from the west of Scotland and South Wales by sea, both for consumption in this part of the Kingdom and for exportation; also from Whitehaven, Staffordshire, Shropshire and Yorkshire.'[78] The commercial wealth of Liverpool in the eighteenth century, as is well known, had accrued to it from the port's interests in the sugar and slave traffic: in the succeeding century, despite the cutting off of these sources of profit with American independence and the abolition of slavery, Liverpool was still able to exploit its geographical advantages for the Atlantic trade. Hence the port was able to channel through its counting houses and wharves the industrial products of Britain for export to the American continent's frontiers of settlement and economic advancement. As a result of this trade Liverpool had a trading community of about fifty commission agents, dealers and others—'usually denominated iron-merchants'.[79] Poole estimated that these merchants, buying and

[76] Addis, *Crawshay Dynasty*, p. 28. In 1822, William Crawshay I asserted: 'We never yet travelled to sell iron, when we do the price must fall . . . ' (Addis, p. 25).

[77] Addis, *op. cit.*, p. 100. Again, as Addis points out, if the two sides of the business were operated as separate entities, then the selling agency could be operated to protect Cyfarthfa during times of depression by bringing up surplus stocks.

[78] B. Poole, *The Commerce of Liverpool* (London and Liverpool, 1854), p. 47.

[79] *Ibid. Dowlais MSS*, V. Higgins to Guest, Lewis and Co., 10 May 1836. 'We

selling iron on commission—charging 10s. per cent—handled exports exceeding 1000 tons each working day. It was claimed that this trade accounted for one-fifth of the total British export of iron and steel products in the early 1850s.[80] This export of more than 300,000 tons of iron from Liverpool's docks came from the following sources:

IRON TRADE OF LIVERPOOL (1852)[81]

Source of Iron		Pig iron (tons)	Bars (tons)	Total (tons
Coastwise	Scotland	78,916	15,534	94,450
	Wales	11,373	126,481	137,854
	England	6,280		6,280
	Ireland	1,064		1,064
Inland	Staffs. Shropshire Derbyshire Yorkshire	1,000	134,000	135,000
	Total	98,633	276,015	374,648

Glasgow at this same time was reckoned to be the greatest stock-holding centre for the iron industry in the country.[82] It was, as we shall see, the centre of a highly-evolved system of marketing and commercial transactions in iron. The Scottish industry was not only the supplier of overseas markets but also to the other metal manufacturing centres of England. The port of Glasgow enjoyed certain advantages in coastal shipping freights over its chief rival, South Wales—4s. to 6s. being the cost of freight charges to Liverpool as against 7s. 6d. to 8s. 6d. borne by the Welsh ironmasters. So, as the above figures of Liverpool's iron trade show, Glasgow was the most important source of supply in the whole country. And Poole's further set of figures, indicating the stocks of pig iron held, the

beg to inform you that it appears to be a general disposition of the Ironmasters to establish Agents here for the sale of Iron. Messrs. Botfields and the Coalbrookdale Coy. have just placed Agents here. The New Works Company at Rumney have their Agents and others we have are doing so, in addition to those already established . . . so that the business here must come to . . . an Agency business.' (Elsas, *Iron in the Making*, p. 133.)

[80] Poole, for the year 1852, gives Liverpool 315,839 tons out of a total export of 1½ million tons (Poole, *op. cit.*, p. 51).

[81] Poole, *op. cit.*, p. 51. In addition Poole estimated that Liverpool possessed 16 foundries using 43,000 tons of pig iron yearly and the shipbuilding yards another 15,000 tons of wrought iron.

[82] Poole, *op. cit.*, p. 48. 'The principal stocks of pig iron held in this country are kept on wharfs and in yards at Glasgow, and on premises belonging to the iron-works in the neighbourhood as Gartsherrie, Dundyvan, etc.'

quantities shipped to coastal and foreign ports, makes it clear that Glasgow rivalled Liverpool's wealth and influence in the iron trade.

SHIPMENTS OF PIG IRON FROM SCOTLAND

	Coastwise (tons)	**Foreign (tons)**	**Total (tons)**	**Stock of pig iron held 31 Dec. each year (tons)**
1846	257,857	119,100	376,951	145,000
7	227,005	143,460	370,465	90,000
8	227,833	162,151	389,984	100,000
9	221,943	153,183	375,126	200,000
50	190,083	134,576	324,659	275,000
1	260,080	192,676	452,756	350,000
2	199,950	224,070	424,020	450,000

Source: Poole, *op. cit.*, p. 49.

This is not the place to trace, if it were possible, the history of Glasgow as a commercial metropolis and as an export port for the iron industry, both during the early days of Carron and during the post-Neilson hot-blast boom; however, we may notice briefly the vital importance of the Glasgow merchants to the successful foundation of the industry itself through their own resilience and enterprise in tapping overseas markets. As Campbell, the historian of Carron Company, shows, time and again they came to the rescue: when the American War of Independence threatened those markets they diverted their sales to the West Indies. This position of dependence upon the Glasgow merchants' acumen was, indeed, acknowledged by the company itself; as Gascoigne told his committee, 'it would be absolutely necessary for the Company to extend the Warehouse at Glasgow and to keep a Stock of Goods there for the export mercat . . . in order that the Merchant may supply himself at the Company's warehouse with every article in the Iron Trade which would be a great means of securing the orders to the Company'.[83]

Apparently, the West Indies market supplied by Carron through the Glasgow agents continued to be important until well into the nineteenth century. Nevertheless, the line of development at Carron demonstrated a reversion to the earlier forms of marketing discussed earlier and as exemplified by Dowlais. In the last quarter of the century, so it was argued, improvements in 'communications and transport had removed the dependence of Carron on well-stocked warehouses'. Accounts of agents were transferred to Carron and the warehouses converted into showrooms. Sales, so the general manager affirmed, could now be better effected by advertising and by the travelling salesman. It would be wrong, of course, to extend this as

[83] Campbell, *op. cit.*, p. 144.

a general tendency in the reorganisation of the iron trade, and, most certainly, the 'new' methods were more applicable to the sales of manufactured goods, like the Carron stoves and so on; for the traditional raw material for the town and country blacksmith (for surely more horses were employed by industry in the towns than by farming) the iron warehouse in each town carrying its stocks of rods, bars, horseshore iron, etc., was the most important distributing centre for the iron trade in Britain.

4 THE COMMERCIAL ORGANISATION OF THE SCOTTISH PIG IRON TRADE IN THE NINETEENTH CENTURY

The organisation of the trade in iron, by ironmasters and merchants, can be said to have reached its most advanced state in the Glasgow market. Scottish makers had concentrated on high outputs at a low cost; moreover, the greater part of Scottish production was for shipment—either directly to countries abroad or coastwise for delivery to the other iron manufacturing districts such as the Black Country *via* the port of Liverpool. As a result of these commercial circumstances there arose a system for storing iron of various grades in warehouses and for the future delivery of this iron to the eventual buyer. Inevitably this system was operated by speculators—ironmasters and jobbers alike—who could make profits on the turns of the market, e.g. by a rise in the price of iron.

Speculation of this kind had appeared in the first boom in the Scottish industry in 1836, but it became a feature of the trade from 1845 onwards.[84] As Campbell explains, it was apparently at Liverpool where it was the practice of the iron merchants to buy large quantities of Scotch pig iron for future delivery and to make 'future' sales—promising to deliver a certain amount of iron at a certain date—that the system had its beginnings. The merchants, like bankers, issued scrip representing more iron than they could lay their hands on at any particular time.

There were certain difficulties, it is true, about these transactions. First, although transfers of the iron could be made by the original purchaser, the contract was not legally assignable and therefore the subsequent purchaser had no legally enforceable rights against the ironmasters. The original purchaser might make arrangements with the maker of the iron unknown to the second buyer; or the ironmaster, in the interval, might be declared bankrupt and the iron be assigned to his creditors. Again, if the bill of exchange drawn for

[84] R. H. Campbell, 'Developments in the Scottish pig iron trade 1844–8', *Journal of Economic History*, Vol. XV, No. 3 (September 1955).

the payment of the original purchase were to be dishonoured, the second purchaser could not demand delivery from the maker.[85]

At this stage, in September 1844, the ironmasters, refusing to run the risks of speculative buying controlling the market, adopted measures to put an end to 'this system of reckless and unprincipalled (*sic*) gambling'.[86] They refused 'to acknowledge any orders upon themselves . . . until they [were] called upon for the actual delivery of the iron itself'. This meant that the ironmasters would refuse to record these transfers of orders in their books; the iron would have to be transferred physically from one speculator to another. That was, of course, unpracticable as there were no suitable central storage yards in Liverpool. To solve the problem a Scotch Pig Iron Association was formed in Liverpool in May 1845. Its aim, among other things, such as the collection of trading data for its members, was to establish stores in Glasgow, where the iron might be warehoused on the most reasonable terms (i.e. close to the furnaces), and such a judicious system as might give the owner of the iron in store all the advantages of the London Dock Warrants.[87] This was a success; certainly, by the end of that month, speculators were taking delivery and housing their iron in Glasgow. The firm which came to be particularly associated with providing these facilities and issuing warrants, charging a rent to their clients for the service, was the well-known house of merchants and shippers, William Connal and Company. The following description of the working of the warrant system is taken from Campbell's article 'Fluctuations in Stocks: A nineteenth century case study':[88] 'Most of the nineteenth century warrants represent 500 tons of good merchantable brands of pig iron (300 tons of No. 1 and 200 tons of No. 3). The warrant could then be sold on the Pig Iron Market, which met daily, morning and afternoon, in Glasgow Royal Exchange. Settlement was made either on a future date, or more generally, and at certain times only, for cash. Alternatively, it was possible to have a warrant discounted at a bank on much the same terms as a first-class trade bill.'

For the next few years, until about 1850, the two methods of sales existed side by side. Some ironmasters, of course, found the old method of promising to make a future delivery more profitable, if risky. As pointed out previously, they could sell more iron than they

[85] Campbell, *loc. cit.*, p. 214. [86] *Mining Journal*, 14 August 1844.

[87] Quoted in Campbell, *loc. cit.*, p. 214.

[88] *Connal and Co. Ltd. 1722–1946*. This said that William Connal, Jun., learning that some Irish clients held the scrip of a Glasgow ironmaster for the delivery of pig iron previously purchased, decided to open a warehouse to provide greater security for buyers of iron (*op. cit.*, p. 10).

could produce.[89] Moreover, some makers gambled by 'selling short', by selling delivery orders when the price was high and buying them back when the price fell. As a result of these facilities for speculation, either by the ironmaster himself or by the broker, there were complaints voiced in the *Mining Journal* of 1849. In a long letter addressed to ironmasters, dealers and others in the pig iron trade of Scotland the writer expressed the opinion that 'the extraordinary system you have latterly adopted for facilitating the transference of pig iron, or of its representative, excites the greatest surprise and distrust amongst the most well-informed mercantile men'. It was alleged that, on the one hand, it enabled ironmasters to raise money without payment of interest . . . storage charges were avoided; on the other hand, the system led to a flooding of the market with scrap at a discount. Thus dealers' expectations were upset. It was concluded that the system led to a depression in the industry since investors were discouraged from purchasing iron.[90] Against this, however, it was contended that the warrant or scrip market could be a simple method to protect railway contractors and others against fluctuations in the price of iron during long-term contracts. In 1852, when the Glasgow iron market was founded with members meeting in the Royal Exchange, business was conducted in pig iron warrants.[91] However, in 1856, legal action was taken against the firm of Connal in an attempt to bring the system to a halt. The issuing of these storage certificates was described as 'an unwarrantable extension of credit', and again it was said that the ironmasters were stimulated to produce more than they ought, by being given access to this form of security which could raise loans from Scotland's bankers.[92] We shall examine briefly the observable influence of the warrants system on the Scottish iron industry; the report of the trial, just referred to, ended with an appeal to keep to morality of commerce pure! At all events, the Scottish market retained this system of warrants. The transactions of the Glasgow iron ring were formalised into an iron trade association in November 1864. In 1890, an attempt to legislate for

[89] Campbell, *loc. cit.*, p. 215, says that in May 1845 the iron available was only 30 per cent of the delivery orders current. In 1852 the Bairds were alleged to have cleared £400,000 in six weeks on these speculative transactions. (*Mechanics' Magazine*, June—December 1852, p. 353, quoting *Falkirk Herald.*)

[90] *Mining Journal*, 1849, p. 481. For an account of the impact of the system on the Scottish iron trade see Campbell, *loc. cit.*, pp. 215–24.

[91] C. C. Brewster, *The Centenary of the Royal Exchange*, Glasgow, 1827–1927 (Glasgow, 1927). For some years before 1845, there was regular dealing in pig iron in the Old Tontine Room in Glasgow, which had been the early home of the exchange.

[92] *The Economist*, 13 June 1857, quoting *North British Daily Mail*, 5 June 1857

the regulation[93] of the system was abortive. A Pig Iron Warrants Bill was read in the House of Commons and dropped with no debate. The system survived until the decline of the Glasgow iron trade, when, in 1916, the iron ring was practically closed down.[94] The system was particularly associated with the Glasgow market for iron; however, through the extension of the activities of the Scottish iron merchants' activities to other iron centres, it was also adopted, for example by the Cleveland and Cumberland districts.

That such a system should be the instrument of speculation and the chief reason 'responsible for the violence and variability of the fluctuations' in the price of the pig iron, is easily understandable. Campbell gives figures of the (average) prices prevailing from 1845 to 1873 in the Scottish pig iron trade;[95] they range from 39s. 10d. per ton in 1851 to 116s. 3d. in 1873. The highest increase in price occurred between 1871 and 1872 when the price jumped from 59s. 10d. to 104s. per ton, the biggest fall in price in 1846–7 when the price slumped from 63s. 11d. to 44s. 7d. It is extremely likely that shorter-term fluctuations were even more severe. Now, as Campbell points out, 'though makers fixed prices independently, they had under normal circumstances to conform to the prices of the speculators—if their iron was of a brand commonly sold in the Glasgow market'.[96] The maker could sell either directly to a customer or to a broker, or put it into store himself. The buyer requiring iron for use could buy from the cheaper source—either from the ironmaster or by buying a warrant and withdrawing the iron from Connal's store. So the makers' prices and the prices in the market had to fall in line.[97]

The speculators, who may have been outside the trade, and who 'gave much of the impetus and character to the activities of the Glasgow market', assumed the financial burdens of storage, as for

[93] Brewster, *The Royal Exchange*, p 31.

[94] *Connal and Co. Ltd.*, p. 12. See also R. H. Campbell's articles in *Glasgow Herald*, 17 and 18 June 1954 and his 'Statistics of the Scottish pig iron trade, 1830 to 1865' (*Journal, West of Scotland Iron and Steel Institute*, Vol. 64. 1964–7, pp. 282–9).

[95] Campbell 'Fluctuations in stocks: a nineteenth-century case study' (*Oxford Economic Papers*, new sei ies, Vol. IX, no. 1, 1957, p. 43.)

[96] *Loc. cit.*, p. 48.

[97] Campbell, 'Scottish Pig Iron Trade 1844–8', p. 215, fn. 6. In April 1845 the meeting of the ironmasters fixed the price at 110s., but warrants were being bought at 120s. When the price fell, the makers were always much more reluctant to lower their price to that of the market. At their July meeting they reduced the price by 10s. to 80s., but speculators were selling at 57s. 6d. At the August meeting the ironmasters conformed and reduced the price to 65s., the price of the speculators.

example when warrants were bought during a depression and held for the expected recovery in price. The intensity of the speculators' activities in the market was determined mainly by the fluctuating Bank of England discount rates; if the rate fell, this led to an increased demand for warrants and a rise in the rate to a slackening of interest. Moreover, the speculators, reflecting a degree of hypersensitivity to the swings in the market, bought indiscriminatingly during periods of prosperity and unloaded their purchases as quickly as possible when the trend went the other way. Indeed, the willingness of speculators to buy iron at improving prices was often an important factor in initiating recovery in the industry.[98] There was almost always someone willing to take up warrants, thus the Scottish ironmasters were protected, in some degree, from the insecurity of the frequent depressions; they could usually sell their iron even if for a very low price. Of course, there was the possibility, considered by Campbell,[99] that such a storage scheme could have been used by the ironmasters in order to stabilise prices. Iron could have been taken into or out of stock to minimise price movements. But this was not so; first, because the storage system functioned only because of the willingness of speculators to hold the warrants, wishing to take advantage of the price changes. The result was, as Campbell concludes, 'it produced even more violent fluctuations than took place in areas where iron was not stored'.

It was thought by one writer that the warrant system had no disadvantages for the Scottish iron industry.[100] But, if price fluctuations were intensified, then this, in the light of many attempts by producers to organise stability of prices and production, would be contrary to this deep-rooted industrial experience. Campbell does in fact admit that, in the short term, the ironmasters found the system beneficial on a short-term view; but the one great disadvantage of speculation was that it distorted the ironmasters' impressions of long-term prospects. In Campbell's words: 'They anticipated that the

[98] Campbell, 'Fluctuations . . .', p. 50. He quotes from *The Economist* of 1863, pp. 1264–5, an article on 'Fluctuations in the Scotch Iron Trade': 'The most important advances in price have been established while stocks are at their maximum, and the greatest falls when they were at their minimum—the principal change occurring at the beginning of each period of prosperity and depression.'

[99] The Scottish ironmasters' way of meeting a depression was always to reduce wages rather than to restrict output. Campbell, however, mentions that on three occasions, 1850, 1860, and 1866–7, were ironmasters' agreements to limit output put into effect. But 'the immediate cause every time stemmed from labour troubles brought to a head through wage reductions as a means of combating low prices' (*loc. cit.*, p. 52).

[100] G. H. Hull, 'Three Systems of Storing Pig Iron', *Cassier's Magazine*, Vol. XVIII, cited by Campbell, *loc. cit.*, p. 53.

extremely high prices were likely to continue at least for a time and failed to realise the insecure foundations on which the prices were based. These undue expectations of future profitability led to more capital investment than might otherwise have taken place.'[101] There was excessive capital investment in the 1840s. For example, twenty furnaces were being erected in 1845, as a result of the high prices of that year.[102] By the end of 1846, resulting from a decline in foreign orders and falling prices, it was difficult for a number of these new ironworks to pay their way. Some of the new works were unprofitable by the time they came into production. There ensued the characteristic battle between the ironmasters and the workers over wages, which the ironmasters, armed with large stocks of coal, iron and Irish labour, won. Even so, the iron industry plunged into the general commercial crisis of 1847—prices came 'thundering down' to 46s. 6d. per ton from a peak of 85s. in the first fortnight of 1846—and the iron companies were only saved from bankruptcy by the credit of their banks.[103] The Ayrshire Iron Company failed with liabilities of £250,000.

Of course, the large-scale flotation of railway companies in 1845, and the prospect of unlimited orders for railway iron, led to a general attitude of over-optimism and expansion of production throughout the industry. Scotland was not alone in its speculative activity. In South Staffordshire, 62 per cent of the blast furnaces went out of blast, as against 31 per cent in Scotland and 22 per cent in Wales.[104] Nevertheless, from the point of view of the industry as a whole, the speculative system in Scotland, together with other illusory advantages such as a flexible banking policy and the issue of paper money,[105] did sharpen regional competition. With such a competition in the home market, it is not surprising that English and Welsh ironmasters should have regarded Scotland as a dangerous invader. Even before the warrant system had been instituted at Glasgow, in 1842 *The Times* had voiced the opinion of many English ironmasters of the evils of this unrestrained expansion: 'Last year,' it commented,

[101] Campbell, 'Fluctuations . . .', p. 54.

[102] Campbell, 'Scottish Pig Iron Trade', p. 216.

[103] *Ibid.*, p. 221.

[104] J. H. Clapham, *op. cit.*, Vol. 1, p. 429.

[105] This is the statement of a Midlands ironmaster before the *Midland Mining Commission*, 1843, Evidence of G. Ward, p. 81: 'I found there (in Scotland) wages were lower and the mineral lying nearer the surface. I came back firmly resolved that Staffordshire was not the place to make cheap iron. I also found in Scotland £1 notes, instead of gold, and above all, I found Irishmen (who had been used to get 7d. a day, and were now getting 2s.) lauding the free shop they had to deal at.'

'Scotland . . . produced an enormous quantity of iron with which she has deluged every part of England without the least regard to the important fact, that the demand all the time was decreasing. . . . What is the result? Prices have been forced down even in Scotland, below the cost of production, the workmen in the meantime being ground to the earth to enable the makers then to continue their insane conflict with each other as to who shall produce the greatest quantity and sell it at the lowest price!'[106] This undoubtedly was an exaggeration, but, as we have seen, the Scottish ironmasters were enabled to continue production even during the depression in the trade generally, and other districts felt the keen blast of this competition, which aggravated the difficulties arising from their own individualism. This erratic course of expansion was, in fact, one of the major problems of the industry during the nineteenth century.

[106] *The Times*, 5 December 1842.

CHAPTER XII

The Ironworkers

> 'Some industries take a relentless hold on all who work in them. They may demand much in courage, skill and team-work. The reasons may be practical, historical or emotional—yet the men become the servants of the machines. The grip may be broken by depression or distress, and the stronger would be the aftermath. Or the hold may grow tighter and tighter, till the victory of the machine is complete, and the individual crushed in the battle.'[1]

This quotation concerns the modern steel industry; yet there is a case for thinking that this relationship reached its extreme limit in the nineteenth century. Today, at least, whilst the individual steel worker is dwarfed by the size and force of the massive scientific equipment he manipulates, and where automative processes are superseding the human element, he is protected from the worst excesses of intensive industrialisation—the attitude that the worker is a mere factor of production expendable at the rate set by the market.

The Industrial Revolution in the iron and steel industry was symbolised by the smoking, country-devouring slag-heap; for the blast furnace—no machine—was a sacrificial fire taking its toll from its servants, the iron-workers, and the surrounding countryside. The field by the stream would be changed very rapidly from a green, rural site with its mantled, soft contours to a harsh, stark gash of grey and black, enlivened only by the lurid glow of fire and the billows of steam. The slow muted rhythm of rural life would be shattered by the unrelenting din of the strike of metal upon metal and the hiss of steam as if stunning, with the rapid reverberating hammer blows, the victims of this onslaught of industry into silent submission. Coal and iron had taken possession. This impact of a new machine civilisation, which if it had a soul, could only be fire, and which we today, accustomed to the pace and noise, still have cause to fear, must have manifested itself in its most terrifying form in the iron works of the Industrial Revolution. Today, if we visit a place like this, which has

[1] *The Times Educational Supplement*, 22 June 1951—Training Steel Makers.

suffered the ravages of industrial corrosion, such as Merthyr Tydfil, only the burnt-out shell of the nineteenth-century iron town remains. We see the amphitheatre of former hills, where nature is trying to re-assert itself with sparse tufts of thin grass on the guttered barrenness of pit and slag heaps, a grim reminder of the volcanic energy of the iron masters who caused the land to be so convulsively upturned. We see what is possibly more depressing, 'man's inhumanity to man'. in the town of crumbling regimented rows of smoke begrimed industrial cottages, whose severity of line and meanness and drabness of appearance deform the impressive features of the landscape, both natural and unnatural, like rows of decaying teeth. The place is like a skeleton; for what is particularly missing is the heartbeat of a live community—the sense of all-pervading activity, of men working. It was this motion of men at work, in the heyday of the town, most obviously and most spectacularly made known, by the throb in the air and the smoke in the daytime and the flames lighting up the skies at night, which forged the Welsh valleys into industrial towns. The result was hardly a community living 'in the light of a tradition which is squarely based on the satisfaction found in daily work', but nevertheless, a society in which the ugliness, the noise and harsh insatiable demands of the blast furnace had a meaning. Thus life seemed bearable.

As a projection of the social background we can first of all examine the conditions of work in the industry. These conditions were the subject of several official enquiries fromthe 1840s onwards, in connection with the employment of children. Detailed first-hand accounts thus exist to draw upon.[2] The Welsh valleys have inspired their novelists, themselves influenced by the elemental example of Emile Zola writing about the coal mines of Northern France; however, the task of documenting the impact of the iron industry on a pre-industrial town like Merthyr Tydfil still awaits the historian. Here, the aim is more limited to describe the conditions of work in the industry.[3]

The labour problem, from the ironmaster's point of view, was that of providing skilled and unskilled men for the new blast furnaces, forges and rolling mills and disciplining them into an efficient body of workers; from the workers' point of view, it amounted to working for a living under conditions which were severe.

[2] For example, *Children's Employment Commission*, 1842. Appendix to 1st Report, Vol. II.

[3] The reader is referred particularly to A. H. John, *op. cit.*, pp. 58–97 *passim*, pp. 169–70, 172, for a detailed examination of labour in the South Wales iron industry.

The skilled men, who could form a nucleus to train raw labour to the operations of the furnace and forge, were throughout the Industrial Revolution brought by the ironmasters from the old-established works. To the examples often quoted, of the Coalbrookdale men employed in the establishment of the Carron Ironworks, and of the migration of the Guests from Broseley to Merthyr, many others can be added. Butler in his tour among the Shropshire ironworks in 1815 came across the manager of Madeley Wood furnaces who had been formerly a puddler at Kirkstall Forge[4]—his name, Williams, suggests too that originally he had come from Wales. John mentions that skilled men were imported into Cyfarthfa in 1816 to break a strike.[5] He is concerned, however, to show that the movement of skilled labour into South Wales was short-distance migration. He says 'the bulk of the long-distance migration of skilled men was probably the result of some special need, rather than a general lack of skilled labour.'[6] But in 1818 there was a movement of puddlers from Tredegar and Nantyglo to Staffordshire and Shropshire, because of an advance in wages there; and conversely, Guest had sent an agent to Staffordshire to seek men.[7] There is no reason to think these migrations of labour were unusual and complaints of ironmasters about men being enticed away from them were numerous.

Having attracted their men, it was the job of the ironmaster to keep them—doubtless difficult at times because of this mobility of labour just mentioned. The more formal method, which naturally tended to disappear with the irregular expansion of the industry, was the bond or contract. As late as 1893, John Davis signed an agreement to serve the Low Moor Company as a puddler for three years upon certain fixed conditions about prices per ton, house-rent and coal free and receiving the expenses of family removal.[8] It is more understandable, of course, that a manager on appointment should sign a contract defining his terms of service. There are several instances of this among the *Dowlais MSS*. In March 1792 an agreement was signed between the Dowlais Company and Robert Thompson, the new manager, with a twenty-one years term (but determinable at the end of seven or fourteen years). Thompson was to receive

[4] *Kirkstall Forge MSS.*

[5] John, *op. cit.*, p. 62. The date of his authority, however, is 15 June 1813.

[6] *Ibid.*

[7] (*a*) *Dowlais MSS.* 'These men have set off, without having given any previous notice of their intention to leave, and I am inclined to think that upon the report they make will depend whether the men here stand out for an advance or not.' Letter to J. J. Guest, 18 February 1818. (*b*) *Ibid.*, 5 June 1818. 'I believe it is known by every ironmaster in the country that I am looking out for men.'

[8] *Souvenir of Low Moor Ironworks 1791–1906*, p. 45.

£250 per annum plus one-sixteenth of the profits, plus travelling expenses, with house and garden rent and tax free, coal, a cow and a horse. A feature of this agreement was a penalty clause in case of default for £2000.[9]

The problems of management were, as was to be expected of these enterprises still suffering from growing pains, diverse and trying. Of these, the disciplining of an ill-assorted body of men, newly brought together, into a co-operative efficient body of smoothly working automata was not the least important. Again, the *Dowlais MSS* reveal many of the problems encountered during the Napoleonic wars. We find several directions to Thompson and Guest, the managers, to get rid of fillers who spoiled the quality of the metal; to see to puddlers who employed boys for this responsible job (the concern of W. Taitt was not, it should be added, out of humanitarian feelings, but for the quality of the bar iron). On the other hand, the company opened a shop to supply necessities to ensure that their men were fed during the famine conditions of 1799. This was bound up with the exploitation of the truck system, nevertheless, at times, the need to obtain food was an overriding consideration.[10] Drink too, perhaps never in short supply, was issued to the workmen; Gilpin reported in 1817 to Dowlais that previous to the recent depression 'they (the workmen) were at 4s. 6d. and 5s. per day and two quarts of ale or porter'![11] The same custom prevailed too in Shropshire and South Wales. But the flow of liquor, supplemented as it was by the activities of South Wales smugglers who came selling brandy in the works, gave rise to another source of inconvenience to the ironmaster—absenteeism. Drunkenness was, of course, symptomatic of the acute social stresses of the adjustment of the industrial worker to the new regime, and was not to be sobered down until strong doses of respectability had given him a new code of behaviour.

If religion in its social teaching on the coalfields was to preach that 'the wages of sin is death', the more immediate payment for the hard work that was wrung from them, which made up most of their life, was of greater concern to the colliers and ironworkers. The question of wages in such a varied industry as the iron trades is complex and, indeed, if sufficient evidence existed, is the subject for a separate study. Wage rates had risen during the war period, as Ashton has

[9] *Dowlais MSS*, 30 March 1792.

[10] In 1800, Evan Evans, employed by the Dowlais Company to obtain barley for the workmen, was mobbed by the hungry men and women working at the Beaufort Ironworks when coming back from Abergavenny laden with the precious food. J. Lloyd, *op. cit.*, p. 183.

[11] *Dowlais MSS*, G. Gilpin—W. Wood, 4 September 1817.

pointed out.[12] The peak year would appear to have been 1811–2, when according to Attwood a labourer could earn 20s. a week—in 1812, this had been reduced to 10s. or 12s.[13] In 1815, in Staffordshire, Butler gives the following wage rates: '4s. for a Gin-driver, 12s. women that Bank; common labourer 18s., some 20s. to 21s. per week, ironstone getters 5s. per day.'[14] Puddlers were paid on output and the wages in this department cost 12s. 6d. per ton. In Shropshire, where Butler observed wages were at least 15 per cent lower than in Staffordshire,[15] the following rates were paid: Colliers 3s. 6d. per day at the most; common labourers 2s. 6d. per day. 'The Keepers and Feeders about 20s. per week, but if they produced more than a certain quantity then they had 6d. per ton extra, as a stimulus to their exertions.'[16] For South Wales, John gives few figures for this date—in 1812 at Ebbw Vale, fillers received 21s. a week with house and firing.[17] However, figures for 1816, when a reduction had taken place, suggest that here the rates of pay had been somewhat higher than in other districts.

Earnings in certain occupations would certainly differ between works even if the same rate prevailed in a district, because of the output bonuses, and wages tended to be higher in the districts in the van of progress. Nevertheless, the fixing of wage rates was not left to the demand in the market. Ironmasters both in their own and other districts conferred with each other to prevent competitive wages being given. In 1808, Gilbert Gilpin, chief clerk at Botfields, one of the largest integrated concerns in Shropshire, was writing to the Penydarren Works in Wales giving them a list of prices of labour at Old Park.[18] These, incidentally, reveal little change from those of 1815. Those making refined iron got 3s. 6d. per ton, those making stamped iron 11s. and puddled blooms 12s. Surprisingly, the price of labourers' wages, in 1808 1s. 4d.–2s. 2d. per day, had increased by about 9d. a day on the average. Inter-regional co-operation among the ironmasters on the subject of wages was, unlike that to attempt price and output control, largely informal; it was highly effective, and played an important part in securing reductions to the wartime wages when depression overtook the industry.

Among the workers, for whom organisation was an act of rebellion or revolution, the Ironfounders' Union had come into existence in 1809 at Bolton, its books being buried in the peat of the moors to

[12] Ashton, *op. cit.*, p. 190.

[13] *Minutes of Evidence against Orders in Council*, 1812.

[14] *Kirkstall MSS.* [15] *Ibid.* [16] *Ibid.* [17] John, *op. cit*, p. 82.

[18] *Botfield MSS*, G. Gilpin—R. Ward, 18 January 1808. Gilpin also revealed that short time was being worked, confining the rolling mill to 8 or 9 hours per day.

ensure the secrecy of its proceedings.[19] However, the severity of the common law and the Combination Act of 1800 kept activities of this kind under cover, so that a Midlands ironmaster declared there had been no combination of workmen in the industry during the Napoleonic wars.[20] Yet, if organised activities by workmen were limited, the workers in the iron industry took part in the riots, upheavals and the strikes which were common on the coalfields. The extreme sensitivity of wages rates to the downward movements of the price of iron, the abuses arising from the ironmaster's exploitation of truck, all went to make the iron districts centres of workers' grievances, agitation and of disturbance. These movements, of course, were of the whole populations in these districts, hence it is difficult to distinguish the part played by the ironworkers. Their neighbours the miners and colliers had certainly a well-deserved reputation for being particularly volatile in these circumstances; and as John suggests, they overshadowed, with their much greater numbers, the efforts of the ironworkers. Nevertheless, the puddlers stand out as a troublesome group to the ironmasters. These were, of course, reckoned as skilled workers—although 'class' differences in times of action to resist reductions could have counted for little. There are several instances, in this period and later, of strikes and illegal activities by puddlers. In 1810 there was a strike at Dowlais for a month against a reduction in wages.[21] Nine years earlier, following the riots in 1800 which had been suppressed, the puddlers presented a petition for higher wages,[22] and there are several references to workers' combinations in the following years. In 1813 the puddlers at Dowlais as well as at Cyfarthfa were associated with the Luddites. Crawshay recommended his son to 'persevere with the puddlers until they agree that their Luddite Oath is more honoured in the breach than the observance'.[23] As a measure to bring pressure to bear upon the Cyfarthfa puddlers, Crawshay was contemplating getting puddlers from Staffordshire or other places.[24] In Staffordshire itself the ironworkers organised themselves into bands, pulling carts behind them, touring the Birmingham district when furnaces were blown out at the end of the war.[25] In Yorkshire, it was the fendermakers at Kirkstall Forge who are noted in Thomas Butler's Diary for September

[19] T. S. Ashton, *op. cit.*, p. 208 fn. 1. also *Centenary Souvenir of the Friendly Society of Ironfounders* (1909).

[20] *S.C. Manufactures, Shipping and Commerce* (1833), W. Matthews, Q. 9819.

[21] C. Wilkins, *History of Iron and Steel Trades, etc.*, p. 126.

[22] *Dowlais MSS*, 1801. Letter Book.

[23] *Crawshay MSS*, 18 May 1813.

[24] *Ibid.*, 29 May 1813.

[25] *Aris' Birmingham Gazette*, 8 July 1816.

1799 as having a 'turnout'. He observed: 'They are a set of ill-natured fellows . . . we have had more trouble with them than in all our Works besides.'[26]

What were conditions like in the ironworks for the boys, youths and men who had to endure for twelve-hour stretches the intense heat and feverish activity? These conditions were the subject of several official enquiries from the 1840s onwards concerning the employment of children. Thus we have first-hand accounts to draw upon. In the old-established centres of the industry the grim hand of industrialism was all too evident. However, even the vast ironworks, such as Dowlais, had not yet assumed the closed-in factory-like appearance of today; but, it was reported, 'the ironworks are generally open and roofed, merely for the purpose of protecting the working people from the rain. The floor of the works is paved, presenting an even and constantly dry surface from the heat thrown out from the furnaces and the hot metal in the process of manufacture.'[27] Thus the works might not seem so oppressive to the new recruit to the industry as the more massive and barrack-like structures, the cotton mills in the North. However, if the workers felt freer within the confines of the yards of the ironworks, the effect on the ironworkers' health was certainly more drastic. In the Bible we read how the fires of the furnace 'consumeth the flesh of the smith', and the intensification of the worst of industrial conditions—excessive heat—in the large ironworks took a heavy toll of the lives and health of those human beings who had to work close to the furnaces. The worst consequences, as the Commissioner pointed out, resulted from the fact that 'constructed as the works are to catch every breeze, the transitions are very great and sudden, occasioning a sudden check to profuse perspiration, which manifests itself in chronic diseases of the respiratory organs and in febrile and inflammatory disease; children and young persons are of course, equally with adults, subject to the same temperature, and in winter this is particularly felt'.[28] One possible reason why the demand for puddlers always outran the supply was that the working life of the puddler was ended by the time he was about forty years old,[29] through close contact with the great heat of the puddling furnace and the great exertions required to work the metal. Another occupation too, that of gagger, whose job it was to 'gag' the jaws of the old-fashioned type of hammer be-

[26] *The Diary of Thomas Butler 1796–99* (London, 1906), p. 325.

[27] *Children's Employment Commission*, 1842, p. 473. Report of Robert Franks on South Wales.

[28] *Ibid.*, p. 473.

[29] *Children's Employment Commission* 1862, 3rd Report 1864, p. xxiii, 'they begin to fail at 40 to 50 years old'.

tween blows, to allow the iron to be turned, was one which meant a life shortened by heat and noise.

The ironmasters, of course, tended to look for these causes of premature old age in excessive drinking, not from the severity of the work or from the impurity of the air from the smoke. It was true, as Bremner pointed out, 'one effect of the severe heat and exertion is the creation of a craving for stimulants, such as beer, which at once cool and support the workmen, and to a certain extent no one would grudge them these, but unfortunately, the craving does not always cease with the work'.[30] Thus the work was responsible, directly and indirectly, for brutalising these men. In 1843, the Mining Commissioner urged the need for three sets of puddlers and mill men to shorten the hours of work, which were long and strenuous—twelve hours, without any break for meals. However, in 1864, the puddlers were still doing six heats per turn (1¾ hours each).[31]

In Wales at this time it was suggested the Welsh system of puddling, using refined metal, allowed more young persons to be employed. Boys and girls under thirteen years of age, and young persons from 13 to 18, who suffered most from the severity of working conditions, and whose lot, from the heavy toil and the dangers of burns, was far worse than that of any child employed in a cotton mill, were a numerous part of the industry's labour force. Of just over 5000 workers at the Dowlais Works, in 1843, over 900 were young persons and children under 18 years of age. There were, in addition, 200 adult women workers employed. In all, among 24 ironworks 6311 young workers were wage-earners. In 1864, when it was said the employment of women and girls in ironworks was peculiar to Wales, it was estimated that 4000 children and young persons were still working in the South Wales ironworks. At the same time, in South Staffordshire there were 600 youths and boys (and 200 women) working at the blast furnaces, 3000 as puddlers' underhands and rollers in the district's 100 mills and forges. What jobs did these young children have to do? Girls were mainly employed outside on the tips removing cinders. It was said, 'their dress is peculiar, their labour hard, and their appearance indicates personal strength'.[32] They needed strength too, for another arduous job they performed, as limestone breakers working 12 hours per day, 7 days a week, for wages of 7s. Girls were also employed in the forges as pilers of the broken pig iron ready for the puddling furnace.[33] Boys generally acted as assistants to the

[30] D. Bremner, *op. cit.*, p. 57.
[31] *Children's Employment Commission*, 1862, 3rd Report, Evidence A.17.
[32] *Children's Employment Commission*, 1842, p. 479.
[33] *Ibid.*, p. 478. See also *Dowlais MSS.* On the Employment of Women and

furnace keepers, puddlers and rollers. Boys and youths of 11 to 15 years filled boxes with charges for the furnaces, performing this duty at night in preparation for the next day; younger children called 'pull-up boys' or 'door-drawers' earned between 2s. and 3s. per week in 1843 opening and shutting the puddling and balling furnace doors; others acted as water carriers, or as scrap gatherers. But the most exhausting jobs of all those they were made to do were those looking after the rolling mills. Here, acting as catchers and hookers-on (to support the iron to the roll), roughers (engaged in light rolling) and as straighteners, lucky was the boy who did not receive bad burns, such was the rapidity of the rolling. Moreover, these unfortunate young lads had to run eleven miles or more during a turn. Yet, despite strongly worded suggestions of the Commissioners to the Government to apply the provisions protecting young persons and children of the Factory Act of 1844 to the iron industry, to stop the two main evils of the system—night work and Sunday work—the ironmasters were able to resist and prevent any legislation. According to the employers' point of view, the abolition of Sunday and night work would mean dismissal for women and young persons, besides causing great loss to the masters, because of the dislocation of the works. As for half holidays on Saturdays, these were quite unnecessary because of absenteeism.[34] As to the number of hours actually worked by the various categories of workers in the industry, since these depended to a large extent upon a system of remuneration based on output, it is difficult to give any average figures. However, there is a small amount of evidence for the years 1850–70, for several districts, based on employers' returns submitted to the Board of Trade.[35] The following examples reveal a great diversity not only between different trades but also as between the different districts.

AVERAGE NUMBER OF HOURS PER WEEK, 1850–70

Trade	District	1850	1860	1870	Remarks
Iron mfr.	Bradford	58	58	58	
Furnacemen	Cleveland	84	84	84	25 per cent of the men work 7×12 hr shifts. The men have each alternate Sunday off. Other 75 per cent work about 63 hrs. per week.
„ fillers	„	63	63	63	
Mechanics	„	61	61	59–61	
Furnacemen	W. Cumberland	60	60	60	

Children in the Ironworks of South Wales: 'Girls are employed generally in piling; this is more suited to girls, because of the stooping and rising (girls being shorter and more active).'

[34] *Dowlais MSS*, On the Employment of Women and Children, etc.

[35] 1890 *Board of Trade Return*—Trade (Hours of Work) (375).

Ironworkers	Derbyshire	63	61	59	
Furnacemen	S. Staffs.	70	70	70	Furnace and puddlers men work day and night shifts, changing in alternate weeks, and working 7 days per week.[36]
Puddlers	,,	48	48	48	
Mechanics	,,	54	54	54	
Furnacemen	Wrexham	72	72	72	
Ironworkers	Dalmellington (Ayrshire)	60	60	60	

Nevertheless, we recognise that organisation and oppression often go together. It is important, therefore, to consider these problems: the numbers of persons employed in the industry, the size of the industrial unit, and the various categories of workers employed at the ironworks. First, the size of the labour force in the iron industry. In 1831, according to the Census of that year, there were 4 smelters and 5882 ironfounders in Great Britain. Ten years later when the detailed classification of occupations made possible the first estimate of the number of persons in the industry, the number accounted for in the Census amounts to 39,031. This figure, however, only includes persons who were tradesmen in the iron and steel industry, and no reckoning can be made of the many unskilled persons employed in the industry when the processes of manufacture were so little mechanized. These men employed for so many important jobs about the furnaces and forges, for instance, charging the furnace from the bridge from barrows wheeled from the furnace bank—later to be replaced by mechanical hoists—were included in the general term 'labourer'. By 1851, the total given for in the Census figures employed in iron manufacture, had doubled, and was 80,032.[37] In 1861, there were 123,430 males and 2341 females employed in the basic trades; whilst in the miscellaneous trades, which were becoming increasingly more important and at the same time inextricably bound up with the expansion of the iron industry, there were 181,999 males and 15,921 females working.[38]

As to the representative size of the unit of output, the Census figures of 1851 give direct evidence.[39] Of 685 employers in the iron industry (another 122 did not submit returns), the largest single group employed between 10 and 20 men. According to the official

[36] In view of this remark, the 48-hour week of the puddlers can hardly be correct.

[37] This included moulders and founders, but excludes the other ancillary trades (such as Pipe, Fender, Grate and Stove makers) included in the 1841 figures.

[38] 1861 Census figures quoted in *Children's Employment Commission*, 3rd Report 1864, p. xxix.

[39] 1851 *Census* Vol. 2, (1), pp. cclxxix and cclxxxi, Tables xxx and xxxi.

figures no more than 68 masters had establishments of over 100 men, and only 8 employed more than 350 men. However, it is apparent that these figures do not include any of the great establishments of South Wales.[40] This evidence, then, is not very satisfactory. However, for the year 1870, when the Factories and Workshops returns were compiled, there is a much more complete set of figures. In that year, in England, Wales and Scotland, there were 761 establishments with blast furnaces and iron mills employing 166,542 men and women. Thus the average for each works was 219 workpeople. As Clapham points out, this figure was reduced by the influence of Staffordshire, where the single blast furnace firm and the small ironmill were common. For instance, the average for South Wales (Monmouthshire and Glamorgan) and Scotland (Lanarkshire) was 360 and 405 respectively.[41] Again, according to Clapham, even these large figures were outclassed by the 26 works of South Wales, each with an average of 650 employees.[42] Although the average for all metal manufacturers (including of course, the many small workshops) was a mere 34·5 employees, the large integrated concerns of Scotland and South Wales were on a larger scale at this date than the industrial units of the cotton industry (with an average of 177 workpeople per works).

However, to concentrate one's attention on the number of workers engaged on the actual and finishing operations is to forget the degree of integration pursued in the industry. The workers engaged on the primary processes were only part of the total labour force coming under the control of the ironmaster. There is a great deal of evidence to indicate that in fact the average ironworks, even in the first half of the century, was a concern something very different from that suggested by the statistics. According to Ashton, even at the beginning of the century, in the Midlands, 'It was by no means uncommon to find 300 or 500 men employed in a single works, apart from colliers and others connected with the supply of raw materials'.[43] Fong suggests that ironworks of 1500 employees each were frequently found, if one counts the miners of raw materials together with other

[40] The tables include 5 in the West Midlands division and 3 in Scotland.

[41] Clapham, *op. cit.* (1932 ed.), Vol. I, p. 116.

[42] Calculations based upon the summary returns published in *B.P.P.* 1871 (LXII) do not always bear out Clapham's conclusions; (the figure of 650 workpeople for South Wales must have been taken from the individual returns); for instance, Clapham gives a figure of 209 as the average for the primary ironworks. Without the actual returns, however, to reconcile cases of overlapping (where the furnaces and mills are counted separately), it is not possible to account for the source of divergence.

[43] Ashton, *op. cit.*, p. 100.

workers.[44] In 1833, the Midlands ironmaster, Matthews, who was the owner of a typical concern, employed 400–500 men,[45] and at the same time one of the large Merthyr Tydvil works had an establishment of 1500–2000 employees. Amongst the larger concerns at this period, in 1839 at Tredegar, Homfray employed 5000;[46] the same number was employed at the integrated concerns of the Stourbridge ironmaster Wm. Foster (who also owned works and forges in Shropshire). Dowlais in 1845 was said to be employing 7000 men, women and children.[47] There was, once more, great diversity in size. In South Wales alone, for which the evidence is very full for the year 1842, the number of workers ranged from 59 in the case of the Tondu Coal and Ironworks to the 5192 of the Dowlais Company. Of the 24 works which can be isolated as being primarily ironworks (although most were, significantly, integrated iron and coal or iron and tin works), only two employed under 100 hands and just over half possessed a labour force of more than 1000 employees.[48] The extent of development of one of these major concerns, both in size and in its integrated departments, is shown by a detailed list of the persons on the Dowlais pay-roll in May 1866:

NUMBER OF PERSONS IN THE EMPLOY OF DOWLAIS IRON COMPANY, 1 MAY 1866

	Under 10 yrs.		**10–13**		**14–18**		**Over 18**		**Total**
	B	**G**	**B**	**G**	**B**	**G**	**Men**	**Women**	
Old works, forges and mills	7		44	5	160	55	1091	67	1429
Ifor works			48	20	98	21	573	15	775
Furnaces			13	16	32	59	559	123	802
Steel works			3		2		39		44
Fitting, pattern and smelting shops / Foundry carpenters / Boiler makers			5	1	78	1	310	2	397
Engines, forge and blast			8		16		130	1	155
Engines, locomotive					3		18		21
,, pits					12		126		138
Collieries			237	16	358	34	1342	37	2023

continued overleaf

[44] H. D. Fong, *The Triumph of Factory System in England*, p. 142.

[45] *S. Committee on Mfres.* 1833, Q. 9774.

[46] *The Times*, 15 November 1839, cited by D. Williams, *John Frost* (Cardiff, 1939).

[47] E. J. Jones, *op. cit.*, pp. 56–7.

[48] *Children's Employment Commission*, 1842, Vol. II, pp. 470 and 594.

Mine works, lime-stone quarries	2	96		122	41	1248	200	1709
Masons, roadmen and navvies		19		2		288	1	310
Church building		2		5		244		31
Farms						19	13	32
Stables		2				82		84
Railway station						3	1	4
Washhouse				1		3		4
Traffic				1		17		18
Weighers, croppers				3		48	1	52
Offices, library				11		32	5	48
Schools			7		22	9	17	55
Cardiff yard				6		58		64
Edge Hill mines		28		58		195	1	282
London houses						23		23
	9	505	64	968	233	6237	484	8500

It will be seen how many activities a large concern such as Dowlais was responsible for at this time. Thus, apart from its agents and clerks in its London, Cardiff and Liverpool offices, a considerable number of workers was engaged on work which was concerned with community building, e.g. church building, in the company's schools, and in the Memorial Library. Industrialisation, too, had not yet severed the ironworks' connection with agriculture, since the company still possessed its own farms and, of course, the stables were an essential part of the local transport system. The actual number of men employed on the furnaces and forges formed only a small proportion of the total labour force. The furnacemen were less than one-tenth of the total. The actual distribution of workers in each concern varied according to the nature of the physical resources at the ironmaster's disposal (i.e. rich mines would require fewer miners and colliers), the degree of integration pursued, and the size of the concern (the proportion of men concerned with the blast furnaces would be smaller in a large concern with many furnaces than in the marginal works), and, of course, the final product. In 1843 it was estimated by a Midlands ironmaster that 157 workers (117 men and 40 boys) were 'requisite to supply with minerals a single blast furnace making 85 tons of pig iron per week'. Only 40 of these were employed in the actual smelting process.[49] Almost twenty years later, it was stated that a mere 26 were employed at the blast furnace and another 210 in the mills and forges, a total of 236.[50] In South Staffordshire,

[49] *Midlands Mining Commission*, 1843, p. 55, evidence of Mr. Best.

[50] *Children's Employment Commission*, 1862, 3rd Report, pp. 1–2. In another place, it was stated that the men worked in sets of 12 (with a day and night set);

however, the practice of sub-contracting was prevalent. According to Gale,[51] the organisation of the working of the blast furnace was in the the hands of only two men who appeared on the pay-roll of the iron-master—'the bridgestocker' who was responsible for handling all raw materials and for hoisting them to the top of the furnace (or bridge) and the charging of the furnace; and the 'stocktaker' who performed the preparation of the pig beds and the handling of the pig iron. These two chief assistants of the ironmaster were paid on output, and they paid the under-workers. However, this system was not always carried out. In Shropshire, and possibly in South Wales too (where direct employment by the ironmaster was more common), the direction of the furnace was in the hands of keepers who, being responsible for both quality and quantity, determined the technical details of the composition of the charge, the kind of fuel to be used, the regulation of the blast and the time of tapping. These men worked in pairs and were paid on output. As it was explained in an early pamphlet on the South Wales district, 'The Keeper has a personal interest in the well working of his furnace and if he is a good and steady workman, will watch all these points with constant care. He is paid on the ton of iron produced—all are generally paid in the same way.'[52] At Stirchley in 1848 the keepers were assisted by a hierarchy of fillers and assistant keepers. The occupation of filling requires no explanation, nevertheless, as Bremner remarked, it 'appears to be a somewhat dangerous one, as the flames at times shoot out upon, and almost surround them'. The assistant keepers 'are employed at the hearth scooping out the slag and cinders with a huge spoon suspended from a crane, and from time to time stirring up the contents of the furnace. This is very severe labour and the faces of the men engaged in it have a half-roasted appearance.'[53] Only the stocker of the bridge and the unloader of the limestone received a lump sum, indicating that he was a sub-contractor. As for the other skilled trades in the forge and mill, the master puddler and master roller, the system of sub-contracting was almost universal, as the underhands were also learners of the art and mystery of puddling and rolling.

What were the factors influencing the growth of this army of workers? The iron industry was an industry in motion, as the changing pattern of the rise and decline of the new and old districts reveals.

then there was an average of 10–20 persons employed in preparing and carrying raw materials.

[51] Gale, *loc. cit.*, p. 22.

[52] Anon., *On the Manufacture of Iron* (1831), p. 9.

[53] Bremner, *op. cit.*, p. 39.

The older districts, South Wales, for example, had depended for their skill on the migration of experienced ironworkers from the Midlands. Now the new districts, Cleveland and the iron industry of the North-east, provided better prospects. Moreover the rise of the Continental and American iron and steel industries attracted men and their familes abroad. Indeed, there is evidence to suggest that migration, although perhaps largely accounted for by short-distance movement, widened its bounds as the edge of economic necessity became keener. As was to be expected, there was always the tendency for the more enterprising of the workers to leave an old-established district and go to a new district where perhaps the ironmasters had not yet come to terms with one another about the price of labour. We have seen how the Staffordshire wages were considerably higher than those prevailing in Shropshire. Redford has pointed out that the migration of ironworkers from Shropshire to the Black Country 'was certainly strong enough to have left its mark on the Census statistics'. His statement finds confirmation in several ways; several of the Staffordshire ironmasters had their origins in Shropshire, for instance Adams of Wednesbury and John Bradley of Stourbridge.[54] Then there is the archaeological evidence to be found in several places of industrial migration—in the names of streets or buildings—in this case, Shropshire Row at Coseley.[55] Also, there is the first-hand evidence of the clerk to the Board of Guardians at Dudley in 1843, 'Our removals are chiefly to Shropshire, the Forest of Dean, Bristol, and some to Wales. We also relieve a great many Irish.'[56] An instance has already been cited of ironworkers being recruited from other districts to break a strike, but puddlers who were on strike in one district would travel to neighbouring districts to look for work; or, as the ironmasters alleged, to 'excite revolt'. Gilpin, in 1817, reported movements of Staffordshire puddlers to Shropshire and to South Wales.[57] The surprising mobility of some of these workers in response to better rates of pay elsewhere almost matched the flexibility of wage rates. When the North-eastern district was being developed, the Welsh technicians were recruited to the responsible positions in the industry. Vaughan, who became a partner of Bolckow, arrived by way of Carlisle, but Williams, appointed to succeed Vaughan as manager in 1864, came from Dowlais. Another candidate for the job, again from Dowlais, became manager of the Bowling

[54] Francis Brett Young's novel *The White Ladies* tells also the story of a migrant ironworker from Salop who became an ironmaster.

[55] W. Leigh, *Cholera at Bilston* (1832), Map. The street adjourned the furnaces of an ironmaster named Baldwin, himself a native of Salop.

[56] *Midlands Mining Commission*, 1843, p. xxiii.

[57] *Dowlais MSS.* 6 and 18 December 1817.

Ironworks.[58] Jenkins of Consett was another prominent Welshman. As to the rank and file, experienced Staffordshire men of 'good character and steady habits' were recruited by being offered 'as inducements, good houses and gardens at a moderate rent, with other advantages'.[59] At Consett, there was a similar report of workers being brought not only from England but also from Wales and Scotland. Here too there was a Staffordshire Row, denoting the origin of its dwellers.[60] Scotland, too, shared a sudden acceleration of activity in the coal and iron trades, with its consequent rapid urbanisation centred upon Monkland, in Lanarkshire, which grew from a town with a population of 9580 in 1831 to 27,333 in two decades.[61] With such a marked growth in population, it might be expected that the currents of migration would be clearly defined. But, according to MacDonald, who has carried out a similar analysis for Scotland to that done by Redford for England, the problem of the origin of Scotland's ironworkers is a difficult one to decide. His conclusion is that 'it seems likely that most employers continued the time-honoured and more or less compulsory policy of importing their labour from England'.[62] He quotes as an example, the case of the Dundyvan and Monkland Works which began the manufacture of malleable iron in 1840. These works were reported in 1842 as having drawn 'a great number of their men amounting, with their families, to about 1000 persons, from Wales, Staffordshire and other parts of England'. The inducements for these people to move, again, included good houses, and for some, pay as much as £1 per day. It seems unlikely that these advantages did much to improve the situation of these workers, however, for 'these people . . . planted in a country of strangers . . . receiving very high wages . . . distinguished themselves even in that country by their excesses and irregular conduct'.[63]

The Irish, too, were an important element in Scotland's industrial growth. Indeed, it has been said that the heavy industries 'owed their rise in a great measure to the supply of cheap labour which was thrown upon the market as a result of the Irish famine in 1847'.[64] But they were quite numerous before this; in 1831 they were one in

[58] J. S. Jeans, *Pioneers of the Cleveland Iron Trade* (Middlesbrough, 1875), p. 288.

[59] *Report of the Mines Commissioner*, 1846, pp. 28 and 29.

[60] *Consett Iron Company Ltd: Description of the Works* (Newcastle on Tyne, 1893).

[61] *Children's Employment Commission*, 1842, Appx. Pt. I, pp. 312–3, and 1851 Census Pt. 1, Vol. II, p. 24.

[62] D. F. MacDonald, *Scotland's Shifting Population* (Glasgow, 1937), p. 68.

[63] *Children's Employment Commission*, 1842, Appx. Pt. I, pp. 346 and 367.

[64] W. H. Marwick, *Economic Developments in Victorian Scotland* (1936), p. 132.

five of the Glasgow population.[65] In the iron and coal industries they were recruited as substitutes for the unskilled grades of work in the mines and in the furnace yards. In Wales, too, the Irish element was considerable, especially after the famine, when a large influx coincided with depression in the trade and caused further falls in wages. After 1849, as John points out, previous antagonism between them and the Welsh was increased by the opinion that the Irish squalor had caused the cholera plague of that year. Twenty years earlier, it may be mentioned, the influx of immigrants from the 1820s onwards caused friction. A Midlands ironmaster who had migrated into Wales recorded in 1840, 'I remember when 20 years ago, Mr. Whitehouse brought a number of Staffordshire men into this Welsh district, the aborigines considered it to be such a wanton unjustifiable invasion of their territory, that many severe battles were the consequence; and at the pays the Staffordshire men frequently went to bed with the fear that they should be murdered before morning.' He went on to say, 'It was now the turn of the Irish to incur the wrath of an "Anglo-Welsh alliance".'[66] At Blaenavon in 1840, the Irish comprised only a mere 1 per cent of the population (the English 38 per cent); at Trevethin, which was nearer the English border, the English and Irish were in equal numbers with the native population.[67] The Irish had penetrated into the Black Country, too. Later, in 1862, it was alleged that mines legislation which forbade the employment of women underground and controlled the employment of children had driven ironmasters to employ Irishmen. They were then the masters' scapegoats for accidents in the pits, being 'reckless and careless'.[68]

Such were the pressures of the sudden spurts of demand that in these boom times the supply of workers with some experience in the industry ran dry.

Complaints from neighbouring ironmasters of men being enticed away from them were very common in this period, and the far-seeing ironmasters like Crawshay and Guest took good care not to lose their men to other masters by standing them off from work, when it could be avoided. Nevertheless in times of prosperity they were not above recruiting other men in their neighbourhood. In 1836 Crane of Ynyscedwyn wrote complaining to Guest that,

[65] W. H. Marwick, *Economic Developments in Victorian Scotland* (1936), p. 132.

[66] *Appendix to the Minutes of Manchester Statistical Society*. A. S. Kenrick (of Varteg Ironworks), 'The Population of Pontypool of the Parish of Trevethin, etc., 1840', p. 12.

[67] *Ibid.*, pp. 6 and 23 (English 44 per cent, Irish 5 per cent).

[68] *Children's Employment Commission*, 3rd Report. Evidence p. 15, para 11.

> 'A person stating he was employed by the Dowlais Iron Company to do so gave out publicly in the chapel yards of Cwm Gier Meeting House . . . viz. the three chapels of this neighbourhood frequented by our workpeople, notice that 200 miners and colliers were wanted by the Dowlais Iron Company and to whom £5 per month would be given, and that that Company would be at the expense of removing the goods and families of any parties who would leave this neighbourhood [and come] up to Dowlais.'[69]

Where these recruiting campaigns failed, then recourse had to be made to the local agricultural districts. In 1846, to satisfy 'the demand for labourers in the iron districts of South Wales which has been very great and still continues' (after the railway boom of 1845, the Mining Commissioner reported that an agent had been sent to Pembrokeshire to recruit workers for the mines and furnace yards, being paid a commission of £5 for 50 recruits.[70] The scheme was not a success because the agricultural labourers were anxious to avoid 'the contamination and moral ruin' of the iron districts. But perhaps of more significance—there was no stern economic pressure to drive them to the iron industry for work; the small tenant farmers of Cardiganshire did in fact enter the industry for the winter season of 5–7 months, but only to pay the rent of their farms where they returned for the rest of the year.[71] Moreover, acting against the pull of the iron industry for labour was the indirect benefit which accrued to agricultural labourers in counties bordering on the coalfields, since they received higher wages than elsewhere on account of industry's competition for labour. Finally, of course, the source of the main part of the iron industry's labour force was found within itself. In common with the population of the rest of Britain, not to mention elsewhere, the industrial workers on the coal and iron fields reproduced themselves at an unprecedented rate. If conditions were unhealthy and insanitary in the towns of the South Wales and the Black Country, so that plagues of cholera raged, killing off both the old and young, then the social conditions also did much to keep up the birth rate. This is not the place to describe the kind of life which the early workers led before social reform and the enlightened efforts of the wealthier ironmasters, such as Guest, made better ways of living possible, but the impression that pervades all contemporary reports is of profligacy and drunkenness. It was the grim necessity of earning a living in a district which presented little choice of a job—only between coal and iron—which tied father and son, mother and daughter, to the industry.

[69] *Dowlais MSS*, G. Crane—T. R. Guest, 18 January 1836.

[70] *Mining Commissioner's Report*, 1846. [71] Kenrick, *op. cit.*, p. 12.

Yet there was a way of escape to prosperity for the most skilled and most venturesome, in emigration. In the 1860s, complaints of the shortage of labour in South Wales were many. In 1866, at Dowlais, it was said: 'Owing principally to emigration workmen have become very scarce. From the same cause, and also from the gradual withdrawal of women from outdoor labour, it is impossible to get work-people to keep the machinery employed.'[72] But the problem was not new; Redford points out that 'in the trade slump after 1825, the new freedom of emigration was eagerly used by workers in the textile and metal trades'.[73] Previously, when it had been illegal to go abroad in this way, the Sheffield ironmasters had passed a resolution of congratulation in 1823 to an ironmaster who had prosecuted a Midlands ironmaster and engineer, 'for his having so illegally taken artisans from England to France so much to the detriment of the iron trade of this country'.[74] Before the *Select Committee on the Exportation of Machinery* in 1841, a witness who was a machinist and engineer at Liège admitted taking puddlers over. 'I gave £3 a week to men who had not been getting more in Staffordshire than £2 a week, and some of them 35s.'[75] This higher rate of pay proved more profitable than paying Belgian puddlers lower wages, because of the bigger yields and better quality which the British puddler could produce. Later the attraction of America proved more powerful; Swank refers to a Welshman, David Thomas, as the 'Father of the American anthracite iron industry'. Another, W. Firmstone of Wellington (Salop), had emigrated to the U.S. in 1835, being one of the first to introduce hot-blast there. Yet another, J. Crowther, born in Shropshire, who had migrated to Staffordshire to become a manager of blast furnaces, emigrated in 1844, and was the first in America to smelt with raw bituminous coal in 1846.[76] These were, of course, pioneers with a knowledge of some special technique or process; however, the rank and file were encouraged to make the trip there too. Just as the Friendly Society of Iron Founders paid travelling expenses to its members looking for work,[77] so this and later societies subsidised emigration. According to a witness before

[72] *Dowlais MSS*, On the Employment of Women and Children, etc.

[73] A. Redford, *op. cit.*, pp. 154–5.

[74] *Jackson MSS*, 16 April 1823.

[75] *Op. cit.*, 1841, Q. 739, evidence of G. Withers.

[76] J. M. Swank, *History of the Manufacture of Iron in all Ages* (Philadelphia, 1892), pp. 361, 368, 373.

[77] During the slump of 1840–2 the Friendly Society of Iron Founders with a membership of 3498 in 1840, spent £21,125 in two years, the greater part in travelling expenses. *Centenary Souvenir of Friendly Society of Iron Founders* (1909), p. 32.

the *Royal Commission on Trade Unions*, workers were being paid £5 a head by the National Amalgamated Association of Iron Workers to emigrate to America.[78] This was in the years just before 1867, when it was alleged that this trade union policy encouraged emigration with the express purpose of thinning the home labour market.[79] In the absence of statistics, however, it is not possible to say how strong was the current of emigration at this period; what we can be certain of is that British workmen taking improved methods abroad were an important factor is developing the iron and steel industries of U.S.A. and the Continent.

We have seen that higher wages were a spur to movement from district to district and to emigration, it is now our task to give some account of wage levels prevailing in the iron industry during the period. This is not a comprehensive and meticulous statistical account, but an attempt to give a general impression.

Examples of the war-inflated wages earned by the ironworkers have been already given. With post-war depression came reductions. In 1817 for Shropshire we have full details of wages earned by the various classes of workers, and also, thanks to Gilpin, some indication of their purchasing power, for he appends a list of prices of provisions: 'Keepers—21s. per week, Fillers—15s., Coke Fillers—15s., Limestone Burners—15s., Cinder Fillers—15s., Bridge Stockers—15s., Engine tenders—15s., Carpenters—15s.–18s., Masons—12s.–15s., Cokers—1s. 6d. per ton on pigs, Mine Burners—2d. per ton on pigs, Colliers—15s., Making Refined Iron—20d. per ton, Making Puddle Blooms—8s. per ton, Shingling Blooms—2s. 3d. per ton, Heating Blooms for Rolling—1s. 9d. per ton, Rolling Bar Iron 2s. 4d. per ton, Slitting Rods—5s. per ton—all long weight. Men are charged 1s. 6d. per week for house, garden and coal. Provisions are now: Wheat—33s. for 230 lbs., Potatoes—1s. 4d. for 90 lbs., Malt—13s. for 36 qts., Beef—4½. per lb., Mutton—5½d., Butter—1s. 2d. for 18 oz., Bacon—10d. per lb."[80]

Wage rates in the iron industry, which pivoted around the puddler's piece rate, were extremely sensitive to changes in the price of iron, particularly to downward movements. The ironmasters often successfully resisted the attempts of the puddlers to force up wages in pursuit of prices. The next set of evidence, this time from Anthony

[78] *R.C. on Trade Unions*, 5th Report, 1867, p. 59, evidence of J. Jones.

[79] *Ibid.*, p. 53, evidence of J. Jones. See also C. Erickson, 'The Encouragement of Emigration by Trade Unions, 1850–1900', *Population Studies*, Vol. III, pp. 248–273, for an account of the roles played by the several unions in promoting emigration of ironworkers.

[80] *Dowlais MSS*, G. Gilpin—J. J. Guest, 12 October 1817.

Hill's Plymouth works, relates to one of those periods when the industry was in the doldrums after a bout of railway prosperity. Miners and colliers were now only receiving 12s. a week, puddlers (were paid by time rates) 20s., refiners also 20s., rollers 18s., and fillers 15s.–16s.[81] The rates which John quotes for South Wales show that ironworkers' wages were rising in the next few years, until they were set back by the depression of 1839–42. Puddlers were getting 30s. in 1837, 35s. in 1839 and 23s. 7d. in 1844.[82] For Shropshire, where there had been a 10 per cent reduction in 1837, puddlers' rates are not given in a list which contains an elaborate list of piece and day rates. Keepers and fillers were paid according to the quality of pig their furnace was producing; loading and unloading ironstone, lime and the pigs were paid by the boat-load; sawing, so much per 100 ft.; labourers were paid at day rates, 1s. 10d. a day, so were the skilled masons and carpenters, 3s. per day.[83] In 1840, at Varteg in South Wales, wages were much higher; according to Kenrick, the skilled man in the industry was earning at least 20s. a week and some 60s. to 70s.; boys from 4s. to 12s. and women 7s. to 8s. In contrast, a farm labourer could earn only 12s. to 15s.[84] While the reports of the Mines Commissioner do not throw very much light on wages in the various districts in the 1840s—an exception is the statement of a Midlands ironmaster in 1843, that men at the blast furnaces could earn from £1 to £1 12s. a week[85]—there is an abundance of evidence for South Wales. The historian of the Monmouthshire iron and steel trade, Mitchell, gives a comprehensive series of figures for the period 1844–50.[86]

	1844	**1845**	**1846**	**1847**	**1848**	**1849**	**1850**
Labourers	11/–	11/–	12/8	13/9	11/–	11/–	10/–
Founders	23/9	26/11	33/8	33/5	35/–	30/4	26/6
Fillers	22/–	25/10	29/10	31/3	32/7	30/4	26/6
Cinder fillers	21/–	19/–	21/8	23/6	22/11	21/1	16/8
Refiners	37/8	39/3	51/2	40/7	48/9	39/6	31/–
Puddlers	23/7	30/6	35/8	32/4	30/11	25/7	23/–
Ballers	26/2	32/7	45/1	38/8	31/7	23/–	18/6
Rollers	33/7	58/10	72/–	80/–	60/9	35/6	32/–
Rail straighteners	26/2	35/5	49/7	47/8	36/8	18/2	21/9

[81] *S.C. on Mfres., Commerce and Shipping*, 1833, Q. 10270–80.

[82] John, *op. cit.*, p. 84.

[83] *Botfield MSS*, Prices of Labour at Stirchley Furnace, April 1837.

[84] Kenrick, *op. cit.*, p. 16.

[85] *Midlands Mining Commission*, 1843, p. 66, evidence of Jas. Boydell.

[86] T. Mitchell, *Monmouthshire Iron and Steel Trade, labour and wages 60 years ago* (Newport, 1904), pp. 10–11. It seems likely that he has transposed the rates for refiners and puddlers.

Before commenting on the movements in wages which this table makes quite clear, this section must be concluded with evidence for the 1860s. This period was a decade of working-class prosperity and this is revealed in the few wage rates that can be quoted here. In 1864, Midlands rates of pay were as follows: shinglers generally clear £3 per week, furnace men £2 10s. to £3 per week, and rollers averaged between £2 10s. to £4—though it was said that 'some rollers earn £5 or £6 per week at certain times'.[87] At the same time, in Yorkshire, at the Low Moor works, hammer men's wages were £2 to £3 a week, puddlers 30s.–50s., rolling-mill men 30s.–80s. and labourers a mere 8s.–16s.[88] Despite the fact the *Royal Commission on Trade Unions* frequently heard the allegation that these bodies had the power to raise wages to an uneconomic level, surprisingly there was little evidence of rates of pay in the industry. However, the way in which regional wage rates were determined was made clear. An ironmaster admitted, 'We have always been in the habit of considering that the South Staffordshire prices regulated the prices elsewhere, and that the initiative for a rise or fall in wages took place there.'[89] The principle of the regulation of wages by the market price of iron had long been implicit in the actions of the ironmasters, but it had been formalised in the Black Country in 1848. The Thorneycroft sliding scale, as it was called, was regulated by allowing 1s. per ton for puddling according to the price of iron. For example, if the price of bar iron was £7 10s., then the puddler received 7s. 6d. per ton.[90] Other wages were altered by 10 per cent for every change of £1 in the selling price of iron. According to the secretary of the South Staffordshire Ironmasters' Association, this agreement only existed up to 1857 or 1858,[91] but the principle of wage adjustment was still broadly retained by several of the district associations. Yet in 1866 the average wage was said to be 35s. a week, when attention was drawn to the very different wage levels prevailing in the other districts; Welsh wages were said to be 40 per cent below, and the South Staffordshire

[87] *Children's Employment Commission*, 1864, 3rd Report, p. 15. para 11.

[88] *Manchester Guardian*, 26 October 1864.

[89] *R.C. on Trade Unions*, 5th Report, 1867, Q. 8271, evidence of J. Kane. One ironmaster did produce a series of wage indices for South Wales of the average rate of wages from 1848, but no definition of that term is made. The series is given for what it is worth:

1848	49	50	51	52	53	54	55	56	57	58	59	60	61	62	63
100	90	83	75	79	83	91	100	90	100	80	72	79	87	78	71

Evidence of G. T. Clark, Q. 10051.

[90] *Ibid.*, Q. 9496. [91] *Ibid.*, Q. 9832.

wages 15 per cent below, the North of England level.[92] What emerges from a random sample of wage rates such as this? First, as with output, the most remarkable feature is the degree of fluctuation of wage rates—both Mitchell's and Clark's figures reveal this most clearly. All classes of workers alike suffered from this extreme sensitivity of wage rates to the market for iron. If we can accept Mitchell's figures as being correct, then the wages of rail-straighteners in 1850 was much less than half they had been in 1846, and similarly for ballers and rollers. However, it has to be noted that not all classes of workers suffered to the same extent; John suggests that this cyclical movement was most marked in those occupations concerned with the production of rails.[93] Nevertheless, all categories were, in fact, indirectly working to meet this demand. What is surprising, however, is that some workers managed to secure advances when others had their wages reduced; for instance, in 1848 the refiners' pay rose by 20 per cent whilst rollers suffered a reduction of 25 per cent. This is probably the results of piece rate earnings. Secondly, whilst there is not sufficient evidence here to make this point obvious, it is clear that regional rates of remuneration differed considerably. We have noted previously Butler's observation in 1815 that the Salop level of wages was 15 per cent less than that prevailing in Staffs. In 1846 wages were said to be 20 per cent lower. It is tempting to ascribe this state of affairs to a survival of that former spirit of give and take between master and man in the old-established districts, but this is unlikely because in Shropshire, as elsewhere, there were strikes and riots over pay; higher rates of wages were paid in districts like the North of England because of the demand for workers. Moreover, higher productivity there, arising from improved techniques and modern plant and equipment, made higher wages possible. However, the importance of productivity arising from greater capital investment must not be over-emphasised for this period, as labour costs represented a far greater proportion of the costs of production than they do today with a highly mechanized and technically integrated industry. How great a proportion of the costs of production it is difficult to say—according to Anthony Hill in 1833 'about three-quarters of the price of (bar) iron was laid out in labour'.[94] In 1867, when ironmasters were getting hysterical about the influence of trade unions, with specious reasoning, they worked out the labour element as 90 per cent of the price of bar iron.[95] Whatever the actual proportion, and this would vary from place to place, this fact was re-

[92] *R.C. on Trade Unions*, 5th Report, 1867, Q. 9520. [93] John, *op. cit.*, p. 84.
[94] *S.C. on Mfres., Commerce and Shipping*, 1833, evidence. Q. 10231.
[95] *R.C. on Trade Unions*, Evidence of W. Williams, Q. 9889 (5th Report).

sponsible for the evils of the truck system. This system of paying wages directly or indirectly in kind—hence providing the opportunity for further profit for the employer—was of long standing in the coal and iron industries. Certainly, the ironmasters had clearly defined attitudes to this trade.[96]

The abuses of the truck system which, as the Commissioner for the Shropshire mining district pointed out,[97] had unfortunate consequences outside the exploitation exacted, also helped to crystallise working-class grievances into action. This development of collective bargaining into organised trades unionism in the iron industry was not the spectacular and well-known process that it was in the textile trades. Indeed in 1924, when Ashton was making his study of the industry in the Industrial Revolution, the evidence of these activities upon the part of the ironworkers was very scanty. It seemed likely that the workers in this industry were too scattered geographically (partly on account of their mobility), too divided, and possibly too well paid, to have developed any class consciousness unity. Thus, it was suggested, they did not develop a coherent union movement. However, this hypothesis was not too convincing. Even if some workers were well paid, their wages were not kept so high and steady as to avoid grievance on their part; further, distance between workers had not prevented the growth of large unions in the allied coal industry; finally, the livelihood of all classes of workers on the coalfield, thrown together by social conditions, was so utterly dependent upon the fortunes of the iron and coal trades that united action was natural. The coalfields were, of course, agitated with the great risings in South Wales in the 1830s and to a lesser degree with the strikes in Northumberland in the 1840s, and these events and their attendant circumstances have been the subject of several studies.[98] In these

[96] G. W. Hilton, *The Truck System* (Cambridge, 1960). This system of payment is more involved than historians used to think. However, it would be possible to document, rather more fully than Hilton has been able to, the prevalence of the truck system in the iron industry.

[97] *Loc. cit.*, 1847, p. 5: '. . . Production is freed too rapidly for the demand; the number of hands brought into the labour market to maintain this production is greatly increased; on a period of slackness occurring in the trade, the smaller capitalist is least capable of resisting it, and is the first to be tempted to reduce prices of wages or both. . . . '

[98] D. Williams, *John Frost* (Cardiff, 1929).
Ness Edwards, *The Industrial Revolution in South Wales* for a lively if partisan description of social conditions which led to the risings.
A. H. John, *op. cit.*, pp. 87–97.
E. J. Jones, *Some Contribution to the Economic History of Wales* (Cardiff, 1928).
R. Fynes, *The Miners of Northumberland and Durham* (1873).

struggles, which were convulsive risings of the whole populations in these districts, the part played by the ironworkers was less prominent than that of the miners and colliers. One gathers the impression from reading accounts of these risings that the ironworkers were less easily drawn into these skirmishes with the military and the magistrates, unless they were intimidated by the miners' threats against blacklegs. Trade unionism was indeed very strong among the coal miners, particularly in the Northumberland coalfield. The Miners' Association, which claimed to have a membership of 60,000 before the strike of 1844, and half that number even after its failure, extended its activities from the North-east into the Midlands, South Wales and Scotland.[99] Such activities could and did dislocate the working of the ironworks, which could be brought to a standstill through the miners' imposing a stranglehold on the raw materials. Some ironmasters, of course, took the precaution of laying in stocks of coal or coke; for instance, Wm. Baird and Co. 'kept a year's stock of coal, ironstone and pig iron enough to supply orders for a whole year'.[100] Others, looking further ahead, saw education as a panacea. At least, it was suggested that

> ' . . . the extreme injury which has been, and in many districts still continues to be, inflicted on their trade by the combinations of their workpeople to raise unduly the price of labour, has given a strong desire to raise the intelligence of the working classes, as the best guarantee against these mutually destructive hostilities between labour and capital.'[101]

The official reports of this period, echoing the sentiments of the masters, are written in a tone bitter with resentment against the injury inflicted on the iron trade by the miners' combinations. Directly, the results of wage-bargaining by the unions were alleged 'to raise the price of coal and iron far above their proper level'; indirectly, all threat of foreign competition and consequent unemployment at home was traced to high costs of production which high wages and restrictive practices necessitated.[102]

A. J. Taylor, 'The Miners' Association of Great Britain and Ireland, 1842–48', *Economica*, Feb. 1955.

Also the Mining Commissioner's Reports of the 1840s.

[99] *Report of the Commissioner for Mining districts*, 1847, p. 10.

[100] *Ibid.*, p. 11. [101] *Ibid.*, p. 9.

[102] *Ibid.*, p. 10. It was alleged by the Commissioner that Scottish firms were losing orders in Germany because of this and that therefore 15 out of about 100 blast furnaces were not at work—a further stimulus to production in Germany and America.

Thus the power of trade unionism was demonstrated to the ironworkers on many occasions during the first half of the nineteenth century, and it would have been very surprsing if they had not imitated the colliers' and miners' example. The account of the workers' activities during the Napoleonic wars, given earlier, has made it clear that the skilled men in the iron industry took part both in the strikes of the period against reductions in wages and in an organised trade union, the Ironfounders' Union, founded in 1809. However, despite this early vigorous activity, there was no development of a well-established and coherent Trade Union movement in the industry until the 1860s; the account to be given here of attempts at collective bargaining is again of short-lived outbursts of protest. As John rightly remarks, 'It was rather a testing period in which the working class obtained experience and discipline in collective action.'[103] Such action as there was, was local and on a small scale; even the Ironfounders' Union apparently continued its existence unnoticed, for in 1833 a Midlands ironmaster could say that during the war unions did not exist and it was only in that year 'they had been recently very extensive'.[104] But this was overlooking much that had taken place on the coalfields.

In 1816 had occurred a violent outbreak of mob violence in South Wales in direct consequence of wage reductions to counter Staffordshire competition. At Nantyglo, Blaenavon and Llanelly the mob put out all the furnaces.[105] W. Crawshay II of Cyfarthfa, similarly affected, wrote to Guest at this time, 'The enemy in too great strength to oppose with any probability of success, have possessed themselves of all our works and wholly stopped them. They are yet exalting in their victory are about to proceed to Penydarren and Dowlais. My spies tell me they threaten hard your shop for they are hungry.'[106] Here it is quite clear that the Cyfarthfa puddlers gave a lead to the disturbance, by striking for an advance of wages in July 1816.[107] In 1817 too, after the previous disturbance had been put down through the intervention of the military, there was a strike at the Plymouth works to endeavour to raise wages once more. In these troubled times, it is worth while noticing that the influence of Methodism, which held a strong power over the most intelligent of the workmen, was divided in its encouragement of the struggles of

[103] John, *op. cit.*, p. 88.

[104] *S.C. Mfres., Commerce and Shipping*, 1833, Q. 9806, evidence of Matthews.

[105] E. J. Jones, *op. cit.*, p. 102.

[106] *Dowlais MSS*, W. Crawshay II—J. J. Guest, 16 October 1816.

[107] A. Aspinall, *The Early English Trade Unions* (London, 1949), p. 220, quoting H.O. 42/152.

labour. The commander of the military at Merthyr Tydfil in 1816 pointed out to the Home Secretary that the Wesleyan Methodists' teachings inculcated patience, resignation and strict obedience to the laws; the fermenters of trouble were the Calvinistic Methodists—'the ranters'—with their 'doctrines dangerous' who had great influence amongst these men.[108]

The clashes between the workers on the coalfields and the ironmasters assisted by the military, although most frequent in South Wales, were also not absent from the districts where something of the paternal relationship between master and man lingered. In 1817 there had been strikes by puddlers in Staffordshire, but before they returned to work 'awed by the soldiery and exhausted of cash', they had gone to Shropshire for support; as a result, the puddlers of Coalbrookdale, who had suffered reductions in their pay, came out in a brief and ineffectual strike. However, in 1821 ill feeling flared up again into a riot similar to those which periodically disturbed South Wales. The affair began with a strike by the colliers of Wellington and Donnington Wood against reductions of wages, but the reports of the magistrates make it clear that even if the ironworkers wished to remain at work they could not do so: 'On Friday morning last . . . we received information that the colliers were assembled in a very large body and were proceeding successively to all the ironworks in this part of the country for the purpose of intimidating those men who were disposed to work at the reduced rate of wages and compelling them to stop the works. . . .'[109] The Horsehay, Coalbrookdale and Old Park works were thus put to a stand, but once more the alliance of master, magistrate and military quickly subdued the rioters.

Several other instances of short-lived combinations, strikes and risings in South Wales in the 1820s are described by John[110] and the more violent outbreaks in 1831; 1839 and 1842 have received much attention.[111] A full account of the circumstances surrounding these outbursts would be out of place here, but the occasions on which they throw light upon the formation of trade unions may be remarked upon briefly. In Wales, union activity, never confined to the iron trade itself, was associated with the local lodges of Oddfellows and branches of the Friendly Associated Coalminers' Union,

[108] In the 1830s T. R. Guest, a devout Wesleyan, drafted a sermon pointing out the sinfulness of combining.

[109] Aspinall, *op. cit.*, p. 368 (H.O. 40/16/15 and 33).

[110] John, *op. cit.*, p. 89, e.g. the revolt of the Clydach ironworkers in 1822; a strike in 1826.

[111] Notably in Ness Edwards, *op. cit.*, and D. Williams, *John Frost*—a study of the Chartist leader.

founded in October 1830.[112] In 1831, partly as a result of a recruiting campaign by a delegate from Bolton, many ironworkers became members, but the ironmasters quickly organised counter measures. As Kenrick of Varteg iron works wrote to Guest: 'If the men are suffered to establish their Union upon a regular system nothing but confusion and violence can be expected and it is better the evil should be crushed in the bud.'[113] This sentiment was quickly acted upon; the Dowlais and Plymouth ironworks by October had discharged 3000 to 4000 men.[114] That ironworkers were among the disaffected is most strongly confirmed by Crawshay in a note to Guest, 'I have this morning stopped one of your Union rollers going by our office with a gun, and a drunken little Unionist too! This now is evidence of battle. . . .'[115] However, the battle had been fought and lost already, in the riots of June; indeed, union activity followed this defeat.

Union activity again is present in 1834, when Owen's Grand National Union raised workers' hopes of the approaching millennium; but again, as John points out, a resolute stand of the masters at Bute ironworks and elsewhere, making the men sign a declaration promising to give up the union, put a halt to the growth of unionism.[116] Where this course of action seemed likely to fail, ironmasters threatened, as W. Crawshay II did in 1834, to blow out the furnaces until the men submitted.[117] His father, alone of all the ironmasters, it should be noted, had looked upon the unions without the fierce antagonism which animated the other ironmasters.[118]

The disturbances of 1839, which made a particularly violent efflorescence of Chartism, were concerned chiefly with the political ambitions of that movement, and did not receive much support on account of any economic distress and misery of the mining population. In fact, it was a period of boom with high wages; according to the Mines Commissioner, 'work was abundant and wages high'.[119] Nevertheless, one of the most obvious ways of the workers venting their feelings and demonstrating their power was to put out the fur-

[112] D. Williams, *op. cit.*, p. 115. In 1831 this union affiliated itself to Doherty's National Society for the Protection of Labour (John, *op. cit.*, p. 91).

[113] *Dowlais MSS*, 30 August 1831.

[114] John, *op. cit.*, p. 92.

[115] *Dowlais MSS*, 5 November 1831.

[116] John, *op. cit.*, p. 93. In 1831 the idea of the document or certificate had been mooted, in order to blacklist all unionists and prevent them from securing employment elsewhere on dismissal from their works (Dowlais MSS).

[117] N. Edwards, *op. cit.*, pp. 91–100. D. Williams, *op. cit.*, p. 117.

[118] *Crawshay MSS*, 7 September 1831, William Crawshay I—William Crawshay II. 'The union of the men to obtain higher wages while they do so by lawful means cannot injure us, I wish them success and shall be happy to pay them better [as] soon as others do.'

[119] *Loc. cit.*, 184 2.

naces of the ironmasters. The Chartist risings of 1842, which were complicated by the intervention of the Anti-Corn Law League, however, had more favourable ground in which to sow the seeds of discontent, for the iron districts had been thrown into depression. One of the centres of discontent this time was the Black Country, where a widespread strike of the colliers against reductions dislocated the ironworks, although at Dudley the ironworkers continued to work so long as they could be supplied with materials, and some acted as special constables.[120]

In Scotland, which never witnessed the bitter struggles on the coalfields as did Wales, the combinations of the workers appear to have been more successful. Possibly this was because union activities, once established, seem to have been able to consolidate their existence. For instance, in 1831 was founded the Associated Society of the Iron Moulders of Scotland.[121] In 1847, this union still seems to have been active and powerful:

> 'The moulders at the various foundries in this part of the country (Lanarkshire) have struck for the last six weeks, to endeavour to prevent their masters taking in such apprentices as they find necessary. The moulders have been making 30s. to 40s. a week for the last 10 years and they are now drawing 12s. a week each from a fund of £5000 which they have been saving up, in order to enter upon this contest with their masters. They want to maintain a monopoly of labour. . . .'[122]

It is perhaps significant that this union, like the English union of ironfounders, existed on the fringe of the basic iron industry among the foundries scattered in towns throughout the country and could be more easily organised on a craft basis than the embryonic and incoherent combinations which came more easily under the sanctions, economic and military, of the alliance of ironmasters. However, there is evidence to show that the Ironfounders' Union suffered from the inflations of membership during periods of agitation, and the after-effects of disillusion or temporary prosperity when its membership fell away to a new low level of apathy. The Union in 1831 possessed 1120 members divided among 44 branches, its entrance fees and contributions were high and its activities restricted to the organisation of the skilled moulder craftsman. After 1842, a year of depression, the membership was a mere 2427, having fallen from 3498 in 1840.[123] Even such an apparently financially strong union as

[120] *Midlands Mining Commission*, 1843, para. 544, p. cxxv, evidence p. 65.
[121] Marwick, *op. cit.*, p. 195.
[122] *Mining Commissioner's Report*, 1847, p. 13.
[123] *Centenary Souvenir*, *op. cit.*, p. 32.

this—it held a cash balance of £18,456 at the end of 1846, an average of £4 2s. 6d. a member—had not the powers of resistance to withstand the severe drain of a slump. After the 1847 depression its funds had dwindled away to £214, with only 1s. 1d. standing to each member's name.

The account of the development of collective bargaining in the iron industry in the first half of the century reveals then the overriding reason for the weakness of Trade Unionism among the workers—the ironmasters were far too powerful and drove the union clubs out of existence by use of the threat of dismissal and the document. Thus in South Wales the men accepted, without a strike, a reduction of wages of 4–5 shillings in the pound; so subdued were they after unsuccessful strikes in 1842–3.[124]

The second half of the century, however, saw the development of the new type of unions, more confined in their aims than the previous Owenite-inspired unions with their grandiose and nebulous schemes. So in the iron industry new unions came into being in the 1850s and 1860s; but their development was at first of the transitional type. For in 1850, at Bury, the central committee of the Iron Trades launched a scheme 'to buy up all the property in the kingdom'.[125] Not only the aim, but the organisation intended to carry out this plan—The British General Labour Redemption Society—and the means, subscriptions of 1d. a week and the profits of co-operative workshops, proclaimed Owenite ideas. The plan received the support of the Glasgow workers, but the purchase of the Windsor Foundry was never completed. There were, however, several other attempts to set up co-operative workshops on the part of the engineers and founders who provided the leaders of socialism among the iron trades. There were ironworks which were organised on the co-operative principle for a short time at Deptford, Southwark and at Wolverhampton, but all were ultimately failures. Yet, as late as 1866, the idea of co-operative production in the iron industry had not lost its attractions. After the six months' strike in the North of England, the Drinkfield ironworks at Darlington was set up with a nominal capital of £100,000, to be run on the co-operative principle. However, the ambitious plan was a failure from the start; only about £5000 was paid up, and when loans had enabled the works to be erected 'on a modest scale', there was a lack of working capital to commence business. 'The company came to a deadlock in July

[124] D. Williams, *op. cit.*, p. 339.

[125] B. Jones, *Co-operative Production* (Oxford, 1894), p. 100.
J. B. Jefferys, *The Story of the Engineers* (London, 1946), p. 268.
G. D. H. Cole, *A Century of Co-operation* (Manchester, 1944), p. 108.

1868.'[126] But to return to the development of coherent trade unions. more concerned to consolidate their prosperity than to transform the economy of production, the National Association of United Trades for the Protection of Industry had included the iron trade in the 1850s;[127] but this did not lead apparently to an extension of unionism in the industry. The critical period of union formation did not come until the few years previous to 1867, when the strength of trade union organisation, not only in the iron trade, frightened the Government into an extensive investigation of their activities by means of a Royal Commission. This was a period when, on the one hand, the employers were trying to meet foreign competition by resisting wage claims, and on the other, the workers' unions, again amongst the skilled classes, exploiting a shortage of workers, did not limit themselves to demanding increases of pay, but aggressively tried to apply the 'closed-shop' principle. Hence since 1864 there occurred a crop of strikes and counter lock-outs in all districts except South Wales, where unionism was said to be unknown.[128] In 1864, in the Leeds district, the men had been locked out through trying to enforce compulsory membership of one man in the Association of Iron Workers (founded in 1863).[129] The following year, 1865, labour disputes were conducted on a national scale, with a strike beginning in North Staffordshire and spreading to South Staffordshire; then the puddlers were supported by the Association of Ironworkers of Brierley Hill and the London Trades Council for a time.[130] The ironmasters countered with sympathetic lock-outs in Yorkshire, Derbyshire and the North of England, in which, for a fortnight, all the furnaces in England were at a standstill.[131] Although this round of the fight was lost by the workers in failing to prevent a reduction in wages, the ironmasters too, despite all the weapons at their command (now including the railway to rush supplies of coke and ore to the affected centres), failed to smash the unions.[132] Again, the succeeding

[126] B. Jones, *op. cit.*, p. 485, quoting *The Co-operator*, Vol. III.

[127] *S.C. on Masters and Operatives* (*Equitable Councils of Conciliation*), 1856 (343).

[128] *R.C. on T.U.s.*, 1867, Q. 10050, evidence of G. T. Clark of Dowlais. There were however 50 benefit societies.

[129] J. Freeman, *The Dispute in the Iron Trade*, 1864 (Civil Engineers Tract Vol. 174).

[130] Anon. *The London Trades Council* (London, 1950), pp. 21–2.

[131] Erickson, *loc. cit.*, p. 258.

[132] There were at this time three unions concerned in these struggles: (1) National Amalgamated Association of Ironworkers with its headquarters at Gateshead, formed of the forgemen. (2) Staffordshire Millmen's Association of West Bromwich. (3) The Association of Ironworkers of Great Britain, with its headquarters at Brierley Hill. *R.C. on T.U.s*, Q. 9416. The National Association

year witnessed another struggle in the North of England which lasted six months (the previous strike in North Staffordshire having lasted almost a year). Yet, despite these exhausting clashes with the iron-masters (who ran an insurance scheme with a reserve fund of £45,000 to combat sectional striking to force up the general level of wages), trade unionism grew stronger in its organisation and in its demand for legal recognition. Even the ironmasters, who were inclined to look upon unionism as 'a corruption of combination', looking at the spread of unionism in all districts with consternation,[133] agreed that trade unions should be given legal recognition and a proper status, in order to bring them under the supervision and control of the Government. This, in fact, was done by the Liberal Government in 1871, although as is well known it was left to Disraeli, in 1875, to free trade unionism from the encumbrances with which the new legal status had been surrounded. However, opinion in the iron industry was looking to the machinery of conciliation and arbitration to attempt to settle its disputes; in 1869, the first board of this kind was set up.[134] Nevertheless, others who recognised the fundamental cleavage of interests between the ironmaster and worker did not put their faith in machinery of this sort; their solution of the problem lay not in improving human relations, but in introducing more and more mechanisation to replace the puddler who, as we have seen, was the centre of agitation for the rights of labour in the industry. The process took a long time to work itself out, but it is indisputable today that the absence of labour troubles in the modern iron and steel industry in the last twenty or so years derives largely from the

---

up to July 1866 had 6500 members (of which 500 were bona-fide paying members); after the strike only one-tenth were paying. Evidence of J. Kane, President, Qs. 8219, 8224. See also *Men of Steel* by 'One of them' (Sir Arthur Pugh), London, 1952, which gives more details of the various unions from the formation of the Amalgamated Malleable Ironworkers Association in 1862 to the eventual organisation of the Iron and Steel Trades Confederation. This book shows that union struggles went on in the 1870s (not least of all in South Wales), but with the decline of the wrought iron trade, membership fell away from as many as 14,000–16,000 to only 5000 in 1878. Eventually, in 1887, a new union was formed, the Associated Iron and Steel Workers of Great Britain. John Kane, the union leader, died in 1876 and was succeeded by Edward Trow.

[133] H. H. Creed and W. Williams, *Handicraftsmen and Capitalists* (Birmingham, 1867), Appx. 1. Extracts from the ironworkers' *Trade Circular and Monthly Record:* 'Newcastle Works—all members unionists but one, Gabriel Parton, a Roller; the members are determined to have no black sheep among them, they anticipate the removal of him by the proprietor ere long.' Union activities included the keeping of a black list of workmen who kept aloof from becoming members (as at Cookley).

[134] *Mining Journal*, 13 March 1869, p. 191.

fact that labour has become less and less a vital element to that industry.

THE IRONMASTERS

'Now, from the very starting of this case, it plainly appears that these men, though few in number, must be men of great substance, great capitals in Trade, and capable of exerting a very dangerous influence, when they find it in their interest so to do. . . . '[135]

'A Furnace is a fickle mistress and must be honoured and her favours not to be depended on . . . the excellency of a Founder is to honour her dispositions but never to force her inclinations.'[136]

'During the last 10 or 15 years immense progress has been made in this country, and a taste for science has been awakened . . . twenty years ago I came a great deal into contact with ironmasters . . . and I certainly found the greatest possible difficulty in getting an ironmaster to look at a new idea, but since that time the Iron and Steel Institute has been established and men who formerly ridiculed the idea of chemical analysis speak now of fractional percentages of phosphorus and sulphur with great respect.'[137]

It is even more difficult to delineate the characteristics of the iron-masters than to describe the attitudes of the workers to their common environment. Still less can one present a balanced picture by a string of quotations.

Ashton, as we know, tackled this problem by discussing the problem of their social origins and their religious affinities.[138]

From what classes were the entrepreneurs drawn? Always aware of the exceptions, he pointed out that—

'Most of the successful men in the industry entered it from the secondary metal trades—successful craftsmen seeking control over the sources of their raw material—or from the merchanting of iron and steel wares.'

He describes the moulding of their lives by religion and by the material limitations of their own working existences.

'Most of them had risen by dint of constant industry and unremitting thrift; their early life and struggles left a definite mark on their minds and outlook; and the narrowing effect of long years devoted to the

[135] *The Case of the Importation of Bar Iron from our Colonies of North America* (London, 1756), p. 7.

[136] John Fuller, a Sussex ironmaster, 1754, cited—G. E. Mingay, *English Landed Society in the Eighteenth Century* (London, 1962).

[137] *Royal Commission on Technical Instruction*, 1881–4, Evidence of C. W. Siemens, p. 141, Q. 1489.

[138] Ashton, *op. cit.*, p. 209.

single object of producing pig iron meant that few of them could feel at home in circles where art and literature and the lighter graces of life were cultivated.'

Perhaps, with a wider perspective of the part played by Nonconformity than was available to Ashton at the time[139]—and in the recognition that the post-Industrial Revolution society of England was far different from that in which the Quaker ironmasters of the earlier century asserted themselves—we might be inclined to give less prominence to this subject. Again, if we agree that 'in the great manufacturers of the eighteenth century the qualities of self-reliance, assertiveness, and adventurous enterprise were strongly developed', we would not necessarily correlate those characteristics in the nineteenth century with religious beliefs.[140]

Finally, Professor Ashton comments upon the reaching out of these industrialists into the growing economy around them, as they found their wealth steadily increasing, and 'they found new ways of making money, particularly in banking'. It is, of course, this interpenetration of hitherto isolated sections of the economy which came together under the stimulus of economic development, which makes any analytical survey of one class of employers of labour complicated and not easily reducible to terms of simplicity.[141] One interesting

[139] For a recent reappraisal of the influence of Nonconformity on English economic life see P. Mathias, *The Brewing Industry in England 1700–1830* (Cambridge, 1959), pp. 289–91. As Mathias says, 'When the study of the religious ethic is supported by investigations on a less abstract level into the familial groupings of the sects concerned, the successes of these persons may be explained with greater cogency. . . . The world of the religious-cum-kinship group provided an environment of mutual trust and confidence within which a private 'invisible hand' could accommodate the advantages of each member with the benefit of all.'

[140] As we shall see, those characteristics of individualism were qualified by several factors—perhaps the most important being the increase in the size of the industry itself; in its structure; in its growing dependence on technical expertise; in the industry's dependence on overseas markets.

[141] One important complication of business enterprise, which has been much written about, the divorce between the owners of the business and its managers 'an outcome of an immanent development of the capitalistic system', has given rise to the distinction used by business historians between the owner-capitalist and the entrepreneur. This question is discussed in F. Redlich, *History of American Business Leaders* (Michigan, 1940). 'For the controlling owner manager there was no difference between output and profit. . . . But so soon as there arises a cleavage between the functions of ownership, control and management, a series of conflicts comes into being. The good and ambitious manager is automatically . . . interested in the productivity of the enterprise, whereas owners may be interested only in making huge profits without any regard as to how productivity may thereby, in the long run, be affected.' Redlich develops a theoretical framework of generations of entrepreneurs, or groups of coevals exhibiting certain characteristics.

framework of reference, however, has been worked out by Fritz Redlich, formerly of the Harvard Research Centre in Entrepreneurial History, for American ironmasters of the pre-Civil War period:

(i) *The pre-classical generation.* The tasks of this generation are technical rather than organisational (e.g. development of puddling or smelting with anthracite). They are technicians rather than businessmen, and their enterprises on the whole remained small.

(ii) *The classical generation.* These men correspond fairly well to the ideal type of the capitalistic entrepreneur, combining in their persons technical insight, organising ability and business acumen. They were able to create and manage big plants and even to combine them into large-sized enterprises in their organisation of production and selling, they point the way to vertical integration.

(iii) *The Carnegie generation*, which Redlich, following Veblen, sees engaged in 'the shift of the prevalence from industry to business'. In the United States, the context of this scheme of analysis, this generation is concerned with the development of transportation and integrating other industries into 'big business'. As Redlich says, 'Having inherited large-scale plants and enterprises from the preceding generation, their problem was to create "big business" with all its implications, thereby automatically organising modern industrial corporations and working out the rules of corporation finance. This required specialists. The leaders . . . therefore cut loose from their plants and became . . . captains of business instead of captains of industry, leaving technical jobs to hired technical experts.'[142]

Nothing has been said here about the spacing of these generations, for obviously Redlich's remarks and his examples belong to the American context. There are several major differences between British and American economic development at this period which prevent any ready application of this scheme to British ironmasters. The development of a transport network, first by canal, then by railway, comes early in the history of industrialisation in Britain; in America, of course, the railway boom of the 1860s and 1870s was the opportunity for the operations of the financier-railroad-steelmasters. It is true that the company formations at this same time, in Britain, bring the rise of the company promoter, and, as we have suggested, the greater power of purely financial interests as technical innovations and the re-organisation of the industry from iron to steel became more pressing. However, in Britain, even from the beginning of industrialisation the lines of distinction are more blurred: would it be accurate to describe the Darby and the Reynolds families as

[142] Redlich, *American Business Leaders*, p. 113. Carnegie was one of the first American business leaders who disassociated himself from the productive process.

technicians rather than businessmen? Was John Wilkinson a greater success as an inventor of technical processes, e.g. the blowing of furnaces by steam power, than as a businessman with a sense of market opportunities, for example, in his exploitation of the Boulton and Watt steam engines? Nonetheless, it must be conceded that Sir John Josiah Guest of Dowlais is undoubtedly a classical figure, a captain of industry. But is there any essential difference between him and John Wilkinson except in the scale of his operations? Yet, more than fifty years divide the two men in age and their entrepreneurial context. And here perhaps we have a clue to the reason for the cogency of Redlich's analysis of the ironmasters as it applies to the United States; and conversely the intractability of the British examples.

In the U.S., puddling and rolling was the first major innovation in the American industry; charcoal was used as the fuel in the furnace until the Civil War, coke was not used in America until about one hundred years after its introduction as a fuel in the iron industry in England. Then, in contrast, in the 1860s and 1870s we see the rapid expansion of the Bessemer Steel.

Thus, where comparable, industrial development followed British experience with a time lag of more than thirty years as in the case of puddling; consequently the whole process of introducing new techniques and evolving new ones was telescoped into a shorter space of time.

It seems plausible to suggest that this shuffling together of innovation and innovator, of technician and businessman, made for sharper distinctions in their functions in a shorter space of time, as if a kaleidoscope were shaken in short rapid spells giving three separate patterns, rather than a gentle rotation, as in Britain.

Already in this book many illustrations have been given of the kind of men who set up business as ironmasters. Indeed, so wide is the range of enterprise conducted by the businessmen with that title that one can only talk about ironmasters as a class if one ignores the fundamental question of economic history: 'How typical?' A glance at the index of correspondents in the Dowlais letters will exhibit the widest contrasts between, say, the humble John Hawkins of Dukinfield Furnace near Manchester, who acted as agent for John Guest in 1782 in trying to sell iron at Chester for the Irish market,[143] and H. W. Schneider—described as 'ore merchant, financier, politician' —one of the founders of the Barrow-in-Furness iron industry. A proper placing of these captains of industry can only be made with the intimate knowledge of their individual background and history. Again, the British industry here reveals another striking contrast with

[143] Elsas, *Iron in the Making*, p. 81.

the U.S. iron and steel industry: these were magnates as we shall see, but the industry was not dominated by three or four Napoleonic 'captains of business' as it was across the Atlantic.[144] Whatever the form of financial framework might have been—partnership or limited liability company with a capital of £1 million—the direction of these concerns depended upon individuals. Indeed, the history of the iron and steel industry could be written in terms of the biographies of the men who began their careers in the furnace yard or in the counting house.[145] Yet the growing scale of enterprise in the industry, the growing wealth of the ironmasters and the subsequent 'rise of the gentry' (a class movement not peculiar to any one century), led to changes. Putting it over simply, one might say the 'captain of industry' disappeared—or, at best, remains a ghostly figure standing behind the managers whose letters and decisions are our evidence on the running of these enterprises. It is clear, too, that the ironmaster, quartermaster, *al factotum* is giving way, or depending more and more on the technician, manager and financier-investor. This is true even of Britain in the last quarter of the nineteenth century.

It is an exceptional man of ability, the ironmaster Lowthian Bell, who, Charles Wilson says, 'can claim to rank very high amongst the great entrepreneurs of the Victorian Age', who points the way to the specialisation of the expert and the division of functions. Yet he was born in 1816, the son of an ironmaster of Newcastle, the leading spirit in the establishment of the Cleveland iron district from the 1840s. As Wilson observes: 'He was an outstanding scientist and he has a fine head for that business to which he directed his scientific talent . . . But to these he added other gifts. He was a first-rate statistician[146] and one of the comparatively few industrialists of the age to see the importance of relating business to economic trends which he analysed so shrewdly.[147] Bell has been represented as the rare

[144] E.g. Carnegie, of course, and H. C. Frick, the organiser of the Carnegie group of enterprises; C. M. Schwab, first President of the United Steel Corporation and later, in opposition, of the Bethlehem Steel Company; E. H. Gary, his successor at United Steel.

[145] Cf. J. S. Jeans, *Pioneers of the Cleveland Iron Trade* (Middlesbrough, 1875). Miss Erickson's book, previously cited, does of course in a more analytical way examine the men behind the Bessemer and open-hearth innovations in the steel industry.

[146] C. Wilson, 'Dorman Long', *Steel Review*, No. 6, April 1957, p. 15. In his *Principles of the Manufacture of Iron and Steel* (London, 1884) Bell drew upon the assistance of a statistician for his data; however, it is indisputable Bell had the statistical sense. This book is a primary source for the economic state of the industry in the 1880s.

[147] Bell's data on the geographical distribution of the iron industry have been used earlier in this book. They were taken from his important Memorandum

example of an entrepreneur with a technical training and as an advocate of a technical education,[148] it is unfortunate that the apparent loss of Bell's business and personal papers makes it well-nigh impossible to make a detailed study of his career as a business leader.[149]

However, to go back to the 'classical' period of the Industrial Revolution, if not of entrepreneurial activity; we are still concerned with the social origins of the ironmasters and the different kinds of interests—technical or mercantile—which came together to establish the pioneer ironworks of the eighteenth century. Indeed, as Miss Erickson has so well demonstrated, this question, with slightly different emphases, is the key to the response of the industry to such major innovations as the Bessemer steel process in the 1860s and 1870s.[150] There is the interesting theme, too, of the rise to fortune and respectability, easier perhaps to document in the case of the success stories of these small industrialists, the lieutenants of industry who became the captains.

It is in South Wales where one can trace most clearly the course of industrialisation as the small native gentry, local landlords, 'adventurers' (e.g. John Guest, Isaac Wilkinson) began to exploit the local resources, first of timber and ore, then, and more decisively, coal and iron.[151] Earlier we have tried to describe some of these enterprises. Another interesting feature of this context of industrialisation is that the social milieu was against aggressive entrepreneurial activity in the eighteenth and early nineteenth centuries, particularly in West Wales, the home of the charcoal ironworks and small tinplate forges. As Minchinton, historian of the tin plate industry says: '... The rationale of capitalism had not been fully accepted, the unrestrained pursuit of profit was not generally recognized as the true end of human endeavour.'[152] It is in Glamorgan and Monmouthshire, nearer to England, where it is more proper to emphasise the economic motive. Indeed, as Hoselitz makes clear in the following tables of the occupational and geographical origin of the entrepreneurs in the South Wales iron industry in this period, 'The entrepreneurs responsible for the great

---

presented to the *Royal Commission on the Depression of Trade* (1886), afterwards published as *The Iron Trade.*

[148] Cf. Erickson, *British Industrialists*, pp. 35 ff.

[149] The fullest biography is to be found in the *D.N.B.* (supplement). Bell died in 1904, two years after his firm had been taken over by Dorman Long.

[150] Cf. Erickson, *British Industrialists*, Chapter VI.

[151] See A. H. John, *The Industrial Development of South Wales* for the most thorough and penetrating account of this.

[152] Minchinton, *op. cit.*, p. 161. I am grateful to Mr. R. Craig for pointing this valuable point out to me.

OCCUPATIONAL ORIGIN OF ENTREPRENEURS IN THE SOUTH WALES IRON INDUSTRY, 1723–1839

Year	Landlord[a]	Merchant	Entrepreneur in other than iron industry (only incl. South Wales)[b]	Iron master	Manager in iron Industry	Misc.[c]	Total No. of Entrepreneurs	Number of furnaces	Operated by known entrepreneurs	Operated by unknown entrepreneurs	Joint stock companies
1723	4	1		2			7	8	7	1	
1750	2	1		6			9	10	10		
1765	1	7		7		1	16	13	13		
1788		9	2	13	1		25	17	16	1	
1796		17	1	15	1	1	35	29	29		
1806		24	2	18	1	2	47	48	48		
1823		18		11	2	4	35	72	67	5	
1830		17	2	16	3	3	41	116	104	6	6
1839		13	2	7	4	4	30	137	110	12	15

[a] The entrepreneurs included under this head may have had other industrial interests than those in the iron industry; they were, however, primarily landlords.

[b] Most often the industries with which they were concerned were tinplate and coal. One works established by four Cornwall tin merchants in 1792 was the only engineering works in the coalfield until after 1830. The entrepreneurs concerned with this works were all by origin merchants (with the exception of the manager to 1800, a local ironmaster) and are not entered in this column.

[c] With the exceptions of the two Bailey brothers (nephews of Crawshay) included in the last three entries and Benjamin Hall (London lawyer, son-in-law of Crawshay, and son of a Welsh bishop) in 1806, these entrepreneurs were all, by origin, local attorneys.

increase in the rate of growth of the industry were English ironmasters and merchants. It was they who erected all but one new works after he first half of the eighteenth century.'[153] This is perhaps not a new

GEOGRAPHICAL ORIGINS OF ENTREPRENEURS IN THE SOUTH WALES IRON INDUSTRY

Year	South Wales	Bristol	London	Mid-lands	Other(a)	Family in South Wales 50 years or more	Total known Entrepre-neurs
1723	5		1	1		5	7
1750	5		1	3		5	9
1765	5	4	2	5		5	16
1788	8	1	5	9	2	4	25
1796	10	4	6	10	5	5	35
1806	11	6	8	15	7	6	47
1823	10	5	5	7	8	4	35
1830	16	5	4	9	7	8	41
1839	14	5	3	1	7	10	30

[a] With the exceptions noted below, the origin of these entrepreneurs was in Cornwall. The exceptions were: In 1788, Cockshutt (Yorkshire ironmaster and secondarily a London merchant) and Atkinson (Yorkshire merchant, 'ex merchant of South Carolina'); Frere (of an old Norfolk family), active from 1793 to 1836; the two brothers Tappenden (bankers of Faversham, Kent), active from 1802 to their bankruptcy in 1814; and the two brothers Bailey (raised on a Yorkshire freehold before transporting themselves to their uncle's ironworks), included in the last three entries.

conclusion, but these tables do give a certain precision to correlating the timing of their entry to (*a*) the broadening of the market for iron, and (*b*) the technological improvements; both of which increased the real rate of return in capital and entrepreneurship.

To quote Hoselitz's conclusions:

> 'Industrial entrepreneurship necessitating the commitment of a large fixed capital in a pre-industrial society, will be carried on by those with either few alternatives for the development of their property, or those

[153] B. F. Hoselitz, an unpublished research report. I am very grateful to Professor Hoselitz for his permission to make use of this material. As Mr Craig has pointed out to me, however, this survey does exclude most of the furnaces west of Swansea which were on the coalfield. Moreover, there is the major difficulty, not completely resolved by Professor Hoselitz, as to what constitutes an ironworks. His criterion, the existence of a blast furnace, excludes the forgemasters and tinplate manufacturers (often combined together). This bias leads him to minimise the influence of a secondary industry to provide 'mercantile opportunity necessary for the training of the entrepreneur and the accumulation of capital'.

who, by dint of other activities or connection, are well secured against loss of capital. The extension of market and improvement in the marginal productivity of capital and entrepreneurship attending the Industrial Revolution overwhelmed these considerations. The later entrepreneurs evinced a predilection to greater risk taking and commitment to wealth than had their predecessors.'

However, in South Staffordshire, a well-developed area economically, and with an active market for iron and iron products, there was a marked difference. Here there were those marginal producers, those anonymous individuals always on the border-line between bankruptcy and modest prosperity, who, in moments of unrealistic ambition, expected 'to be at least a Bailey, a Guest or a Crawshay'.[154] Their emergence was explained by Gilbert Gilpin, himself described as an ironmaster by Miss Elsas, but in reality a managing clerk and one of the first industrial correspondents of the Industrial Revolution, informally keeping all the industrialists in touch with each other and passing on news by letter. 'In Staffordshire', he explains, 'landed property is very much divided and naturally all the proprietors [are] desirous of turning their coal and iron mines to immediate account. Hence, there is a colliery in almost every field. As there is no sale for such an immense quantity of coal and limestone, several of these little proprietors unite together and build furnaces; clerks from the neighbouring manufacturers are taken in as partners to direct the concerns; the tradesmen of the towns in the vicinity who can raise a 100 or 200 pounds form part of the firm, and it is in this way that the Ironworks have been multiplied in that country. The proprietors embark all their property and *all they can borrow* in these establishments.'[155]

The slender capital resources of the small Staffordshire ironmasters obliged them to dispose of their iron at any price. 'For', to quote Gilpin again, 'blowing out their furnaces would only be the prelude to their being blown up themselves by the demands of their creditors.' This situation led to small profits, if not losses, and this poverty engendered still more desperate measures, until these small men gave up the unequal task. The small ironmasters of the nineteenth century were, it appears, in a far worse position than their predecessors. They had a struggle to keep their heads above water; but now the tides of competition were heightened by the superior efficiency and greater financial resilience of the large-scale concerns of the ironmaster magnates, such as Guest and Baird. When there was such

[154] *The Times*, 21 June 1843.
[155] *Dowlais MSS*, G. Gilpin—J. J. Guest, 3 October 1819.

a gulf between the wealthy ironmaster, with an economic and social influence extending beyond his industry,[156] and the struggling one-man business tied to the rack of his enterprise, the only common denominator between these men was having a financial stake in one or several of the many processes and trades of iron making, manufacture and selling, and of being an employer of labour. (This, from the point of view of the working classes at that time, would have been sufficient for class discrimination—employers large and small were alleged to have the same attitudes to labour, if not to each other.) So, as the metamorphosis of the iron industry into the modern steel industry took place, so the class of men known as 'ironmaster' became extinct. We have reached the stage when it is possible for Miss Erickson to trace the function of the metallurgist, the manager or accountant: and the director, representative of financial interests.[157]

A representative figure among the new class of ironmasters was Charles Geach (1808—54).[158] A man primarily of finance, he is known as the founder of the Midland Bank. He had come to this position of importance after an early career as a junior clerk at the Bank of England in London, later being transferred to Birmingham in 1826. Ten years later he commenced his real operations as a financier when he was engaged in the establishment of two local banks at Birmingham, of one of which, the Birmingham and Midland Bank, he became manager in August of that year. Amongst its first directors was Samuel Beale, to be associated with him, later, in the Park Gate Iron Company, and Robert Webb, the Bank's first Chairman.

In 1838, by way of bank business, in the negotiation of a loan began Geach's connection with the iron industry. Two years earlier, a company with a capital of £2000 had been formed at Wednesbury, Staffs, to exploit the patent of a local Independent minister-mechanic, named Hardy, for making an axle of laminated layers of iron upon a central core. However, as often happened, the first flush of enthusiasm having evaporated, the company had to approach Geach's

[156] One important feature of industry in the eighteenth century, as Peter Mathias observes, was that 'the business world lived very much in a personal way, almost as a vast straggling club' (Mathias, *op. cit.*, p. 295).

[157] Cf. the opinions of Stanley Baldwin, a former managing director of a public company in the steel industry, through his family connection with an old-established iron firm in Stourport, Worcs. 'Since the days when private industry gave place to the joint stock company there have battened on to joint stock companies large numbers of men connected with the management, and directors, who are parasitical and nothing but parasitical' (cited in Erickson, p. 188).

[158] W. Edwards, *Personal Recollections of Birmingham Men* (1877), pp. 125–31; *Birmingham Post*, 31 January 1938.

bank for extra working capital, the original partners being unwilling to advance further money. On Geach's initiative, a new company, retaining some of the old partners, the Patent Shaft & Axletree Company, was formed. By 1844 Geach had bought out all the partners. Thus he came into sole possession, in time to enjoy the great demand of the railway boom of 1845; hitherto, the business had not been a success and no dividends were paid. In 1844 the works consisted of a small forge and 10 furnaces: in 1854 with 4 forges and 4 mills, it employed 12 hammers, 100 furnaces and 800 men.[159] Then for Geach followed the period of investment and enterprise further afield. It is said that he bought a controlling interest in the distressed Park Gate Iron Company for the price of an 'old song', though it appears that Geach's part here in the negotiations was perhaps less spectacular than his biographer suggests. In 1840 the works had come into the possession of William Scholefield and in 1845 he admitted Geach and Samuel Beale (a director of the Birmingham and Midland Bank, it will be remembered) into co-partnership. Geach and Beale carried on when Scholefield retired in 1847.[160]

A concessionnaire in several Continental railways and director of the Manchester, Sheffield and Lincs. Railway and the Shrewsbury and Birmingham Railway Companies, as Crick and Wadsworth suggest, he doubtless advanced the fortunes of the Park Gate Iron works, which concentrated upon an output of rails.

In connection with these linked enterprises in railways, it is interesting to note that Geach is said to have assisted financially in the establishment of the Gorton foundry of Beyer Peacock and Co., which soon became famous for its railway locomotives.

In 1851 Geach retired from the active control of the bank, and chose iron as being more profitable than credit.[161] In that year, too, having served in municipal administration as Mayor of Birmingham in 1847, he rounded off his career with a seat in Parliament (M.P. for Coventry), though unlike Chamberlain who followed his example, he did not make his mark there.

Of course, Geach was a banker, and the connection of banking with iron is too well known to need elaboration; however, the flow of investment was now the other way. Another example of the banker industrialist in the first half of the nineteenth century was William Thompson, M.P., Alderman of the City of London and controllling shareholder in the Penydarren Iron Company in South Wales. This

[159] Crick and Wadsworth, *op. cit.*, p. 421; J. F. Ede, *A History of Wednesbury.*

[160] Information supplied by T. W. Edwards, Director of Park Gate Iron and Steel.

[161] Crick and Wadsworth, *op. cit.*, p. 421.

'paragon of Victorian virtue' died worth nearly £1 million.[162] He had been director of the Bank of England, director of the Globe and Marine Insurance Companies, on the board of Lloyds as well as of several railway companies.[163] He had large financial interests in overseas copper mining companies supplying copper ore for the Welsh works. Here then is a link between the establishment of the coke-iron industry with the help of merchant capital during the Industrial Revolution, and the investment of Manchester cotton magnates, company promoters and financiers in the steel industry of the 1860s.

A different illustration of this impersonalisation of industry, through a larger scale of operation, more specialised management, and the '*rentier* interest' in the industry, is the path of respectability trod by the ironmaster as he gained wealth and local influence. The career of Sir Ambrose Crowley in the early eighteenth century had shown that the iron industry was an avenue to wealth and title.[164] However, he and his enterprise were exceptional; and even if Ashton overdraws the picture of the arduous, harrowing and confined life of the eighteenth-century ironmaster, it is true that the ironmasters of the first generations of the Industrial Revolution appear to have had few social pretensions. It was an age of strict economy in order to accumulate reserves of capital. (As we have seen, incidents in the life of Sir John Josiah Guest as a young man, when he was warned against dancing, illustrate the austerity of the older generation; and, certainly, the conflicts of the elder Crawshay with his son William II are a perfect example of this generalisation.)[165] Certainly the early industrialist had to concentrate upon building up his enterprise. The pioneering ironmaster was a busy man, we can be sure, and William Baird, the successful ironmaster of Gartsherrie, in Scotland, only found time in his later life to give an account of his energetic daily round.

> 'The routine of my day's work', says Baird, 'was to be in the works before breakfast; then, in the forenoon, to ride round the ironstone

[162] Lady Charlotte Guest was not so eulogistic: '(Thompson was) the Alderman in every sense, (but) has not the uprightness which I should have been inclined to give most City Merchants credit for . . . " *Journal of Lady Charlotte Guest*, p. 41.

[163] *Banker's Magazine*, March 1854, p. 219, and A. B. Beaven, *The Aldermen of the City of London* (London, 1913).

[164] Cf. Flinn, *Men of Iron.*

[165] Sir Jeremiah Homfray was, so far as is known, the only ironmaster to obtain a title during this period (in 1810). He became bankrupt soon afterwards (John, *op. cit*, p. 34). His brother was High Sheriff of Monmouth in 1813, and M.P. for Stafford 1818–20.

pits, always some of them in the process of sinking, and at some of them the erection of engines going on under my directions. Some time between 2 and 4 o'clock I came home and had dinner in the "Coal Hole"—a two-storied dwelling house, in which all the brothers, at one time or other lived. The afternoon was spent about the works, attending to what was most necessary to be done; and although the hour of shutting the office was seven, my work not infrequently did not cease till 10. . . . " Afterwards there was often a spell of "toddy drinking" and receiving business associates in the house until midnight.'[166]

However, with the stabilisation of these enterprises the greater ironmasters outgrew the harassing ties of active managership. It was possible to engage a certain type of clerical worker with sufficient intelligence to master the practical side of the business; which experience he might think of using for his own enterprise later on. The ironmaster might be freed from many of the small and trivial tasks of day-to-day management necessitating residence in the house overlooking the furnace yard, where supervision could go on, even when the master was at meals or spending the evening over his books—meditational or financial.[167] Naturally, the expression of social ambition and achievement was the ownership of a country estate, or a newly-built mansion, to confer status for a political career as Member of Parliament or Sheriff or deputy Lieutenant of the county. This rise of a new gentry, which, surprisingly, did not accelerate in the nineteenth century as a result of industrial fortunes,[168] was a permanent feature of the English social scene. Thompson uses as an illustration the career of John Barker, a South Staffordshire ironmaster who purchased Albrighton Hall in the adjoining county in the 1830s. He became Sheriff of Staffordshire in 1851.[169] But other examples might be added: in Shropshire, not the Darby family, who branched out into banking and the iron and steel business elsewhere, but the Botfields of Old Park and Stirchley furnaces, owning collieries in the Clee Hills, who cut fine figures as country gentlemen, riding in only to collect their profits.[170] According to Gilpin, in 1807, these amounted to £13,000 for one partner alone: 'This gentleman will take no part in any business. . . .'

[166] D. McGeorge, *The Bairds of Gartsherrie*, p. 66.

[167] For an account of these changes in management see my description of the fictional treatment of the history of the Darby and Reynolds business in 'Coalbrookdale, fact and fiction', *Edgar Allen News*, Sept. 1956, pp. 198–200.

[168] F. M. L. Thompson, *English Landed Society in the Nineteenth Century* (London, 1963), p. 120.

[169] *Op. cit.*, p. 128.

[170] The drawing up of a family tree by the College of Heralds was a necessary step towards this social aspiration. The Botfields obtained theirs, 'Stemmata Botevilliana', in 1853.

It might be mentioned in passing that the aristocracy, the ancient landed proprietors (as opposed to the *nouveaux riches* of the Industrial Revolution), had not given up completely their long-standing interests in the industry. As late as 1861 the Earl of Dudley owned furnaces at the Round Oak Works in Staffordshire; Earl Granville eight at Shelton, North Staffordshire. This is not to say that managers and agents did not look after these ironworks, but these peers were not wholly sleeping partners or proprietors drawing an income from these businesses, established to exploit the natural resources of their landed estates. Thompson records that Granville, presumably busily engaged in public life, complained of the tedium of having to go down to the country to look after his iron works—'a necessary but tiresome operation'.[171] However, as Miss Erickson makes it clear, the Cavendishes, the Lindsays, the Wards, and Leveson-Gowers, who retained nominal control as Chairmen of the board of directors of their firms until well into the twentieth century, did not take an active part in the day-to-day management of these firms.[172] They continued their diplomatic, political or military careers.

Finally, it is curious that Dowlais should have provided the only two examples from the iron industry in the nineteenth century of what may be termed 'industrial peerages'. In 1873, the Hon. H. A. Bruce, Home Secretary since 1868, became Lord Aberdare; he had been appointed co-trustee of the Dowlais Iron Company in 1852 on the death of Guest. The resident trustee of the Dowlais estate, however, was G. T. Clark, engineer, archaeologist and antiquarian. In 1880, Sir John Josiah Guest's son was elevated to the peerage as Lord Wimborne.[173]

This elevation into the aristocracy, rare and certainly untypical, had resulted from the union of the grandson of one of the founders of the firm with the daughter of an impoverished peer, the ninth Earl of Lindsey, in 1833. This marriage symbolised the recognition of industrial success by the aristocracy. This again, it is fair to add, was not typical, although, of course, Peel, the grandson of a cotton manufacturer, was soon to be Prime Minister. It was at any rate the culmination of social acceptance for these South Wales ironmasters. In 1815 William Taitt, Guest's uncle, had been 'pricked' Sheriff for the county of Glamorgan. Four years later John Josiah Guest himself became Sheriff. A year or two earlier, as Justice of the Peace he

[171] Thompson, *op. cit.*, p. 174.

[172] Erickson, *British Industrialists*, pp. 51–52.

[173] As a result of Guest's death in 1852, it is at Dowlais where we see most clearly the emergence of managerial responsibility for a big business. John Evans became manager 'just when Sir John's health was failing'. He made 'detailed reports for Sir John . . . '. In 1855 his salary was £1300.

exerted some influence in local political affairs. Wyndham Lewis, a Dowlais partner, was 'elected' to Parliament in 1820 for the Glamorgan Boroughs; in 1826 he was nominated to the pocket borough of Honiton; in 1832, John Josiah Guest became M.P. for the newly enfranchised borough of Merthyr, the industrial town created by Dowlais, Cyfarthfa and the neighbouring ironworks. All this perhaps was an example of industrial feudalism, but the story is not as simple as that. Certainly Guest's influence, a local paternalism, being greatly responsible for the livelihood and welfare of the inhabitants of Merthyr and Dowlais, owed much to the co-operation of his wife, Lady Charlotte. She undoubtedly brought to the alliance a strong sense of aristocratic duty and influence; she also lived down any squeamishness about 'Trade', the sordidness of commercial life, as well as about poverty, squalor and disease. Her own active intelligence and interest in human affairs and industrial management inevitably drew her into the works, the mines, the schools, churches and homes of the Dowlais workpeople. Indeed, it is thanks to her 'Diaries' that we have this intimate knowledge of this outstanding industrial family business.[174] Dowlais, with some justification, was described in the first half of the nineteenth century as 'the largest manufactory in the world'.

The Guest family had been associated with the Dowlais Works for over half a century, as managers and partners. When in 1815 his uncle, William Taitt, who had been largely responsible in promoting the wealth of the company, died, John Josiah had been jointly responsible for the management with Alexander Kirkwood, another nephew of Taitt, since 1807; however, Kirkwood died in 1814. Thus Guest was left the largest partner—William Taitt had held eight shares out of sixteen which came to J. J. Guest; in later years, in 1837, his brother left him another share.[175] Apart from his brother, Thomas Revel Guest, a Methodist with strong religious inclinations,[176] who

[174] Miss Elsas, *Iron in the Making*, being a selection of the available *Dowlais MSS*, does contain a few letters by J. J. Guest. The *Dowlais MSS* are essentially, however, a collection of letters coming into Dowlais. They certainly give a unique cross-section of the working of the iron industry in the first half of the nineteenth century, and enable the historian to describe the growth of the firm for the early period from 1782 onwards; but, unfortunately, we have no remarkable or significant statements of policy or attitude here from John Josiah Guest.

[175] Elsas, *Iron in the Making*, Chapter VIII. The other family interest was that of William Lewis, of Pentyrch forge, a local landowner, and original partner. Wyndham William Lewis, his great-nephew, was bought out by Guest and his nephew, Edward John Hutchins, for £200,000 in 1848. After this transaction Sir John Josiah Guest held fourteen shares; in 1851, Hutchins' two shares were purchased for £58,000, leaving Guest in sole control.

[176] See T. R. Guest, *Plain Address to Unionists* (1831) printed in Elsas, *Iron in*

acted as the company's agent in Ireland and directed Dowlais during John's absence, the Lewises were sleeping partners.

Guest was well trained for his duties, since besides receiving a Grammar School education, he had served an apprenticeship in the works. Even in later life, he prided himself on his practical knowledge of every process in the works. Lady Charlotte Guest makes it quite clear in her *Journal* that he wished his wife to be conversant, too, with the running of the works.

As owner of the second largest works in the country (Crawshay's Cyfarthfa by output being the largest, and Dowlais having, in 1815, five furnaces turning out 12,500 tons a year), as a country banker, with branches in Cardiff and Merthyr, Guest soon became a local magistrate and played a prominent part in bringing some kind of order and justice to the teeming town of Merthyr Tydfil. From a village of a few hundreds in the middle of the eighteenth century Merthyr had swollen to a town, with a population of some 26,000, by the 1830s.[177]

It is important to remember, when thinking of the stature of these captains of industry, the absolute dependence of an industrial community such as Merthyr upon the ironmaster, not only for its livelihood but for the very tone of its social life. In fact, so great was the power of the ironmaster in these circumstances that the fate and character of a town depended upon his sense of responsibility, his beneficence and good sense.

In 1818, following his uncle's example before him, his position in the county was recognised. As we have seen, he became High Sheriff of Glamorgan. In the same year he was elected a Fellow of the Geological Society. This, perhaps, was no academic recognition of work in the scientific field; however, throughout his career as an ironmaster, he was always ready to experiment with new methods and to give inventors of promise facilities to demonstrate their patents.[178] Moreover, as a friend of scientists (for instance, of Michael Faraday),[179] and later, in 1830, in becoming a Fellow of the Royal

the Making, p. 612, '. . . Is not this "Union" a confederacy that brings you into those places where, as men fearing God, you should not be found. . . ?'

177 Gwyn Williams, 'The making of Radical Merthyr, 1800–1836', *The Welsh History Review*, Vol. I, 1961, No. 2, gives a vivid and detailed picture of social and political life in Merthyr at this time.

178 E.g. he was experimenting in 1815 the use of cinder in blast furnaces and later in 1822, with David Mushet: *Dowlais MSS*, David Mushet—J. J. Guest, 21 October 1822.

179 In this connection it is interesting to see that Faraday at this time (1819) had been experimenting in metallurgy. M. Faraday—J. J. Guest, 28 August 1819: '. . . I am also somewhat anxious to know the fact of the new change in

Society, he was following the tradition established by the eighteenth-century ironmasters and engineers.

Guest was Vice-President of the Cambrian Society in 1824; he was a shareholder in the University of London and helped in the formation of King's College, London. It is evident he was anxious to promote education, scientific discovery and its practical application.

Dowlais during this time was prospering indeed. As early as 1819 it was reckoned that profits amounted to more than £20,000 per annum and it is not surprising, especially since his partner Wyndham Lewis had been a Member of Parliament since 1820, that Guest should have felt the urge to indulge himself in parliamentary politics. Doubtless through the influence of Lewis, he was adopted as member for the pocket borough of Honiton in 1826. The year previously he was reported as having expressed his intention to retire from the active management of Dowlais.[180] In 1832, as we saw, he was returned to Parliament as Member for Merthyr. In politics, Williams describes him as 'orthodox, if slightly Whiggish liberal'—in favour of free trade, against corn laws, slave trade and monopoly.[181]

A portrait of Guest at this period in his life reveals him not as a ponderous man of business but rather as a prosperous country squire with a face showing no sign of hard struggle, but of an easy and confortable living.

His next step into the ranks of society was his marriage to Lady Charlotte, a daughter of Albemarle, ninth Earl of Lindsey, in 1833. This was not just a marriage arranged to seal an alliance of an impoverished noble house with one of the *nouveaux riches*. Anyone who has read Lady Charlotte Guest's *Journals*[182] will be very much aware she was a woman of character and intelligence. Knowing that a brilliant match in the ranks of the aristocracy was closed to her, the prospect of being partner to the 'largest manufacturer in the world' was not unattractive, despite the prejudice to be overcome of marrying a 'man in trade'. By then, Dowlais with its eighteen furnaces and supplying the railways all over the world gave her one kind of preeminence. Later she confessed that she had become a thorough business woman, and added '. . . whatever I undertake, I must reach an

---

the Blast Furnace and of the coffers or boxes for the heating Furnaces and also of any other trials that may have been made. There can be no doubt but that unremitted exertions like those you are continually making must produce some important and valuable result in the manufacture of iron.'

[180] *Crawshay MSS*, 9 June 1825.

[181] Williams, *loc. cit.*, pp. 180–1. By 1837 Guest's views had advanced towards zeal for the ballot and he was a sharp critic of the New Poor Law.

[182] *Op. cit.* (ed. Earl of Bessborough, London, 1950).

eminence in. I cannot endure anything in a second grade. I am happy to see we are at the head of the iron trade. Otherwise, I could not take pride in my house in the City and my Works at Dowlais and glory (playfully) in being (in some sort) a tradeswoman.'[183] Perhaps one false impression of Lady Charlotte ought to be corrected. It has been suggested that she acted as secretary to the business for her husband. Certainly, as she says, she played an important role during her husband's absence in conducting negotiations. However, although she may have acted as private secretary for Guest's private concerns and parliamentary business, she at no time acted as a book-keeper for the business.

Henceforth Guest's ties with the source of his wealth were relaxed. Life was spent between a country house in Wales and later, in 1846, a mansion at Cantford 'improved in the Gothic manner by Sir Charles Barry'. In London the Guests met fashionable society in town houses at Spring Gardens and in Bedford Square. It is true that he now represented the newly enfranchised borough of Merthyr Tydfil (since 1832), and his interests in the South Wales railways[184] and periodic visits took him to the family house overlooking the Dowlais furnaces, but the City was now the focal point of his interests. In common with the other large merchant houses in iron, e.g. the Crawshays and the Carron Company, he had offices in London in Token House Yard and, in 1839, in the City, at Lothbury. Moreover, Guest had transferred his banking interests from the local sphere of South Wales to the first new London joint stock bank, the London and Westminster.

The natural summit of a career like this, of a great industrialist, financier, and as an assiduous if undistinguished parliamentarian,[185] was of course a title. In 1838 Guest 'was elevated, if so I must call it, to the rank of Baronet'. And although Lady Charlotte thought little of the honour, saying 'I consider it a paltry distinction and was much averse to him taking it . . . I shall not rest till I see something of more value bestowed upon him',[186] Sir John Guest was more impressed. He, not without justification, looked upon it as recognition

[183] *Journals* p. 89. She also engaged in literary work on the translations of the Mabinogion; she also contemplated a 'History of the Iron Trade'. However, she does not appear to have got far with it. She found, she tells us, the contrast between her 'musty antiquarian researches, and the brilliant fêtes and the plodding counting house' a great delight.

[184] E.g., the Taff Vale Railway of which he was chairman.

[185] He was evidently a valuable financial support to the Whig Party, paying the election expenses not only for himself but for his nephew at Southampton and his wife's cousin at Carlow in 1841.

[186] *Journal*, p. 71.

'for my successful efforts to advance the commercial interests of this great country'.[187]

In the years immediately following, Sir John Guest was at the zenith of his fame and fortune, and held an established position in the social world. Trips to the Continent were undertaken, though not only for pleasure but for carrying out the many contracts for the European railways which flowed into Dowlais, and in return, such was the fame of Dowlais, like Soho before it, that from many parts of the world came visitors, to be impressed by the sights and to talk business.[188]

Dowlais too, it should be noted, had shared in the prosperity of its works. Sir John Guest was perhaps no more enlightened than his contemporaries on the larger issues of the struggle between capital and labour. However, his marriage with Lady Charlotte, who was a natural reformer, proved fruitful to Dowlais in many ways.[189] A school had been built before her reign there in 1819, but she improved and extended it. As the historian of this scheme claims, 'From the standpoint of established schools for the education of the working classes, the Guest family of Dowlais was probably the most important and most progressive not only in the industrial history of South Wales, but of the whole of Britain during the nineteenth century.'[190] Lady Charlotte provided for a library and a church and did much, through her influence, to alleviate the toll of squalor, disease, ignorance and poverty, which seemed an inevitable part of industrial life at that time.[191]

The last few years of Sir John's life, when he was weakened by ill-

[187] C. Wilkins, *op. cit.*, p. 106.

[188] Miss Elsas, p. xii, gives a list of distinguished visitors to Dowlais between 1792 and 1860. They include scientists, technicians, scholars, diplomats and princes.

[189] A works surgeon had been appointed in 1813. By the 1840s, along with the other Merthyr ironworks, the Dowlais workmen had a contributory medical scheme, paying 1½d. in the £ on their earnings per week. The doctor attended a man's family, but did not attend to confinements without a fee. In 1860 the stoppages were 4d. in the £: 1½d. for the doctor, 2d. towards a sick fund and ½d. for schooling—this was subsidised by the company (*Dowlais MSS*, Elsas, pp. 71–4). A Savings Bank for the men at the works was not started until 1853.

[190] L. W. Evans, 'Sir John and Lady Charlotte Guest's Education Scheme at Dowlais in the mid-nineteenth century', *Journal of National Library of Wales*, Vol. IX, No. 3, Summer 1956.

[191] As late as 1854 cholera swept through Merthyr. John Evans to Lady Charlotte Guest, 20 September 1854: 'It attacks all the dirty courts and gulleys and the overcrowded houses . . . we shall remove a lot up to the Barn today, and further provision is being made for the reception of a larger number. It is very distressing to see the state of excitement which prevails amongst the men. . . . The men are leaving the place altogether . . . '. (Elsas, p. 73).

ness, were taken up with the vital struggle with the Marquis of Bute for the renewal of the lease of the land upon which the works were situated. Indeed, this struggle between the two astute men of business to obtain the wealth of Dowlais forms a central thread to the story which is unfolded in Lady Charlotte's diaries.

Guest had much at stake; his works covering an area of 48 acres, with 18 blast furnaces pouring out over 70,000 tons of iron a year. In 1847 Dowlais employed about 7000 men, women and children, bringing in a profit of £50,000 a year.[192] However, Guest almost succumbed to the attractions of a wealthy retirement, free from responsibilities, and spoke several times of giving up Dowlais. As early as 1837 he had contemplated a partial retirement, in view of the expiring lease and the competition of neighbouring works.[193] In 1850, doubtless, this resolution was strengthened by a heavy loss made in the business.

Nevertheless, six blast furnaces were blown in again, for Guest fortunately had a feeling of paternal responsibility to his 14,000 dependants living at Dowlais.[194] As the Reverend Mr. Jenkins, the local parson, somewhat eulogistically put it when the crisis was over, 'God had blessed him with abundant means, and he might have retired in the enjoyment of a princely income, free from the immense responsibility of carrying on such large works. . . . But in mercy, more especially to this populous district, he acted otherwise and in so doing he was influenced by strong feelings of compassion and kindness to the thousands he had collected together from North to South, from East to West and who looked to him as the only one likely and able to carry on the largest ironworks in the world.'[195] Guest resisted the temptation to leave Dowlais to die and the lease was renewed in 1848 at a very high rent.

Nevertheless, the association of the Guests with Dowlais was drawing to a close. In 1852 the partnership of this well-matched couple came to an end. It was left to Lady Charlotte to sign the new lease and to carry on. But, although she tried to rule Dowlais single-handed for a time, paying much attention to the welfare schemes for her men, and, indeed, taking pride in the profit of £43,000 made during her year of control, she soon found the job too much for her. Dowlais was henceforth administered by a board of trustees. Guest's son, Ivor, did not take up this hard uncongenial work of making money.

[192] *Journals of Lady Guest*, p. 199. [193] *Ibid.*, p. 60.

[194] Elsas, *Iron in the Making*, p. 73. In 1847 Guest proposed to set up a soup kitchen, 'If there should be much distress when the works stop'.

[195] *Wilkins*, op. cit., p. 115.

Such was the attraction of the profession of ironmaster; which, apart from its pecuniary appeal, also satisfied ambitious men's desire for power over large establishments of men and over the fiery forces of nature imprisoned in the furnaces. The grandeur of the furnace districts at night, when the flames belching forth lit up the skies for miles around, had impressed many travellers from the eighteenth century onwards. Indeed, such observations became commonplace, but there was, too, about the position of ironmaster, this pride in one's power. The greatest ironmasters of the day thought of themselves, not just as businessmen, nor as manufacturers, but as rivals for the 'Iron Crown'.

Part III

CHAPTER XIII

The Process of Supersession: The Age of Steel

THE PRE-BESSEMER ENGLISH STEEL INDUSTRY[1]

The modern English steel industry may be said to have had its beginnings in the early seventeenth century when the cementation process of converting bar iron into steel was perfected in this country.[2] Until the sixteenth century, England had been an importer and user of steel, indeed, from as early a period as that of the Celts of Anglesey, 150 B.C. to A.D. 50,[3] but, apparently, there was no native production of steel until 1523.[4] In that year, according to Wolf,[5] a steel furnace was built in this country. Dr. Schubert has described the efforts of forty years later, of Sir Henry Sidney to establish steel forges at Robertsbridge and Boxhurst in 1565.[6] How-

[1] This chapter is a shortened version of the article by M. W. Flinn and A. Birch, 'The English steel industry before 1856, with special reference to the development of the Yorkshire steel industry', *Yorkshire Bulletin of Econ. and Social Research*, Vol. 6, 1954. See also S. Pollard, *History of Labour in Sheffield* (Liverpool, 1959), Chapter III, 'The Steel Industry before Bessemer', pp. 78–85.

[2] Patent No. 33, 1614. Ellyot and Meysey. It is very likely that the cementation process was 'invented' upon the Continent. In 1613 two armourers in Maastricht received a charter of monopoly for the production of steel made in this way. (See the investigations of Brownlie and Baron de Lavelaye in the *Journal of the Iron and Steel Institute*, 1930, Vol. CXXI.)

[3] H. R. Schubert, 'Early steel', *Steel News*, Vol. 3, No. 3, December 1952. The early product was steely iron rather than steel itself, although it is quite possible that this may have been made in England during the Middle Ages, as Schubert suggests; the evidence he brings forward points to the use of steel rather than its manufacture in this country. However, he gives an interesting account of the process of cementation from the legend of Wayland the smith. This involved feeding birds with iron filings mixed with their meal, smelting the dung, filing and feeding it again to birds 'kept hungry for five days', and a second smelting.

[4] Rhys Jenkins, 'Notes on the early history of steel making', *Trans. Newcomen Soc.*, Vol. III, p. 17.

[5] A. Wolf, *History of Science, Technology and Philosophy in the XVIth and XVIIth centuries* (London, 1935), p. 495.

[6] H. R. Schubert, 'The economic aspect of Sir Henry Sidney's steelworks at

ever, this enterprise founded with the help of foreign migrant labour, which was at first successful, was undercut by the import of cheaper steel from the Swedish Baltic ports, and the works appear to have come to an end in the 1570s.

The method of manufacture at this time was doubtless fundamentally the same as that practised upon the Continent right up to the nineteenth century. Since steel is iron with approximately 1 per cent of carbon in its composition, its manufacture may be approached in two ways. The Continental and earlier English method was to start with pig-iron containing 4 per cent carbon and to refine it in a small furnace similar to the finery for making wrought or bar iron. The method which was to supersede it was to start with wrought iron, which was practically free of carbon, and by a process of cementation to add the requisite amount of carbon. It was this process, patented by William Ellyot and Mathias Meysey in 1614 and 1616[7] (the latter for use with pit-coal in the reverberatory furnace), that established the English steel industry.

It may be as well at this point to attempt to clarify the position regarding the many different varieties of steel which occur in contemporary sources. References have been found to raw steel, blister steel, shear steel, faggot steel, doubled steel, bar steel, crucible and cast steel, English, German, Spanish, Turkish and Flemish steels. Several of the terms are, of course, self-explanatory, denoting the country of origin. Spanish steel in particular, from Toledo, was highly regarded in England. In the sixteenth century, according to Thomas Fuller, it was thought a fortunate circumstance that Englishmen had mastered the art of casting iron cannon to counter the advantages of Spain with their steel. 'Some observe that God hath so equally divided the advantage of weapons between us and Spain, that their steel makes the best swords, our iron the most useful ordnance.' English steel referred to that made by the cementation process; it was also known as blister and bar steel. Raw or natural steel was the product of the Continental finery process.[8] The other names describe raw or blister steel which had been subjected to further mechanical or smelting processes in order to give the metal greater homo-

Robertsbridge in Sussex and Boxhurst in Kent', *J.I.S.I.*, 1950, Vol. 164, pp. 278–80. See also E. Straker, *Wealden Iron* (1931).

[7] A. Wolf, *History of Science . . . in the XVIIIth Century* (London, 1938), p. 633. The inventors were not successful in putting their theories into practice despite having been granted a monopoly to prohibit the import of foreign steel. Other experimenters appear to have perfected the process soon afterwards.

[8] Described in some detail by C. Sanderson, 'On the manufacture of steel as carried on in this and other countries' (*Journal of the Royal Society of Arts*, Vol. III, 1854–5), pp. 450–2.

geneity and purity. Raw steel was inconsistent in quality since much depended upon the skill of the workman in working the iron in the furnace; likewise blister steel was suitable for cruder pieces of metal work, but not for taking a fine cutting edge. Or, as Fuller put it, 'It is more hard and brittle . . . useful for the making of English knives, sithes, scissars, shears, and etc.; but fine edges cannot be made thereof as lancets for letting blood, incision knives. . . . I have been informed that Sir Basil Brooke (the great steel maker in this country viz. Gloucestershire) his patent to prohibit the importing of foreign steel was revoked on this account, because that no artist could make the aforesaid instruments of English steel, but must have it from Damascus, Spain, Flanders, and etc.'[9] Brooke, incidentally, had come into possession of the patent as early as 1618, but was dispossessed by another contestant, Doctor Fludd, two years later.

According to Rhys Jenkins, it was Brooke who took Ellyot and Meysey's patent and started actual production some time after 1635. Thus it was from the Forest of Dean that the cementation process seems to have spread by way of the Midlands to the North of England.[10]

Already, before the end of the seventeenth century it was stated, 'The English steel is made in several places in England, as in Yorkshire, Gloucestershire, Sussex, the Wild (*sic*) of Kent, and etc. But the best is made about the Forest of Dean . . . if it be well wrought and proves sound it makes good edge tools, files and punches.'[11] An important centre of the industry, not mentioned by Moxon, was the district around Stourbridge. One reason for the establishment of the industry here was that this district was the source of the refractory clay which was indispensable for the chests or pots for conversion. According to Le Play this Stourbridge clay withstood six heatings.[12] Dud Dudley in his *Metallum Martis* refers to the use of coal for the making of steel[13] and Dr. Plot's account of John Heydon's works at Bromley near Kingswinford is one of the earliest giving details of the method carried out. This was sometime before 1686.[14] However,

[9] T. Fuller, *A History of the English Worthies*, ed. P. A. Nuttall (London, 1840), Vol. 1, p. 547. He described steel as 'the elder brother of iron'.

[10] Rhys Jenkins, 'Early history of steel making', *Trans. Newcomen Soc.*, Vol. III, pp. 16–40.

[11] Moxon, *Mechanick Exercises* (2nd edn., 1693), p. 54, quoted by Jenkins, *loc. cit.*

[12] Le Play, 'Mémoire sur la fabrication de l'acier en Yorkshire', *Annales des Mines*, 4me série, Vol. III, 1843, p. 644. Huntsman, incidentally, used this and Dutch clay, which he found was difficult to obtain.

[13] *Op. cit.* (1855 edn), p. 35.

[14] R. Plot, *Natural History of Staffordshire* (1686), p. 374.

perhaps the most important manufacturer of steel here was Ambrose Crowley, senior, the father of Sir Ambrose. He owned a 'steel house' in Stourbridge as early as 1682. But it is rather strange that evidence of his activity in this industry in the shape of a letter in which he describes experiments upon the cementation process dates from more than ten years after Plot's account of this method of steel-making at John Heydon's.[15] Of course, it is quite possible that he was experimenting to improve the construction of the furnace and the making of the chests in the cementation process which had been already long established. What is more important is that it is quite probable that it was through the Crowley family that the process was introduced to the North of England. The exact date of its introduction is not known, but by 1701 there is clear evidence that Ambrose Crowley, junior, was producing steel at Winlaton[16]—'when the Steelmaker shall want iron to sett the furnace, the ironkeeper is to lett him pick out of the sorts appoynted to make steele'. Crowley's Council Instruction No. 41, also of 1701, carefully details the procedure for sorting the bar iron and putting it into the furnace. Another steel maker in the North of England was Dan Heyford of Roamley, near Pontefract, who supplied the Shotley Bridge sword-makers with steel.[17] Isaac Cookson, who established the first blast furnace in the district, at Whitehill, about 1704,[18] was also a converter of steel. His concern was engaged in trade with Peter Stubs of Warrington, the file-maker, as late as 1799.[19]

The Sheffield district was apparently backward in the development of steel conversion, especially in view of its important cutlery trades. It is true that Samuel Shore had set up a furnace in 1709 and, in 1719, this was followed by the similar enterprise of George Steer.[20] The Cutlers' Company, which in many ways was antagonistic to the home producers of the town's raw material, also in 1730–1 investi-

[15] Sir Ambrose Crowley's Council Instruction No. 41, 11 December 1701, MS volume in the library of Newcastle-upon-Tyne Society of Antiquaries. I am indebted to Mr. M. W. Flinn for this reference and others relating to Crowley.

[16] *Crowley MSS*, Council Instruction No. 39, 4 November 1701. Verse 151, Steel Account.

[17] E. Hughes, *North Country Life in the XVIIIth Century* (London, 1952), p. 60.

[18] *V.C.H. Durham*, Vol. II, p. 290.

[19] T. S. Ashton, *An Eighteenth-Century Industrialist* (Manchester, 1939), p. 48. He quotes a letter from the firm advising the use of German or spur steel instead of Huntsman's cast steel. It is most probable that spur steel should be shear steel. Stubs, it should be noted, also obtained steel from another Newcastle firm, Emerson, Hulme and Co., and from Hull (this might have been imported). Cookson's steel had to be sent to Oughtibridge, near Sheffield, for tilting.

[20] G. I. H. Lloyd, *The Cutlery Trades* (London, 1913), p. 74.

gated the possibility of making steel for itself, but it was not until 1759–72 that it embarked upon manufacture of steel. A furnace was rented in Sheffield and the enterprise begun with £100 capital contributed by the Cutlers' Company and £580 of loan capital. Iron was obtained from London and Joseph Sykes of Hull—a few tons was bought from Attercliffe Forge in 1767. The most successful period in the history of this enterprise was 1759–62, when a profit of £212 was made under the management of a Mr. Ibberson. But afterwards, a committee of management was set up, with almost annual changes in the managers. The concern was minutely controlled in theory in accordance with the object to supply steel at a price 'if possible something below the common market, and yet to bring gain to the Company something more than equal to answer the expense of the trust, and the interest of the capital stock or fund'. But the enterprise was a failure, despite the introduction of £260 extra capital, and was given up in 1772. The furnace, which had in the meantime been built for the company, was let to another Sheffield firm of converters at £20 a year.[21]

The Sheffield area, with its wealth of water power and its nearby market for the steel, was of course closely associated with many of the later stages of preparing steel, e.g. tilting, but until the end of the eighteenth century the industry was scattered in many places. Localisation and concentration had not taken place to any marked extent. The main consideration was the availability of Swedish bar-iron, and as this was well distributed from London and Hull, the largest importing centres, cementation furnaces could be set up wherever there was a local demand for steel from the tool-making and tool using industries.[22] Thus it is not surprising to find evidence of steel furnaces at Coalbrookdale and Bristol.[23]

There are several references to steel-making in Birmingham itself, which like Sheffield was a town with steel consuming industries at its doorstep. The Birmingham steel converting furnaces were run, as a rule, in conjunction with the manufactories of the jewel and toy makers. In the late 1730s about 220 tons of first quality Swedish iron were annually converted to steel in the Birmingham neighbourhood and a plan of the town, dated 1731, shows two steel houses there. One gives its name to a street which still survives—Steelhouse

[21] R. E. Leader, *History of the Company of Cutlers in Hallamshire* (Sheffield, 1905), Vol. I, pp. 174–5.

[22] Steel was frequently used for plating. That is, welding a steel cutting edge to a wrought iron implement.

[23] Raistrick, *op. cit.*, p. 286. He quotes a lease dated 7 June 1734, mentioning a steelhouse which had been lately converted into a malthouse.

Lane.[24] In 1738 the cementation furnaces of Birmingham attracted the attention of a French traveller who reported upon the process there.[25] Some years later, Matthew Boulton, the famous industrialist, in partnership with a 'toymaker', also owned a steelworks in Birmingham.[26] As a result of this expansion of the industry, which was perhaps encouraged by the protective duties placed upon foreign imports at the end of the seventeenth century, England began to export steel of her own making for the first time. Between 1719 and 1735, 1096½ tons of English steel went abroad, as against 207½ tons of foreign steel re-exported.[27]

One important task before the historian of the early steel industry is to try to clear up the question of the innovation of making shear steel in this country, by methods similar to those for forging wrought iron. It seems fairly conclusive that the method originated in Germany, where it followed the finery process. The operation, as it was described by a nineteenth-century steelmaker, is as follows: 'The raw steel bloom is drawn into bars, one or two inches wide and ½ inch thick, or less; a number of these are put together and welded; these bars are thrown into water, and they are broken into smaller pieces to examine the fracture, those bars which are equally steelified are mixed together.' Then a packet of these small pieces was sandwiched between two whole bars, two to three feet long. 'This packet is then placed in a hollow fire, and carefully covered from time to time with pounded clay, to form a coat over the metal, and preserve it from the oxidising influence of the blast. When it is at full welding heat it is placed under a hammer, and made as homogeneous as possible; it is again cut, doubled together, and again welded.'[28] The product of this forging was known as German steel, and when Gabriel Jars visited the Winlaton works of the Crowleys in 1765, the methods for making shear steel there were identical with those employed by Styrian steelmakers. Sir Ambrose Crowley has, in fact,

[24] R. A. Pelham, 'The West Midland Iron Industry', quoting *J. H. C.*, XXII, p. 854, and citing Westley's *Plan of Birmingham*. The steelhouses were Kettle's and Carlesse's.

[25] J. U. Nef, *The Rise of the British Coal Industry*, Vol. I (London, 1932), p. 246, refers to a French MS in the Archives Nationales (O¹1293). In 1754, Angerstein, the Swedish representative of the Board of Mines, also visited a steel manufactory in the town.

[26] H. W. Dickinson, *Matthew Boulton* (Cambridge, 1937), p. 203. See below for a description of the processes carried at Boulton's works in 1770.

[27] Scrivenor, *op. cit.* (1841 edn.), p. 346, for figures of exports; Le Play, *loc. cit.*, pp. 628–9, remarks upon a sevenfold increase in the duties upon foreign steel in the seventeenth century.

[28] C. Sanderson, 'On the manufacture of steel, as carried out in this and other countries', *Journal of the Royal Society of Arts*, 1854–5, Vol. 3, p. 453.

been credited with the introduction of the manufacture of shear steel into this country at some time between the beginning of the century and as late as 1730 (when he was more than ten years dead!). However, Flinn, the historian of the Crowley business, has traced evidence for the making of shear steel at the Teams steel furnace in the 1750s. Only a small proportion of the output of steel was shear steel, which had a higher cost of production.[29]

At that time shear steel was only made to order, according to Jars, and it seems that the adoption of the process was a slow one. It is, incidentally, a Sheffield tradition that shear steel got its name as the result of local demand from the Yorkshire cloth shearers, who specified this special steel for their tools of trade. But, if this is so, then Sheffield must have continued to import this steel from Germany and Newcastle (as Lloyd in fact suggests, up to 1775); it was not until 1767 that the manufacture of shear steel was taken up by Sheffield Steel-makers.[30]

Shear steel here came into competition with cast steel made according to the Huntsman process. In 1799 Cookson and Co. of Newcastle were advising Stubs to use 'spur steel' in preference to cast steel, because of the frauds practised by the crucible steel-makers in using inferior Russian iron.[31] Even today, opinion seems to be divided as to whether cast steel was superior to genuine double sheared steel. At all events, following the disclosure of the secret process, perfected about 1740, the Walkers of nearby Masborough set up a 'fire-hole' furnace in 1750; then, in later years, the manufacture of crucible steel began to be more widespread. By 1774 at least two other steelmasters were making cast steel.[32]

We need not be concerned here with the details of Huntsman's career and the involuntary sharing of his trade secret with his competitors in the trade.[33] However, besides his fellow countrymen,

[29] Flinn, *Men of Iron*, pp. 187–8. German steel (shear steel) sold at £50 per ton as against £27–28 per ton for cemented steel.

[30] Wolf, *op. cit.*, pp. 633–4. Cf. Lloyd, *op. cit.*, p. 74; forges were erected in 1785 for this purpose.

[31] Ashton, *Eighteenth-century Industrialist*, p. 48. It is to be noted that Sanderson, writing in 1853, indicated that only because cast steel was cheaper was it preferred to shear steel.

[32] Bolsover and Co., and John Marshall.

[33] See E. W. Hulme, 'The pedigree and career of Benjamin Huntsman', *Trans. Newcomen Society*, Vol. XXIV, pp. 37–48. The agent of Huntsman's undoing was a goldbeater named Waller. It is curious, however, that Broling, a Swede, who investigated the cast steel industry in 1797–9, came across another claimant to the honour, with the very similar name of Walter. However, as Beck says, no English writer has given his claim any serious consideration (Beck, *Geschichte des Eisens*, Vol. 3, pp. 277–8).

foreigners were also interested in the process which Huntsman had tried to keep for his own benefit. Since the Swedish forges supplied Sheffield with best quality bar iron for conversion, the Swedish ironmasters did not need to seek out that secret of the quality of Sheffield steel. But one can well imagine that the more enterprising of them were anxious to experiment with the manufacturing processes of cast steel in their own country. In 1761, Ludvig Robsahm made a visit to the English steel manufactories.[34] He visited Crowley's works at Swalwell, and at nearby Newcastle, the steel furnace of a manufacturer named Hodgson. His main goal, however, was to inspect Huntsman's establishment, two miles outside Sheffield.

Robsahm's account of these cast steel works precedes that of Gabriel Jars by a few years, hence his observations are of considerable interest. It is quite obvious that he understood that the secret of Huntsman's success lay in the making of the crucibles from a clay capable of withstanding the high temperatures necessary to melt the blister steel. Huntsman lived up to his reputation as being secretive. He declared that even if he were offered £50 he would not show anybody the method of making the crucible, but he would show his Swedish visitor the finished article and, apparently, he allowed Robsahm to inspect his establishment. The most important secret that Robsahm discovered was that some clay came not only from Stourbridge but also from Holland. Huntsman, apparently, was having difficulty in obtaining this raw material from abroad, but could at that time, with his two assistants, work up 8 tons of steel yearly. This very small capacity, Huntsman said, could be increased to 12 tons, if he could employ more workmen.

Robsahm's visit did not have the practical results he wished, but it was not many years before another ironmaster from the same country paid a call upon Huntsman in 1766–7. As a result of this visit his secret was again revealed and Quist Andersson was able to set up the first cast steel works in Sweden, upon his return there.[35]

Huntsman was a self-effacing man, and in 1792 it was remarked in a *Report on Huntsman's Cast Steel by the Engineers to the Prince of Wales and the Duke of Clarence* that, 'during the course of more than 30 years of time devoted to the manufactory of it, he has so much neglected his own interest and credit, as never to give the public, through any general or circular medium, any account whatever of his steel.'[36]

[34] S. Rydberg, *Svenska Studieresor till England under Frihetstiden* (Uppsala, 1951), pp. 189 ff., quoting L. Robsahm's Dagbok (MS Journal in Kungliga Biblioteket, Stockholm).

[35] Rydberg, *op. cit.*, pp. 295–7.

[36] *Op. cit.* This report is printed as an appendix to Sir R. A. Hadfield's 'Early

The advantages of the processes were twofold. In the first place, with later developments, it was possible to mix steels and ingredients to produce special alloy steels. However, it may be mentioned in parenthesis that William Reynolds of Ketley had introduced and patented a method of introducing manganese into steel as early as the 1790s.[37] But, as Le Play pointed out in the 1840s, it was the forging processes to produce shear steel which accounted for the greater part of the cost of high quality steel.[38] Thus the Huntsman process enabled costs to be reduced and, eventually, larger outputs to be obtained. Despite these advantages, however, it seems that the cutlers were prejudiced against the cast steel, and, as Professor Ashton has shown, most of the product was sold abroad.[39] Among Huntsman's customers was Matthew Boulton, who purchased cast steel for the dies used in his mint at Soho.[40] This is a curious circumstance because, as we have seen, Matthew Boulton owned a steelworks of his own at Birmingham, and a report made by Robert Erskine, later to be an American ironmaster, in 1770, describes the casting process. When James Watt saw this steelworks in 1767, he wrote afterwards in his *Memoir* of Boulton, that he had 'a steel house for converting iron into steel, which was frequently employed to convert the cuttings and scraps of the chapes and the other small ironwares into steel which was afterwards melted and made into cast steel for various uses'.[41] Three years later Erskine gave a full description of the establishment.[42] The coverting furnace held 'about 8 Tons'. The method of working was as follows: 'The Iron is put in at holes in the wall, at the Top of

---

history of crucible steel', *Journal of the Iron and Steel Institute* 1894, No. II, pp. 224 ff.

[37] Reynolds' Patent No. 2363 of December 1799. See also a letter from Thomas P. Smith of Philadelphia to the *Conseil des Mines* (Archives Nationales F14 4261), dated 1803. This is an account of iron and steel making in England based largely upon his observations at Reynolds' works at Ketley. ' . . . *Steel.* To prepare the iron for making steel the refined iron is placed in the puddling furnace as soon as it is softened to the state of paste 40 lbs. of Exeter manganese finely powdered is added for every 270 lbs. of iron. Immediately a violent effervescence takes place. After this it is operated on precisely in the same manner as for common bar iron. . . . ' He also describes the production of cast steel, as practised at Sheffield: 'The blistered steel produced by the above operation is broke to pieces and melted in crucible, each containing about 25 lbs. in blast furnaces. The Crucible is kept constantly well covered with coke. Eventually in about 3 hours it will be melted and is then run into iron moulds. The blooms thus produced are drawn or rolled into bars of the desired sizes. . . . ' (I am indebted to Dr. Chaloner for this reference.)

[38] Le Play, *op. cit.*, p. 631. [39] Ashton, *op. cit.*, p. 58.

[40] Ashton, *op. cit.*, p. 59. [41] Quoted in H. W. Dickinson, *op. cit.*, p. 203.

[42] *Erskine MSS*, Robert Erskine—R. Atkinson, 11 October 1770 (Library of the New Jersey Historical Society, Newark).

the Chests, and taken out steel at opposite holes, all which holes while the operation is performing, are stopt with Brick and mortar. The fire is kept up six days, and then it is left to cool, to such a degree, that a person can go into the furnace and lift and push the Steel Bars out of the Chest.' Unfortunately Erskine does not say much about the casting process. The first preliminary, however, was that the blister steel should be 'twice converted', then, 'the steel in order to be melted is broke into small pieces and put into a crucible about 5 inches diameter and 15 inches high. The crucible is placed in a furnace of live coake cinders.' The furnaces themselves were small affairs, '3 feet deep' and 'about a foot and ½ square Each', so that several of them were placed together in a row. The whole structure had a chimney 30 feet high.

Boulton, is seems, was in the forefront of technical progress, for it is possible that his Huntsman establishment preceded the adoption of the process in Sheffield.[43] Huntsman himself is said to have been urged to settle in Birmingham and it may be that Boulton was behind this. The fact that Huntsman preferred the Sheffield district is probably irrelevant to the subsequent development of the steel industry in the Midlands. Yet even in the early nineteenth century, when Sheffield had already established itself as the predominant district, Birmingham's steel production seemed capable of expansion. In 1815 Johann Conrad Fischer, a Swiss metallurgist who claimed to be the inventor of cast steel, visited the steel works of 'Herr P.' But operations were only upon a small scale. There was only one Huntsman furnace melting 20 lb. at a time; however, Fischer was told that it was proposed to increase the capacity four-fold by building another furnace.[44]

However, the district did not expand its production; steel forges and mills persisted near Stourbridge[45] as late as 1832, and the Brades works of William Hunt and Sons supplied the edge-tool and steel pen works in the locality, but manufacture of steel did not achieve large proportions until the coming of the Siemens open-hearth process.[46]

[43] The first recorded evidence of the adoption of the process in Sheffield is dated 1760, when John Love and Thomas Manson entered into a partnership to erect a furnace 'for fluxing or running steel' (quoted in Walton, *op. cit.*, p. 179).

[44] J. C. Fischer, *Tagebücher* (Zurich, 1951), p. 109.

[45] Scott, *History of Stourbridge* (Stourbridge, 1832), pp. 595–7. William Jackson, the founder of the French steel concern at Assailly and St. Seurin, where the Bessemer process had been pioneered, had been a steelmaker in Birmingham. In 1815 he emigrated to introduce the Huntsman process to France (*Jackson et Fils*).

[46] Hence it is very surprising to learn from Lord Aberconway's *Basic Industries*

The history of the pre-Bessemer steel industry's development in Sheffield, apart from the circumstances surrounding the Huntsman invention, has been strangely neglected by economic historians,[47] although the cutlery trades themselves have been studied because of their interest as a survival of the domestic system of manufacture. Perhaps the main reason is that, like their associated trades, the primary processes of converting and casting were in the main backyard activities. When Fischer visited the Huntsman steelworks in 1814 he was very surprised to find the access to the business through a living-room, and he was led into the yard, where the storehouse and furnace were situated, past the Huntsman family.[48]

Since Fischer was himself a skilled manufacturer of steel, and although it was his chief mission to compare Huntsman's technique with his own, he says very little about the establishment. He found that his cast steel had the same general qualities as Huntsman's, which itself far excelled other cast steel he had seen, namely that of 'P. & Co. ', 'the so called Marshall steel' and that made by the Walkers of Masborough. Walkers must, incidentally, have been the largest producer of cast steel in the district, for Fischer saw there twelve melting holes, from which the iron moulds for casting the ingots were teemed. The cast steel was sold by its appearance and quality. The best quality sold for as much as £120 a long ton.[49]

The history of the steel industry in Sheffield in the first half of the nineteenth century is of the building up of small-scale concerns in the backyards and orchards of the houses in the town. There was a great proliferation of small concerns specialising in the business of converting and refining. The tilting was a separate branch of the

---

of Great Britain (London, 1927), p. 290, that South Staffordshire as late as 1865 had as many as 45 furnaces for making blister steel. His authority, not a very good one, it seems, seems to be Timmins, *Products of Birmingham . . .* , p. 70. The figure given of 'connecting furnaces' (*sic*) is 'not above 45'. There is another erroneous statement in the same work (p. 212) to the effect that steelmaking had not been carried on in Birmingham for more than seventy years.

[47] It is true that Professor Ashton has discussed the trading relations of the Huntsmans and Stubs connections with the Sheffield steelmakers. More recently the history of a family business (S. Pollard, *Three Centuries of Sheffield Steel*, Sheffield, 1954) has appeared.

[48] Fischer, *op. cit.*, pp. 151–8. Huntsman did not set up as a steelmaker at Attercliffe upon a commercial basis until 1751. Thus the Walkers were first in the field, although even here there is a difficulty with the evidence. According to H. G. Baker, a local authority upon the subject, the Walkers had given up Grenoside by 1750, when the Huntsman furnace was supposed to have been built.

[49] Fischer, *op. cit.*, pp. 158–9. See also W. O. Henderson, *J. C. Fischer and his Diary of Industrial England, 1814–1851* (1966), pp. 64–8.

industry conducted in the mills upon the streams surrounding the town. From the directories of Sheffield, which are the chief source of information for the growth of the industry, it appears that in 1787 there were eleven firms making cast steel and nine which specialised in the process of conversion.[50] By 1856 there were 135 steel makers in the town.[51] The rapid expansion of steel-making, which continued in this type of industry until the 1860s, can be gauged by the following statistics of furnaces and melting-holes in Sheffield:[52]

	Conversion furnaces	**Melting holes**	**Estimated production of blister steel**
1835	56	554	14,000
42	97	774	20,500
46	105	974	26,250
53	160	1494	40,000
56	206	2113	51,500
62	205 (?)	2437	78,720

As Mary Walton, the historian of Sheffield, says, 'The majority of the steel makers, then, were cutlers or edge-tool makers.' And it is significant that the captains of the later steel industry built up their wealth from small beginnings in the firms of the city which carried on these small workshops and furnaces. Thus Mark Firth, Thomas Jessop, Charles Cammell, Samuel Fox and John Brown, who were later to establish large-scale steel works in the Brightside area in the 1840s and 1850s, all either came from parents engaged in a small way in the industry or started their industrial career by entering small partnerships themselves.[53]

One set of accounts which has become available to illustrate the rise of the small man in the steel industry, those of Daniel Doncaster and Co., would hardly seem to indicate more than a very modest

[50] Gales and Martin, *A Directory of Sheffield*, 1787, quoted in Ashton, *An Eighteenth-century Industrialist*, p. 37.

[51] Mary Walton, *Sheffield, its Story and its Achievements* (Sheffield, 1952), p. 178.

[52] Hunter, *History of Hallamshire* (1882 edn.), pp. 214–16. The estimate of production is based upon an output of 250 tons per furnace. In 1846, according to a statistical report of the consumption of coal in Sheffield, 32,760 tons were actually produced. In 1856, there were reckoned to be another 54 conversion furnaces elsewhere in the United Kingdom, bringing the total of blister steel production to 65,000 tons. 3000 tons of steel were also made by a direct process.

[53] See J. H. Stainton, *The Making of Sheffield, 1865–1914* (Sheffield, 1924), and Walton, *op. cit.*

level of profit.[54] The first Daniel Doncaster set up business as a file-maker in 1778. The process of steel converting soon became a necessary adjunct to the business. Nothing is known of this period until 1829, when the second Daniel Doncaster joined his brother to take over the business left by their father upon his death in 1819.[55] The joint capital in 1829 was £2000 contributed in equal shares, and was added to by family loans made at 3 and 4 per cent interest in the following years. The partnership came to an end in 1833, when Daniel took over the whole concern and concentrated upon the converting and refining of steel. The sum of £2000 was spent upon new premises and plant and this was the beginning of a very steady if modest growth of the establishment. Between 1834 and 1840 the average profit was only £289 a year, but by 1855–61 it had mounted up to £1399.[56]

By contrast, the firm of Marsh and Shepherd, whose history has been related by Pollard, had a balance of capital and profits, in July 1810 of £1013. During the Napoleonic wars the firm flourished and in 1816 the capital had been increased from profits to £7233. Even during a period of depression and crisis in England, gross profits for the next three years averaged 20 per cent. By 1828, thanks again to still rising profits (now 23 per cent), the total capital had increased to £29,000.[57]

An account of this small-scale industry was given by the Frenchman, Le Play, at the beginning of the 1840s.[58] He estimated that the capacity of the furnaces had been increased eight fold in the century and a half following the beginning of Sheffield's development as a steel centre. His figures of output bear out to some extent those already given. In 1842 97 conversion furnaces had produced in 33 weeks some 8250 tons of blister[59] steel. This, however, was a time of depression, and Le Play suggests that output could have been doubled. There were at that time, too, 51 cast steel works, which had a weekly production of about 80 tons.[60] The cementation furnaces were usually to be found two to a works. According to Le Play, five

[54] P. C. Garlick, *The Sheffield Cutlery and Allied Trades and their Markets in the Eighteenth and Nineteenth Centuries.* (Unpublished M.A. thesis of Sheffield University, May 1951).

[55] William Doncaster had evidently carried on the business himself for ten years.

[56] Garlick, *op. cit.*, citing Daniel Doncaster's Cash Book. Doncaster's survive in Sheffield today as the only firm making blister steel there.

[57] Pollard, *op. cit.*, pp. 10–11.

[58] Le Play, 'Mémoire sur la fabrication de l'acier en Yorkshire', *Annales des Mines*, 4me série, Vol. III, pp. 583 ff.

[59] *Ibid.*, p. 621. [60] *Ibid.*, p. 639.

furnaces was the greatest number worked by any one establishment. The process was very economical of labour, for three men could look after two furnaces, and were paid upon a commission basis. The owners of the furnaces agreed to convert the bar iron at an agreed price. Hence it is not surprising, as Le Play says, that these small masters were very susceptible to depression of trade,[61] for they evidently required little capital to operate and therefore had nothing to fall back upon. The cast steel works, too, were no larger. The size of the establishments ranged from those with four to ten melting holes, and a staff of only eight persons was required to look after ten holes. The work was very arduous in these works, especially the task of treading the Stourbridge clay to make the pots—Le Play says he had not seen anyone in England or Europe who could have kept up this work as these men did.[62] However, the cast steel works usually only operated in the day, and again, short-time working, caused by depression, often reduced the working week to three days.

Nevertheless, a change was beginning to take place in the structure of the steel industry in Sheffield. Already, between 1823 and 1826, the first instance of the integration of the processes of steel manufacture and cutlery production had taken place, when William Greaves and Sons centralised all the stages of production in their works.[63] Similar cases followed in the 1830s and 1840s.

Associated with this integration was the move to Brightside and the introduction, later, of puddling and iron forging into the Sheffield works.[64] One very compelling reason attracting the steel makers to this newly developed district was the Sheffield-Rotherham railway opened in 1837, which opened up both sources of raw materials from the Rotherham coalfield and a railway connection to the London market.[65] The railway, too, was responsible in another way for the expansion of the industry in that much of the demand for the iron and steel work for the locomotives and rolling stock was met by such manufacturers as Charles Cammell and John Brown.

[61] *Ibid.*, p. 624.

[62] Le Play, *op. cit.*, pp. 652–5.

[63] A. Gatty, *Sheffield: Past and Present* (Sheffield, 1873), pp. 212–13.

[64] D. L. Burn, *Economic History of Steel Making* (London, 1940), p. 235. As Burn puts it: '. . . by the 'sixties the crucible steel traditions of the district had been utilised by such firms as John Brown's in the creation of a high-grade heavy steel industry alongside the old light-steel manufactures.' However, in the 1850s there was a prosperous trade in wire for crinolines; and even today there is for umbrella frames. At Birmingham in the nineteenth century the steelmakers specialised in making ceremonial swords.

[65] The land adjoining the railway, too, seems to have been very cheap. When Cammell set up his new works there in 1845 he only paid 2½d. a yard for the site.

The latter is reputed to have made his fortune with his invention of a conical spring buffer.[66]

In 1854 Brown entered into a £14,000 partnership with J. D. Ellis and William Briggs of Birmingham and two years later the Atlas Works were established. Soon afterwards Brown turned his interests to iron and set up a forge with puddling furnaces and two Nasmyth steam hammers. It is said that the Atlas Works were expanded to three times their original size in three years at this time in the late 1850s and this may well be so since, in the years 1859–61, ten 35-ton cementation furnaces and 48 more puddling furnaces were added. This rapid expansion, which was shared by Cammell's and Firth's, was, it is important to note, still derived from wrought iron. The first armour plate was rolled in 1853 or 1854 for the floating battery *The Terror* for the Crimean campaign at the nearby works of the Parkgate ironworks, and in 1859 Brown took up the production of this important branch of the armaments industry. Henceforth the wealth of the heavy Sheffield steel industry was to depend to very great extent upon the duel fought between the newly developed artillery, firing armour-piercing shells, and wrought iron (later, steel) armour plate. Both Brown's and Firth's developed the manufacture of these two contesting materials of modern war. The railways and wrought iron had brought the first powerful impetus to the expansion of Sheffield; the weapons of war and Bessemer steel were to establish the town as the centre of large-scale metallurgical industry. It is noteworthy that Bessemer's first motive in developing the pneumatic process of steel-making in the converter was to build a gun powerful and strong enough to fire his self-rotating shot. And John Brown was the first Sheffield steelmaster to adopt the Bessemer process in 1860.

It was, of course, Alfred Krupp who played the most decisive part in shaping this deadly role for steel. It is certainly curious that he should have visited Sheffield in 1839 in order to learn the secrets of England's supremacy in the manufacture of steel.[67] His letters written to the firm and the War Minister at Berlin during the period from 1827 onwards throw many interesting sidelights upon the dependence of Britain upon Swedish wrought iron as the necessary raw

[66] Stainton, *op. cit.*, pp. 306–9. P. L. Payne, *Rubber and Railways in the Nineteenth Century* (1961).

[67] *Letters of Alfred Krupp, 1826–87*, edited by W. Berdrow (London, 1930), p. 53. In September 1838 he wrote to Hermann Krupp from Paris: '. . . I look upon a trip to England as much more important than my business here, on account of the improved methods of manufacture of which I hope to get hold.' He was anxious to obtain Swedish wrought iron from Sykes and Sons of Hull, who were said to be 'the sole importers of the Danemora iron which the English steel factories work up into their best product'.

material for its high-quality steel; upon the competition between English and German steel in the Continental markets; upon the development of the weldless steel tyres for locomotives; and, inevitably, of the steel artillery gun. In 1827 the Prussian Mint was buying steel from Marshall's in Sheffield, and although the Krupp steel was said to be as good as this, shortly afterwards the German firm was soliciting for orders to establish the 'success of the solitary Prussian steel factory which is so useful to the state'.[68] Alfred Krupp was confident even at this time that 'instead of the English steel which is imported in such large parcels, mine will gradually oust the former and find a widespread sale'.[69] Indeed, one reason he gave for visiting England was that he expected to be able to sell his steel here.[70] Krupp was always striving to outdo the English steel manufacturers, and, in 1851, after hearing that Turton's of Sheffield were to exhibit a 27 cwt. piece of crucible steel, he wrote back to direct the casting of an ingot from 98 crucibles. He was very scornful of a monster casting of some 2400 lb., and boasted, 'I have been saying we make little bits like that every day, and I would send its grandpapa.'[71] The firm of Krupp, incidentally, put on show a large casting at the Great Exhibition, which won him a bronze medal. He also exhibited a cast steel gun; for as early as 1844 he had come to the conclusion: 'There is no profit in crucible steel in bars.'[72]

[68] *Letters of Alfred Krupp*, pp. 11.
[69] *Ibid.*, p. 14.
[70] *Ibid.*, p. 41.
[71] *Ibid.*, pp. 127–8.
[72] *Ibid.*, p. 82.

CHAPTER XIV

The Development of Mild Steel

In the 1860s the iron and steel industry experienced revolutionary technological innovations—the Bessemer and open-hearth processes which created the heavy steel industry. From being an expensive and indispensable raw material of the cutler and to be used in small quantities, steel, or rather the new product 'mild steel', was transformed into the subject of mass-production techniques and was used as the ubiquitous source of strength for the architect, the engineer, the shipbuilder and armaments manufacturer. Even today, its indispensability and universality are only slightly challenged by aluminium, necessary for man's technological exploits in the air.

Bessemer, surveying these innovations on the threshold of the era of their impact, in 1861, asserted: 'Steel, to the engineer, has hitherto stood in much the same relation as granite to the builder.' Its qualities of toughness which, as we have seen in the preceding chapter, were becoming more and more vital to the railways as well as to the makers of cannon and armour-plate, were recognised; but two major difficulties stood in the way preventing the metal's use on a massive scale. First, the cost and risks involved in making steel by the available techniques—of co-ordinating a melting of many small crucibles of cast steel, itself the costly refined product of expensive wrought iron, for large metal forms. The consequences of an unfortunate extrusion of impurity in the metal could render a casting weak and useless. The second disadvantage was the difficulty in having it worked or shaped by the forge-hand or engineer. Although cast steel was regarded as a pure and homogeneous metal (and on that account superior in strength to the scoriae-grained wrought iron which often peeled off in layers), it was until recent improvements considered by the smith as hard and brittle, and incapable of being welded together. It is significant that as certain improvements, particularly those introduced on the Continent by Fischer in Switzerland and Krupp in Prussia,[1] could be utilised to make the metal 'milder' and weldable, steel became to be used more and more, particularly in high-class

[1] See below, pp. 318–9.

machinery. These then were the inducements to inventors to produce a steel stronger yet more pliable than cast steel and, of course, cheaper to make by dispensing with the dependence on bar-iron as the raw material for the conversion process.

Attempts to make steel by a direct process had, indeed, attracted the efforts of inventors for a long time. In 1800 David Mushet, whose son was to play an important role in the perfection of the Bessemer process by the reintroduction of the use of manganese, had taken out a patent. His efforts to produce steel at the Calder Works in Scotland and to satisfy the requirements of Peter Stubs with his cast steel, have been described in Ashton's *An Eighteenth-Century Industrialist.*[2] The primitive Catalan process was the only successful way of accomplishing the task of by-passing the blast furnace, but this only operated on a very small scale. There were several attempts made during the century, but all were unsuccessful. In 1825 Charles MacIntosh, the Scottish industrialist, who was more successful in other fields of industrial chemistry, experimented with an ingenious method of using carburetted hydrogen gas to effect the conversion.[3] The aim of this was to improve upon the ordinary methods of cementation; however the process did not work upon a large scale. 'It was found impossible to keep the chambers in which the bars of iron were suspended air-tight, at the very high temperature to which it was necessary to raise them.'[4]

With many of these patents it is difficult to decide whether the principles they contained were valid, and more so, to decide if the subsequent development owed anything to the claims of these early experimenters. A case in point is that of Joseph Heath. In 1839, as a result of some metallurgical researches in India, he patented the use of 'carburet of manganese' in the conversion of iron into cast steel. The following year he is said to have gone to Sheffield and have 'made a valuable improvement in the quality of the steels for which Sheffield, even then, was famous'.[5] His patent was quashed in the law courts and Heath was finally ruined by long litigation against the Sheffield steel manufacturers who infringed his patent. He died in unfortunate circumstances, just before making a last and desperate demonstration of his contribution to the industry by exhibiting at the 1851 Exhibition. To some historians of the iron and steel industry he

[2] *Op. cit.*, pp. 48–50. In 1867, R. Jordan listed sixteen different processes of steel-making practised in Europe (Jordan, *Revue de l'industrie du fer en 1867* (n.d.), p. 281).

[3] J. Percy, *Metallurgy: Iron* (London, 1863), p. 773; G. MacIntosh, *Biographical Memoir of Charles MacIntosh* (Glasgow, 1847), pp. 94–7.

[4] D. Mushet, *Papers on Iron and Steel*, p. 671.

[5] F. M. Osborn, *The story of the Mushets* (London, 1952), p. 58.

thus belongs to that trinity of metallurgical martyrs—Cort, Heath and Mushet—who were sacrificed by the greed of their fellow ironmasters to that fate which overtakes the unsuccessful, poverty and bitterness. Yet, all the same, some of the claims which have been made upon their behalf seem somewhat exaggerated. It is claimed, for instance, that 'Heath was thus the author of an invention conferring commercial profits to be reckoned in millions. . . .'[6] Such statements only obscure the real significance of these contested inventions.

One method of making steel, which became important as a competitor to the pneumatic process, was the puddling of steel by the ordinary methods as patented by Cort, except that the pig iron was not decarbonised to the same extent as when making wrought iron. The stumbling-block would, of course, be the question of deciding at which stage in the process a new manufacture came into being. Steel, it should be remembered, had no specific identity to be indicated by a definite carbon content, at that time. The puddling process, strangely enough, had been confined to the Westphalian steel industry for roughly twenty years before it was taken up by the Low Moor Company in 1858.[7] Thereafter its output appears to have been considerable, but no statistics are available to measure its production in this country. This metal, of course, suffered from the same imperfections as natural steel; indeed, in 1854 Sanderson noted that its price was only £14 per ton as compared with £18 per ton for charcoal natural steel.[8]

However, until Bessemer's process got under way in the late 1860s, the steel industry was still based upon the cementation and cast steel processes. Moreover, although the amount of British bar iron which was being used was increasing, the greater part of this raw material of Sheffield's industry came from Sweden. The import of Swedish wrought iron in the 1850s was as follows:[9]

Year	Tons	Year	Tons
1845	18,607	1850	28,096
1846	30,840	1851	35,467
1847	28,264	1852	23,817
1848	20,438	1853	23,540
1849	26,605	1854	24,436

[6] *Ibid.*, p. 58. Bessemer, however, also paid tribute to Heath's experiments with manganese in a lecture given to the Cutlers' Company, London, in 1880, 'On the manufacture and uses of steel'.

[7] J. Percy, *op. cit.*, p. 793.

[8] C. Sanderson, 'On the manufacture of steel, as carried on in this and other countries', *Journal of the Royal Society of Arts*, 1854–5, Vol. III, p. 454. Not less than 15,000 tons were made in 1854; and it was largely used for railway springs. Puddled steel was made by Firth's.

[9] *Ibid.*, p. 451.

The Swedish iron was absorbed by the many small converters, who still specialised in the old process of steel-making, despite the competition of the large works in Sheffield, at the end of the 1860s. One of the largest of these, Vickers, indeed, had no cementation furnaces; it concentrated upon the making of cast steel in its 384 melting holes holding two crucibles each. Each crucible held about 40 lb. of blistered steel. But, of the other large-scale steel makers mentioned by Jordan in his review of the British Steel industry made between 1862 and 1869,[10] Turton's possessed 13 cementation furnaces, each containing 22 to 23 tons of bar iron for conversion; John Brown's in 1867 had 18 cementation furnaces. It is apparent that the manufacture of crucible steel was booming at this time. Indeed, between 1851 and 1861 the production of crucible steel at Sheffield was expanded from 18,500 tons to 51,500 tons.

It was in the technique of casting from the crucibles, however, that the greatest advances had been made. None of the Sheffield cast steel makers concerned with bell-casting, among other things, came quite up to the level of achievement of Krupp. Nevertheless, thanks to the military discipline of the organisation, Vickers could teem ingots weighing 20,000 kilograms from 500 crucibles.[11] Krupp, on the other hand, with an output of about 62,500 tons in 1866, was a year later making ingots weighing 40,000 kilograms.[12]

One point, however, which had not greatly concerned the British manufacturers, was the possibility of making cast steel direct from bar iron without the intermediate process of cementation. In this Britain lagged behind the Continental steelmakers, especially Johann Conrad Fischer of Schaffhausen. We have already noticed his account of the Huntsman establishment in 1814, and he made several more journeys to England to compare technical progress here with his own experiments. His diaries give details of several visits to

[10] S. Jordan, *op. cit.*, Vol. II, pp. 297–304. He notes that Vickers also possessed Siemens gas-heated furnaces. The steel industry also survived in Newcastle (at the Newburn works of John Spence and Son) and in Scotland (at Hawksworth's cast steel works, Linlithgow). There are also brief details of several of the Sheffield works, particularly from the point of view of the labour employed, in the 4th Report, *Children's Employment Commission* (1865, C. 3548). Brown's and Cammell's together employed about 6000 works people.

[11] It was this feat of organisation which pointed to the need for a major innovation making large-scale production possible. As Pollard, *op. cit.*, p. 160, says: 'There can have been few moments in industrial history when a technical revolution, to create a new basis for large-scale production, had become as urgent as in steel making in the 1860s, when "a little army of men" had to be co-ordinated to heat and then to teem 672 crucibles into one mould, one every half second within five minutes to maintain a constant flow and produce a faultless ingot.'

[12] Jordan, *op. cit.*, pp. 300 and 307.

Sheffield, and to the Huntsmans in particular, for he was on the best of terms with the family.[13] Fischer's own experiments—with a hot blast furnace to fire crucibles of improved ceramic material to produce cast alloy steel from wrought iron[14]—also brought him in contact with the great scientist Michael Faraday in the 1820s. Faraday was the first of British scientists to give his attention to metallurgical problems, and his experiments in collaboration with Stodart attempted to produce a synthetic meteor iron by adding minute quantities of other metals.[15] Fischer was working upon the same problem independently and achieved some measure of practical success; Faraday's experiments were the product of disinterested curiosity, although he did consult John Josiah Guest of Dowlais. Hence the work of the scientist is to be regarded rather as a precursor of metallurgical research. Fischer was working towards a more scientific method in the manufacture of steel,[16] and it is significant that in January 1846 he was supplying the Royal Mint at London with his special steel.[17] Competition of this sort was to be very dangerous to the British steel industry. Nevertheless, despite the competition of Alfred Krupp and experimenting foreigners, the British steel industry was in the ascendancy for the next forty years. This was due in the first place to the invention of the pneumatic process of making steel.

The story of Sir Henry Bessemer's sensational paper read to the British Association at Cheltenham in 1856, which seemed to herald the approach of the age of steel, is well known. Nevertheless, the part actually played by Bessemer has been frequently misunderstood, and what is more, there is a need for the invention to be put into perspective against the development of the wrought iron industry, which was to be rendered obsolescent by steel. The adoption of the Bessemer process was a much slower process than is sometimes recognised.[18]

There is no authoritative modern full-length biography of Sir Henry Bessemer (1813–98) to state objectively the exact course of events in the development of the pneumatic process, and it is to the

[13] J. C. Fischer, *Tagebücher* (Zurich, 1952), pp. 537 and 592 ff.

[14] *The Metallurgist Johann Conrad Fischer (1773–1854) and his relations with Britain* (Schaffhausen, 1947), pp. 12 and 28. Fischer was also experimenting with nickel steel. W. O. Henderson, 'J. C. Fischer, a Swiss industrial pioneer', *Zeitschrift für die Gesamte Staatswissenschaft*, April 1963, Tübingen, pp. 366–70.

[15] See C. R. Fay, *Round About Industrial Britain, 1830–60* (Toronto, 1952), pp. 201–2, quoting Sir Robert Hadfield's *Faraday and his Metallurgical Researches* (1931).

[16] He exhibited at the 1851 Exhibition in London a model of a furnace, which acted upon regenerative principles.

[17] *The Metallurgist J. C. Fischer*, p. 36.

[18] Cf. Erickson, *British Industrialists*, pp. 141 ff.

Autobiography[19] that one must turn to get a first-hand account. The account of the invention there, written by Bessemer in his old age and when he was resting upon his fame, is almost certainly less than objective.[20]

Born at Charlton, near Hitchin, in January 1813,[21] of a father who earned a living as a typesetter but who was both an engineer and inventor, the young Bessemer, very early in life, decided to follow his father's footsteps. In 1830 he went to London to seek his fortune, and in the course of the next twenty or so years his inventive genius worked upon several successful enterprises. A mere list of his many activities—including electroplating, producing perforated dies for the Stamp Office, using plumbago waste to make lead pencils, inventing a type-setting machine, producing machine-made imitation Utrecht velvet, and work upon sugar-refinery machinery as well as plate-glass manufacture—is evidence at any rate of the scope of his inventive mind. Yet he was not always the successful entrepreneur, as in later life—perhaps his business sense was not so well developed at that period. For instance, he claims to have saved the Government £100,000 with his dies, but received no reward for his contribution to its administrative efficiency.[22] On the other hand, one of the most profitable of his ventures was his secret process for making gilt powder, from which he realised £5 a pound from copper alloys costing only a few pence. It was this source of income which financed the experiments which ultimately realised his ideas of making steel in a converter.

It was the stimulus of warlike demand which spurred on the development of the steel industry. During the Crimean War Bessemer was working upon artillery guns. Already in 1853 he had patented an automatic breech loading gun to be operated by steam under pressure. This apparently was only an idea on paper. After being

[19] *Op. cit.* (London, 1905). Unfortunately, it has not been possible to trace his papers. It has been suggested that Bessemer issued instruction for these to be destroyed upon his death, but his will contains no clause to this effect. Bessemer's *Autobiography* (published posthumously) may be usefully compared with the substantial memorial lecture by James Dredge to the American Society of Mechanical Engineers, 1898.

[20] Bessemer's claims have been shown to be grossly inaccurate and exaggerated in connection with his inventions for glass making.

[21] There is no lack of accounts of his life and his inventions. See *D.N.B.*, Vol. XXII (Supplement), and an excellent recent article by E. N. Simons, 'Pioneers of the modern steel industry', *Murex Review*, Vol. I, 1950. Bessemer figures, too, in Dr. Chaloner's *People and Industries* (London, 1963).

[22] It is perhaps unnecessary to repeat the reservations with which all his claims should be treated and any biographer of Bessemer would have to check these by independent evidence.

spurned by or himself ignoring the War Office, he had taken his invention—a self-rotating shot to be fired from a smooth-bored gun —to France to the Emperor Napoleon III. However, it was suggested at the trials at Vincennes that his guns needed to be made of a stronger material than the iron then available. In order to perfect his artillery invention, Bessemer turned to the problems of metallurgy.

Inside three weeks he applied for a patent in which he anticipated the Siemens-Martin process, using the open-hearth furnace, in which blister steel was fused in a bath of molten pig iron to produce a tougher kind of iron.[23] Throughout 1855 he was occupied with the problems of casting steel and malleable iron made by the new process for cannon.

His experiments had left him with traces of decarbonised iron, the carbon having been burnt out by the blast during the smelting operations. According to R. D. Allen, his brother-in-law, who had been in charge of the practical operations at Baxter House, London, they had found difficulty in generating sufficient heat to melt the pig iron, and Bessemer had directed 'the nozzle to be put in, to try and convert that which was melted'.[24] Bessemer, it should be noted, at this time was trying to make malleable iron, but it was during these experiments that the basic principle of the converting pneumatic process was discovered. Noticing some pieces of apparently unmelted pig iron, and discovering they were 'merely thin shells of decarbonised iron', he deduced that air alone was capable 'of converting it (pig iron) into malleable iron without puddling'. Bessemer claimed this discovery gave a new direction to his thoughts. In October 1855 Bessemer patented the use of air to remove impurities from melted iron.[25] By this time, as Dredge claims, 'Bessemer appreciated the end in view and the general way of obtaining it, though his mechanical details were still crude and imperfect'. In December of the same year he patented what was in essence the tilting Bessemer

[23] Patent No. 66, 10 January 1855, 'Improvements in the manufacture of Iron and Steel'. It will be remembered that this was an ancient Continental method of steel making. Dredge confirms this was not the Bessemer process, 'only the first practical step toward it'. Bessemer claimed his metal had the appearance of steel and, according to Bessemer, he was then on the point of building a furnace at Ronelle, France, with the encouragement of the French Government.

[24] Presidential Address of R. D. Allen, *Journal of the Iron and Steel Institute*, 1890.

[25] Steam, then air, was forced into the iron in this process melted in crucibles with a perforated pipe down the centre. Dredge quotes Bessemer: 'Air is used to complete the operation . . . the metal . . . will rapidly brighten up, and increase of flame will be observed, and a rapid increase in the temperature of the metal will take place.'

converter. Now, the molten pig iron was run from the blast furnace into a large vessel, into which the air was introduced by tuyères, and this could be tipped for receiving and ejecting its charge. The characteristic blowing of the charge in the converter is described in the specification of this patent: 'During the operation the metal undergoes ebullition and increases in temperature; the appearance of the flame, sparks and slag issuing from the top of the crucible indicate the state of the metal. After the operation has continued for about half an hour the flame gradually diminishes, and this indicates to the workman that the process is completed, and that the crude metal has been converted into a nearly pure malleable iron.' Earlier, Bessemer had tried using a fixed converter, possibly adapted from his glass-furnace experiments.[26]

In the next year, in January and February 1856, patents were taken out for improved and new mechanical details of the process. As Bessemer wrote to Krupp, an eager seeker of information, at this time: 'In our daily experiments every trial gives rise to some suggestions for improvement of the process.'[27] In March he added another patent to his name for returning the necessary amount of carbon to 'the burnt metal' to produce the quality of steel required. On this point it should be mentioned that, as Bessemer was at pains to point out in his claims, the idea of blowing air or steam into molten iron to refine it was not new. Bessemer, indeed, acknowledged his debt to James Nasmyth, the inventor of the steam-hammer, who was one of several working upon this plan, and offered him one-third of the value of his own patent.[28]

In America too, by a strange coincidence, the pneumatic process had also been evolved by a Kentucky ironmaster, William Kelly, between 1847 and 1851, and his biographer puts forward suggestions that Bessemer travelled to the United States to steal Kelly's secret.[29] There does not, however, appear to be much foundation for these stories.

[26] W. M. Lord, 'The development of the Bessemer process in Lancashire', *Trans. Newcomen Soc.*, Vol. XXV, 1945–7.

[27] *Krupp MSS*, H. Bessemer to A. Longsdon (Krupp's agent in London), 25 January 1856.

[28] Nasmyth had taken out a patent for steam puddling in 1854.

[29] J. W. Boucher, *W. Kelly: A True History of the So-called Bessemer Process* (Greensburg, Pa., 1924). For a recent appraisal of the Kelly contribution see Philip W. Bishop, 'The Beginning of Cheap Steel', United States National Museum *Bulletin* 218, 1959, pp. 42–4. Again there is the familiar situation of the difficulty of reconciling his biographer's claims with the few ascertainable facts. The successful operation of the first 'Bessemer' plant in the United States at Wyandotte owed a great deal to the information on the French application of the Pneumatic Process brought over from St. Seurin by L. M. Hart.

Bessemer claimed not only to make refined iron, but to convert crude pig iron into steel or malleable iron *without the use of additional fuel*; and almost certainly it was this feature of his invention which attracted so much attention when, overnight, after the reading of his paper at the Cheltenham meeting in August 1856, the invention was taken up by the press and Bessemer became a nine days' wonder.

Nor was the eager reception confined to journalists; Thomas Brown of Ebbw Vale offered Bessemer £50,000 for the outright purchase of his patents, which was refused. Within a month he had received £27,000 from ironmasters for licences.[30] Among them was the Dowlais Iron Company, upon the following terms: 'On condition of the Dowlais Iron Company undertaking to at once erect the necessary apparatus for carrying into practical operation Henry Bessemer's new process of manufacturing malleable iron, and on the Dowlais Company paying to Bessemer and Longsdon the sum of £10,000 and 10s. per ton up to 20,000 tons.'[31] Thus, it will be seen, expectations were high on both sides.

Then came failure which threatened to discredit Bessemer, who had appeared like 'a brilliant meteor that had flitted across the metallurgical horizon, dazzling a few enthusiasts, and then vanishing for ever in total darkness'.[32] At Dowlais, experiments had been conducted to attempt to perfect the process, for Bessemer himself had not at this time secured consistent results nor discovered how to control the chemical changes in the process. The course of the events was described by E. P. Martin, who later migrated from Dowlais to Bolckow Vaughan's and played an important part in the perfection of the later Thomas process. 'When Mr. Bessemer came to Dowlais to continue the experiments, a convenient refinery happened to exist opposite the furnace making cinder and pig, and the iron from this furnace was by a most singular and unfortunate mischance employed for Mr. Bessemer's trials.'[33] Thus working with this inferior quality pig iron Bessemer's process would not repeat what it had successfully

[30] Cf. Erickson, *British Industrialists*, pp, 141–2. 'Bessemer himself planned to grant such licences to one firm in each area at a reduced royalty.' W. H. Chaloner, 'John Galloway, 1804–1894', *Trans. Lancs. and Ches. Antiquarian Society*, 1954, noting that this engineering firm, which was closely associated with Bessemer in his experiments and in perfecting the converter, makes the suggestion that even as early as 1855 an ingot of Bessemer steel—the first in the world—had been made at these works. Galloway, in fact, took out a licence before Bessemer announced his discovery to the world in August 1856.

[31] *Dowlais MSS*, agreement dated 27 August 1866, reciting the earlier terms.

[32] Boucher, *op. cit.*, pp. 67–76, quoting a letter of Henry Bessemer to Sir James Kitson, 10 September 1890.

[33] Presidential Address of E. P. Martin, *Journal of the Iron and Steel Institute*, Vol. LI, 1895, p. 22.

done before. His experiments showed that 'only from the very best descriptions of foundry iron could wrought iron be produced which had at all the character of ordinary bar iron . . . with ordinary forge pig—nothing could be done with it . . . it crumbled to powder and dust or else the bar was as brittle as glass.'[34]

A similar sorry experience was repeated at the Mersey Forge in Liverpool; at Butterley in Derbyshire; at the Coats Iron Company and at William Dixon's Govan Ironworks in Scotland.[35]

This failure came to Bessemer like 'a bolt out of the Blue',[36] for where he had been working with Blaenavon pig iron, almost free of phosphorus (which was the unknown cause of the trouble), he had not yet realised the special qualities of pig iron necessary for success. Baffled and frustrated, all that Bessemer could do was to begin new experiments (and employ a chemist, which was not a great help) to discover the cause of the failure. Meanwhile, the British Association refused to print his paper on the grounds that so great a fallacy ought not to be encouraged by its publication in their Report. The *Mining Journal* in November of 1856 wrote: 'The excitement of the Bessemer process, by which we were to have had malleable iron at the price of pig iron, has now nearly subsided.'[37] However, as Dredge says, if Bessemer was temporarily discredited, he did not accept the verdict passed on his invention. He knew 'the defects which had presented themselves did not touch the principle on which the invention was based. . . . He addressed himself to overcome the difficulties of detail.' 'Furnace machinery and apparatus were constructed at the Baxter House works, to be pulled down and rebuilt or altered; thousands of pounds were spent in experiments, and some two years passed away in incessant preliminary work.'[38]

The first step towards getting the pneumatic process perfected was to discover which kinds of pig iron would work and give con-

[34] E. Riley, *On the Manufacture of Iron* (Society of Engineers, London, 1862), p. 79.

[35] Erickson, *British Industrialist*, p. 142. It would be interesting to know exactly how many licences were taken out in the first place. Allan Nevins, *Abram S. Hewitt* (New York, 1935), p. 129, discussing an early trial at the Trenton Ironworks in December 1856, mentions that five firms had become licensees and all had met with failure. In Germany, Krupps' efforts to take out the Bessemer patents for making malleable iron in Prussia from January 1856 were unsuccessful on account of the priority given to Nasmyth's and Martien's processes. Krupp did not enter into an agreement to become licensed to use the Bessemer process until August 1861. ('Alfred Krupp and Henry Bessemer', *Stahl und Eisen*, Heft 13, July 1962, pp. 911 fn.)

[36] Bessemer, *op. cit.*, p. 170.

[37] *Loc. cit.*, 29 November 1856.

[38] Dredge, 'Henry Bessemer, 1813–1898', *loc. cit.*, pp. 922–3.

sistent results in the converter, and then possibly to find out the chemical cause of failure. Bessemer, had, in fact, been buying his raw materials 'simply as pig iron' without ever apparently suspecting that iron from other sources was so different.

Eventually he discovered that the haematite pig irons, and iron from the Forest of Dean, Weardale and Blaenavon, best suited the process, but it was the success of his Swedish licensee, Göransson, in 1858, after many trials with Swedish charcoal iron, which pointed to the cause of failure—the presence of phosphorus and sulphur.[39] As Carlberg claims, Göransson 'put Bessemer's ingenious invention on a sound industrial footing, thereby inaugurating the age of steel'. Göransson brought over to Sheffield, where Bessemer had recently commenced his steel works in partnership with W. D. Allen, Robert Longsdon and others, fifteen tons of steel ingots made at Edsken in the Converter. This steel was successfully hammered and tilted and made up into artillery, tools, etc. Göransson initiated Bessemer into the Swedish way of blowing; as a result the tiresome and profit-consuming Sheffield process of granulating the steel and remelting it could be eventually abandoned. Bessemer and Co. bought 100 tons of the Edsken pig iron for conversion and on 18 June 1859 W. D. Allen was able to record in his diary, 'First made Steel direct.'[40]

Thereafter, Bessemer himself made steel, again according to his own estimates which were disputed, worth £50 or £60, from Swedish pig iron costing only £7 a ton. Bessemer proclaimed the end of the age of iron: 'I could now see in my mind's eye, at a glance, the great iron industry of the world, crumbling away under the irresistible forces of the facts so recently elicited. In that one result the sentence had gone forth, and not all the talent accumulated in the last 150 years . . . no, nor all the millions that had been invested in carrying out the existing system of manufacture, with all its accompanying great resistance—could reverse that one great fact.'

But that is not the end of the story, for in reality the development, of the process was not just the work of Bessemer and his associates to perfect and realise a brilliant idea, but rather of a series of complicated and not yet completely unravelled set of circumstances in the contemporary metallurgical world. Moreover, although Bessemer recognised 'great resistance' of the iron industry with the large

[39] E. F. Lange, 'Bessemer, Göransson and Mushet, a contribution to Technical History', *Mem. and Proc. of Manchester Literary and Philosophical Society*, 1912–13; Per Carlberg, 'Early Industrial Production of Bessemer Steel at Edsken', *J.I.S.I.*, Vol. 189, July 1958.

[40] Lange, *loc. cit.*, pp. 13–17, quoting a letter from Göransson to Professor Richard Ackerman, dated 6 November 1879. Carlberg, *loc. cit.*, p. 204. Dredge, *loc. cit.*, p. 928.

amount of capital tied up in its puddling furnaces and forges, he too easily dismissed the difficulties and obstacles to a large-scale change-over to steel.

We have already noted the coincident invention of Kelly, who used the same principle as Bessemer, but who used different machinery for the conversion. The American courts, incidentally, gave Kelly's patent priority over Bessemer's. But although metallurigists themselves have recognised the value of the contribution of Robert Forester Mushet, the son of David Mushet, the famous metallurgist of the first half of the century, historians have too often been dazzled by the limelight which Bessemer threw upon himself. This section is therefore intended to consider briefly the part played by Bessemer's rival in metallurgical invention.

At the outset, it can be said that Mushet, a steelmaker in the Forest of Dean, was in possession of the knowledge and experience which, given co-operation between the two inventors, could have solved Bessemer's problems sooner than he did. In fact, Mushet, who had been given some of the samples available at the Cheltenham meeting by Thomas Brown, of Ebbw Vale Iron Works, was able to forecast Bessemer's difficulties, and what is more, he had been able to improve bars of metal made by a Bessemer converter in South Wales into steel which was forgeable.[41]

What was the secret of Mushet's success? One cause of the metal crumbling and being unsuitable for forging was the superabundance of oxygen in the metal. Mushet, who had been using a manganese compound, 'spiegeleisen', since 1848, to remove the oxygen as an oxide in the slag and at the same time to replace the carbon which had been burnt out, was able to produce ingots of steel equal in quality to cast steel.

The ensuing story of the relations between Mushet and Bessemer—Bessemer's approach to Mushet to share his secret with him; the consequent rivalry upon Mushet declining the offer and the lapse of Mushet's patents, which finally gave Bessemer the use of the spiegeleisen secret, need not be related here.[42]

[41] See my article on the first steel rail in *Engineering*, 8 February 1952. This rail was rolled and laid in 1857.

[42] See F. M. Osborn, *op. cit.*; Lange, *loc. cit.*; and Bishop, *loc. cit.*, pp. 33–5. The clarification of Bessemer's position over the rival patent claims of Joseph Martien to purify pig iron by forcing currents of air under it, which were taken up by Brown of Ebbw Vale is also a difficult matter. Eventually, Bessemer was glad to pay £30,000 for the Ebbw Vale patents and Ebbw Vale took out a licence for the manufacture of steel by the Bessemer process, 'which from the peculiar resources they possess they will be enabled to produce in very large quantities.' (Bishop, *loc. cit.*, p. 37). Lange's comment is appropriate: 'This agreement

In 1858 the problems of production having been solved by the use of Swedish phosphorus-free iron and Swedish techniques, practical trials of the product having been put to the test at Galloway's works in Manchester, the next step was to establish a steelworks at Sheffield, the headquarters of the steel industry. As Bessemer told the Institution of Civil Engineers in May 1859, 'It was then decided to discontinue for a time all other further experiments, and to erect steel works at Sheffield, for the express purpose of fully developing and working the new process commercially.'[43] According to Dredge, the resources of the firm were very small—less than £12,000 in all. Messrs. Bessemer and Longsdon subscribed about £6000, Messrs. W. and J. Galloway £5000 and W. D. Allen £500. 'These were the narrow means with which to enter on so great a struggle. . . .' According to Bessemer they were 'determined to beard the lion in his den, and to undersell the trade, until we forced them in self-defence to take a licence under my patents and carry on my process'.[44]

In June 1859 *The Mining Journal* remarked that Bessemer tool steel had become a recognised article of manufacture; the Bessemer works using the new methods could sell the best quality steel at 'little more than two-thirds the normal price'.[45] Accordingly, 'it would appear almost impossible for success to be wanting to the seller'. In the making of rails, too, the new firm introduced stiff competition, cutting its prices by £10 per ton. Bessemer was now probably justified in asserting: 'This soon brought the trade to a proper frame of mind.'

In 1860 John Brown, the leading Sheffield heavy steel maker, took out a licence; Charles Cammell's followed suit and began rolling rails in 1861.[46] By that time the output of Bessemer steel was estimated at 3000 tons a week and the process was adopted in Liverpool, Crewe, South Wales, Manchester, Cumberland and in the North-east, besides Sheffield.[47]

Of course, the Bessemer process was by no means perfected, even then. Indeed, the basic problem remained, to convert steel from pig

---

removed the last barrier to the quiet commercial progress of Bessemer's invention throughout Europe and America, and showed Bessemer in the light of a very astute business man. In fact, the Ebbw Vale royalties eventually amounted to between £50,000 and £60,000. This was Bessemer's last battle.' (Lange, *loc. cit.*, p. 23). In America, Kelly's patent stood in the way of the exploitation of Bessemer licences and steelmakers had to take out licences from both interests.

[43] Dredge, *loc. cit.*, p. 927.

[44] Letter of Bessemer (no date cited), Dredge, *loc. cit.*, p. 929.

[45] £2 4s. per cwt. as against the usual price of £2 15s. to £3 5s.

[46] See Erickson, *British Industrialists*, pp. 143–5, for an analysis of these firms.

[47] L. Grüner, *The Manufacture of Steel* (London, 1872), p. 49.

iron made from phosphoric ores. As Miss Erickson says: 'None of the existing firms of iron or steel makers had either the staff, the ingenuity or the foresight to try to solve the problems that Bessemer's original idea posed.'[48] But, apart from that line of development, there were problems to be overcome arising from the rapid rate at which the linings were consumed by the intense heat—Bessemer used a locally available ganister; the tuyères needed to be replaced; the waste of metal during the explosive chemical reaction amounted to as much as 18 per cent. These technical problems were the subject of discussion at Sheffield at the meeting of the Institute of Mechanical Engineers in 1861, when demonstrations of the largest 4-ton converter at the Atlas works and at Bessemer's ironworks began to silence critics.

Significantly enough journalists seized upon the potentiality of Bessemer steel as making it possible for English steel cannon to outnumber 'the vast batteries of the same material already mounted on the Continent'—at one-third the cost. Bessemer himself spoke of an 18-pdr. gun being cast at 11.20 in the morning and ready for the boring mill at 7 in the evening. He claimed the cost of his furnaces and the plant to make forty such guns would not exceed £5000.

This was unrealistic, perhaps, but certainly his appraisal of the position in the world of iron and steel showed a greater sense of the future. It is significant that he could now set out the specifications of the new industrial material mild steel: 'The problem we have now before us, is how to produce cast steel that will take any form in the mould, or under the hammer; that will yield quickly and readily to all our present cutting and shaping machines; will retain all the toughness of best iron with a much greater tensile strength; and all the cleanness of surface, beauty of finish and durability, that so eminently distinguishes the harder and more refractory qualities of steel in common use.'[49] This steel had been already used for marine engine shafts, cranks, propellers, anchors and the like; Bessemer urged its use in girders, bridges and viaducts. 'The manufacture of cast steel has only to produce at a moderate cost, the various qualities of steel required for constructive purposes to ensure its rapid introduction, for so certain is it that the age of iron superseded that of bronze, so will the age of steel reign triumphant over iron.'

We shall be considering the impact of the Bessemer steel process on the location of the British industry in the next chapter; here we may examine in a little more detail the subsequent history of Besse-

[48] Erickson, p. 143.

[49] H. Bessemer, 'On the manufacture of steel and its application to constructional purposes', *The Engineer*, Vol. 12, 1861, pp. 62–3.

mer's own Sheffield works. As a business interest, this was distinct from the royalties accruing to the inventor and his partner, Robert Longsdon, from the firms taking up the Bessemer process. In fact, in 1862, a one-quarter share of the interests in the patent royalties was acquired by a syndicate of fifteen members, including the Galloways and John Platt of Oldham, for the sum of £25,000 immediate cash and £25,000 in deferred payments. This agreement was to run until 1884, covering the period of currency of Bessemer's many patents relating to metallurgical processes and the techniques of working the metal into manufactured goods.[50] According to Dredge, even though this was a high price paid by the firm for its extra capital, it enabled it 'to pass through the last struggle'. Hence, it is unfortunate that we know so little about the fortunes of the firm in Sheffield and what exactly this 'last struggle' was. However, it is clear that, unlike the industrial giants of Sheffield, Brown's and Cammell's with their £1 million capitals, Bessemer's remained comparatively small. Miss Erickson suggests that the firm did not use its profits for large-scale expansion: in 1895 its paid-up capital was only £96,000.[51] However, during the 1860s, when it began to operate at a profit, these were modest until 1867 when some £28,622 profit was made.[52] In 1877 the firm was converted into a public limited company; by then, according to Bessemer, 'including the selling price of the business at the end . . . the profit on the original capital throughout the fourteen years [from 1862?] averaged cent per cent every two months.'[53] This would mean an average profit of £72,000 a year.

Bessemer also amassed a large fortune through the payment of royalties. Up to 1869, if we can infer the scale of charges from the Dowlais agreement drawn up in 1866, the charges were £2 a ton for small ingots up to 5 tons and £5 per ton for bigger ones. There was an allowance of 50 per cent for losses and waste, and as an encouragement for the manufacture of large ingots, Dowlais could claim £15 drawback on each casting.[54] It is perhaps a sign that these charges

[50] Throughout the 1860s Bessemer followed up the 'original' steel patent of 1856 with inventions for the manufacture of malleable iron and steel. His last in this respect was taken out in 1880. Dredge lists 37 Bessemer patents between 1854–80 relating to iron and steel and a further half a dozen for the manufacture of iron and steel.

[51] Erickson, *British Industrialists*, p. 143.

[52] Dredge, *loc. cit.*, p. 930. He gives a table of 'profits' (not defined) for the period 1858–67. The fixed capital of the firm in 1868 was worth £34,000.

[53] Bessemer, *Autobiography*, p. 179. In 1890, according to W. D. Allen, the firm declared a dividend of 25 per cent on a capital of £90,000.

[54] *Dowlais MSS*, agreement dated April 1866. Carr and Taplin, *History of the*

were regarded as excessive, for it was at this time that Bessemer began to pay Mushet a pension of £300 a year in order to create a body of favourable public opinion. He admitted, 'The press at that time was violently attacking my patent and there was the chance that if any of my licensees were thus induced to resist my claims, all the rest might follow the example.'[55] Certainly Bessemer's friend and biographer Dredge did not conceal the fact that the Bessemer patents were a remarkable financial success. He observes that Bessemer received in royalties no less than £200,000 a year during some of those fruitful years up to 1870. By then royalties paid amounted to one million pounds.[56]

This great wealth flowing into the Bessemer bank account, of course, represented the inventor's toll on a vast empire of industrial enterprise brought into being through the exploitation of the mass manufacture of steel and the wholesale application of the metal to new and old uses.

British Steel Industry, p. 24, say Bessemer repurchased all the original licences to make malleable iron and subsequently stiffened his terms.

[55] Bessemer, *Autobiography*, p. 294, cited Bishop, *loc. cit.*, p. 41.

[56] Dredge, *loc. cit.*, p. 946. He points out, however, that one-quarter went to the syndicate and the remainder was shared with Longsdon. Bessemer in his *Autobiography* speaks of the partnership paying £3000 a year income tax. This would presume an income of £180,000 a year. In 1869 the royalty dropped to 2s. 6d. per ton and Bessemer had, in Dredge's words, gathered 'the golden harvest', This reduction in the royalty charge, representing a saving of about £1 per ton, gave a great impetus to the production of steel rails.

CHAPTER XV

The New Factors of Industrial Location: Ores for Steel

1 THE CLEVELAND DISTRICT

'Before the railway system was developed to its present extent,' observed the former chief engineer of the Dowlais Iron Works, 'it was always considered that the ironstone should be carried to the coal, but partly by increased facilities of communication, and partly by improved manufacture, it has in many cases been found more advantageous to carry the coal as coke, to the iron ore. . .'.[1] This change in the balance of the factors of location, noted by an intelligent observer of the state of industry in his own day, is most strikingly illustrated by the rise of the Cleveland District of North Yorkshire.

This was a rich area of mineral wealth extending about 200 square miles, capable of yielding at its richest points at least 50,000 tons of ore per acre,[2] yet it lay unexplored by prospecting iron-masters for the first four decades of the century. The explanation of this period of inactivity lay in geological factors. As Lowthian Bell, an ironmaster who played an active part in the rise of the iron industry of the North-east and who later wrote its history,[3] says: 'The iron metallurgy of the North . . . owes more of its importance to the existence of any of the ores in those measures which belong more immediately to the coal formations.'

The early history of the exploitation of the Cleveland ores in the attempt to introduce this phosphoric ore to the Tyneside smelters in the Newcastle area from 1815–20 onwards, by means of summer

[1] E. Riley, *On the Manufacture of Iron* (Society of Engineers, London, 1862), p. 66.

[2] W. Fordyce, *A History of the Coal, Coke, Coalfields, Iron, etc., of Great Britain* (London, 1860), p. 141.

[3] I. L. Bell, 'On the manufacture of iron in connection with the Northumberland and Durham coalfield', *Industrial Resources of the Tyneside* (London, 1864), p. 73 ff.

shipments from the Whitby coast, can only be summed up as insignificant.[4] Indeed, the conditions under which it was mined only served to throw the ironstone into disrepute and retard the development of the industry. Moreover, such was the effect of the competition of Scottish pig iron, used extensively on account of its cheapness in the North-eastern malleable industry (which, in contrast, had grown at a great pace until in 1850 there were about 300 puddling furnaces with a total capacity of 150,000 tons of finished iron a year), that the supply of Whitby stone ceased. This had become more regular, in fact, since 1836 after the laying of a railway to the coast. But, as Bell pointed out, the selling price of the Scottish product was well below the cost at Birtley in Durham, even when the cost of transport was taken into account. Thus, 'there was no inducement for a practical man to sink capital for a few years' prosperity with the knowledge of the Scottish superiority'. Up to 1843 there were, according to Bewick, only five old blast furnaces at which Cleveland ironstone was used.[5] It is apparent that the value of the ironstone was not recognised for, a year earlier, the firm of Losh, Wilson and Bell had built a furnace at Walker-on-Tyne to use accumulated stocks of mill cinder from the ironworks already on the site, not with a view to exploit the Whitby ironstone.[6]

The nearest ironworks to the mining area had been set up in 1839, when Bolckow and Vaughan formed a partnership with a capital of £15,000 and established a works to manufacture bar iron and ships' engines from imported pig iron.[7] This was on the bank of the River Tees at the then desolate mud flat which was to soon become the booming iron and steel town of Middlesbrough.

This partnership was an excellent example of the combination of technical skill with merchant capital. John Vaughan, of Welsh upbringing, had previously been manager of the Walker ironworks at a salary of £250 a year,[8] Bolckow was the son of a Mecklenburg corn merchant, and before embarking upon the enterprise of being an ironmaster had been a commission merchant in Newcastle-on-Tyne. The firm prospered in its early years, enjoying the advantages of a cheap supply of Scottish pig iron from Fifeshire at the extremely low cost of 35 shillings per ton, and coal from the Durham coalfield by way of the extension to the Stockton and Darlington Railway.

[4] See J. Bewick, *A Geological Treatise on the District of Cleveland* (London and Newcastle-on-Tyne, 1861).

[5] Bewick, *op. cit.*, p. 24.

[6] Charles Wilson, 'Dorman Long', *Steel Review*, No. 6, April 1957.

[7] *Romance of Industry*, Bolckow and Vaughan (privately published).

[8] J. J. Burton, 'Some notes on the early history of the Cleveland iron trade', *The Teeside Chamber of Commerce Journal*, December 1929, p. 106.

It is said that Pease, the railway magnate, whose family was to be prominent in the North-eastern steel industry, encouraged Bolckow and Vaughan in their enterprise, and was a leading promoter of the company which bought up a large area of Middlesbrough, on which the main plant of Bolckow and Vaughan was built.[9]

The 1845 railway boom in the iron trade proved that the wrought iron business was not so advantageous as first thought, especially when pig iron reached £6 per ton.[10] Hence in that year Bolckow and Vaughan established furnaces at Witton Park to smelt the coal measure ores, situated 30 miles west of Middlesbrough, in an attempt to reduce costs of production. Yet this was not the exploitation of the Cleveland ores; although, in fact, the owners of Witton Park, who made 20,000 tons of pig iron in their first year, did import several thousand tons of Crosmont iron ore.

Nevertheless, the North-east soon assumed the appearance of a new Glamorgan. Such was the energy of the nineteenth-century entrepreneurs, in coal, iron railways and other industries, that they thought little of summoning new towns into existence in order to provide a labour force for their works. Thus, when in 1846 'two furnaces (were) in blast and two others with rolling mills and forges nearly ready', the Mining Commissioner reported that 600 or 700 cottages were 'to be built by speculators' to house the men who were being imported from Staffordshire by Vaughan.[11] At the nearby Consett Iron Works, 'reputed to be the second largest in the Kingdom' with 14 blast furnaces, rolling mills, a forge and a foundry, there was a new settlement—one could hardly term it a community—of about 10,000 folk 'all brought together from the different parts of England, Wales and Scotland'.[12] Other works which, on a smaller scale, were being established at this time were that of the Weardale Company (1845), purchased by Attwood upon its construction, and the Tow Law furnaces also owned by him. These, incidentally, were later to be run as an integrated concern with the Tudhoe ironworks making wrought iron.[13] All these concerns, and others unmentioned,

[9] *Romance of Industry*, p. 4. See D. L. Burn, *op. cit.*, pp. 253 and 273, for details of the firm of Pease and Partners.

[10] J. T. Gleave, 'The Tees-side iron and steel industry', *Geographical Journal* (January–June 1938).

[11] *Report of the Mining Commissioner*, 1846, p. 29.

[12] *Ibid.*, p. 28. The establishment of the Consett Iron Company, the result of a rash speculation by a banker, has been commented upon earlier. This, too, was a large integrated concern with branch forges at Bishopwearmouth and a tinplate mills.

[13] Burton, *loc. cit.*, p. 108. R. W. Hidy, *The House of Baring in American Trade and Finance* (Harvard, 1949), p. 349, reveals that from 1846 onwards these bankers and financiers controlled the Weardale Iron Company.

however, relied in the first place on the local Durham coalfield iron ores.

It was fortunate, as another writer suggests, that the discovery of the main seam of the Cleveland ores in the Eston Hills came into the hands of ironmasters who already had had some experience of smelting this kind of ore—Bolckow and Vaughan—and this led to its rapid exploitation. On 8 June 1850 the solid exposed face, 16 feet deep, was discovered. On 13 August quarrying began; a few weeks later a light railway was laid. By the end of that year 4041 tons were despatched. The first furnace at Middlesbrough was erected during the next year. Such were the revolutionary consequences of the discovery and the exploitation of this orefield that ten years later there were 50 blast furnaces on the Tees-side alongside the fuel artery—the Stockton and Darlington Railway; and in 1871 the furnaces numbered 122.

Moreover, there were many technical innovations in the working and construction of the furnaces, in particular, closed tops for the combustion of the gases and greater height added to the furnaces, so that, where in the 1840s a furnace in this area was making only 57 tons a week, in 1869 there were giants 85 feet tall at Bolckow's producing 350 tons a week[14] or 16,000 tons of pig iron a year.

It was on the Tees-side that the industry in the North-east (Northumberland, Durham and Yorkshire) tended to concentrate. In 1869 there were 52 furnaces in blast in the immediate neighbourhood of Middlesbrough[15] and there had been a shift of the industry from the old-established centres near Newcastle-on-Tyne. The Bedlington ironworks, established in the eighteenth century, stopped production in 1856; the Crowley works at Swalwell were sold, possibly in the 1850s, and Winlaton in 1863.[16] Before the end of the century the iron industry on Tyneside had faded out altogether.[17]

The malleable iron trade was important long before the rise of the smelting industry; however, here again there was an amazing spur in production. In 1864 there were 188 puddling furnaces with a capacity of 100,000 tons; in 1872 there were 1000 with an output of

[14] *Mining Journal*, 12 June 1869. See also *Select Committee on Coal*, 1873 evidence of I. L. Bell, Q. 6212.

[15] *Ibid.*

[16] M. W. Flinn, *Men of Iron* (Edinburgh, 1962), pp. 91, 93. 'The Peace of 1815 appears to have been the first serious reduction in the scale of the firm's activities . . . (it) meant the reduction of Admiralty contracts for anchors, chains and the miscellaneous small ironware which made up their valuable naval trade . . . The Winlaton Mill was sold to Messrs. Pow and Fawcus on 10 February, 1863, for the pitiful sum of £780.'

[17] S. Middlebrook, *Newcastle-on-Tyne* (Newcastle, 1950), p. 231.

440,000 tons a year.[18] This side of the industry was centred upon Stockton-on-Tees. Fortunately, for this district we have the statistics of output of the wrought iron industry from 1872, which was the turning-point in this trade when at last rails, which had been the chief product, lost their advantage of comparative cheapness against those of steel. As a result there was a short-lived boom in wrought iron plates for shipbuilding, as these figures show:

PRODUCTION OF MANUFACTURED IRON IN THE NORTH-EAST[19]

Year	Total (tons)	Rails (tons)	Plates (tons)
1872	610,253	299,075	177,857
3	613,862	324,440	165,593
4	583,846	265,020	178,272
5	562,522	246,218	173,417
6	421,175	107,832	172,374
7	395,640	36,750	214,723
8	421,544	21,646	233,964
9	290,209	6,769	173,700
80	508,437	27,417	316,722

Bolckow and Vaughan had adopted the acid Bessemer process in 1871, and pioneer work was carried out by this firm to perfect the basic Thomas improvement, which was seen to hold the key to the future exploitation of the Cleveland resources of iron. In 1881, when this new process was being introduced, the works were one of the most modern in the country with a planned flow of pig iron from three blast furnaces, working at a temperature of 1000°F[20] and each giving 500 tons of iron a week to the converters (4 × 8 tons) via pig-ladles, through to the ingot stage, the reheating with a Siemens gas furnace, and finally to the rolls.[21] In May 1881 the works had an output of rails to the extent of 3600 tons a week (1000 tons of which were made from Cleveland steel). One new factor, incidentally, in this mass production was the use of electric light in the iron and steel

[18] H. G. Reid, *Middlesbrough and its Jubilee* (Middlesbrough, 1881), p. 91.

[19] *Royal Commission on the Depression of Trade and Industry*, evidence of I. L. Bell, pp. 322–3. Bell points out that in 1864 the comparative price of iron rails was £7 a ton as compared with Bessemer steel, which cost £17 10s. The number of puddling furnaces in the country had increased to a remarkable extent between 1860 and 1875—from 3642 to 7575. The capacity then was between 4½m. and 5m. tons. By 1870, however, there was only £3 difference between steel and iron rails, hence the fate of the wrought iron industry was sealed.

[20] There were 6 hot-air stoves to each furnace.

[21] Reid, *op. cit.*, pp. 95–7. There was only one Siemens o.h. furnace in the district at the Roseberry works making 15–20 tons weekly.

works. Reid reports that the modernisation of the plant included the fixing of seven Siemens lamps of 3000 candles each.[22]

Finally, at this date we may note the rapid growth of the town of Middlesbrough. Again, a local historian has provided a table which is extremely valuable for showing not only the substantial increase in population, but also the even more remarkable exploitation of the mineral resources which gave rise to the town's origin in the first place.[23]

Year	Output of Cleveland ironstone (tons)	Production of pig iron (tons)	Exports of finished iron and steel (tons)	Population of Middlesbrough
1851	187,950	24,300	—	7,631
61	*c.* 2,000,000	543,000	unobtainable	18,892
71	4,855,000	1,813,000	666,670	39,284
81	6,538,000	2,634,000	1,211,857	55,934

The scene was set for the enactment of the drama in a steel town, the life of a young community dominated by the demands of the furnaces. This was the everyday life and work of the Middlesbrough steel workers and their families so faithfully described by Lady Bell in her social document *At the Works.*[24]

2 THE NORTH-WEST: CUMBERLAND, LANCASHIRE AND CHESHIRE

Both Lancashire, in the Duddon valley and the Furness peninsula, and Cheshire had been flourishing centres of the iron industry from the fourteenth century, if not earlier.[25] In the Furness area was situated the old-established Backbarrow Iron Company, whose early history illustrates in many particulars the significant features of the industry.

It was, as Raistrick suggests, 'right to say that the ironmasters of the early and middle eighteenth century were the intermediaries by whom the industry was nourished and organised to a condition in

[22] *Ibid.*, p. 101.

[23] A. D. Stuart, *The Cleveland Iron Industry* (1936), typescript in Middlesbrough Public Library.

[24] Lady Hugh Bell, *At the Works* (London, 1911).

[25] See G. H. Tupling, *loc. cit.*, for a brief survey of the Lancashire iron industry and B. G. Awty, 'Charcoal ironmasters of Lancashire and Cheshire, 1600–1785', *Trans. Hist. Soc. of Lancs. and Cheshire*, CIX, 1958.

which it could carry the weight of the first stages of the Industrial Revolution'.[26]

Thomas Rawlinson, a small landowner and steward of an estate, came into possession of a forge—the Force Forge—in 1681. In 1711 it was transferred to the Backbarrow Company. This was a partnership consisting of William Rawlinson, John Machell and two others owning a one-third share between them. A furnace was built in 1711–12 in connection with which Abraham Darby tried to interest Rawlinson in his coke-smelting method; however, this was not adopted.[27] Another furnace was built at Leighton Beck and various forges—four in all—were operating in conjunction with these furnaces.

In 1753 the business was extended by the addition of the anchor smithy and in 1770 a new furnace replaced the old one. The company also traded in fruit and provisions—selling apples, cider, butter and fish to West Yorkshire and Westmorland dealers and grocers.

There was also the usual penetration upon the part of its partners into other iron works, which reflected their growing prosperity as well as a concerted policy of the integration of processes. In 1750 the neighbouring furnace at Cunsey was leased in this way;[28] the year following the Nibthwaite furnace, founded by a manager of the Cunsey Company, Richard Ford, was also leased. In 1784 the Low Wood Company's properties were joined to the Backbarrow concern. But this extension was local and perhaps is not extraordinary, yet as early as 1727 the possibilities of unlimited supplies of charcoal in the Scottish Highlands had attracted the enterprise of the Rawlinsons to this distant spot. This was, however, a not altogether successful venture, for the Invergarry concern only lasted for nine years.[29] The subsequent and revealing history of these ironmaster colonisers

[26] Full details will be found in Raistrick, *Quakers in Science and Industry* (London, 1950), pp. 95–107. This brief account is based largely upon this work.

[27] J. D. Marshall, *Furness in the Industrial Revolution* (Barrow-in-Furness, 1958), pp. 19–41. He suggests that Darby wished to go into partnership with Rawlinson, the latter's concern being worth, in 1715, not less than £6500 as against the £5000 Coalbrookdale works.

[28] Raistrick lists the following furnaces and forges as having connections with the Rawlinson and Fell families:

Furnaces: Backbarrow, Leighton, Cunsey, Lowood, Newland, Nibthwaite, Duddon, Invergarry, Lorn, Penny Bridge.

Forges: Force, Backbarrow, Dunsey, Coniston, Hackett, Cartmel, Burblethwaite, Stoney Hazel, Spark, Newland.

[29] See, G. R. Morton, 'The Furnace at Duddon Bridge', *Journal of the Iron and Steel Institute*, Vol. 200, Pt. 6 (June 1962), for a detailed examination of the Duddon Furnace and works. The furnace was in use until 1867.

who lived and worked in the remote woodlands of Cumberland is well summed up by Marshall. 'Its chief proprietors', he says of the Fells, Rawlinsons and Jacksons, 'learned the art of combination against the wood owners. They disposed of their pig and bar iron to merchants in Severn ports, who doubtless resold it to industrialists in South Wales and the South-west Midlands; but as charcoal became more expensive and national competition keener, they were obliged to export haematite [ore] for mixing with poor-quality ores. After its early progress, the local industry tended to stagnate, and to be run on traditional lines by established families of gentry. Only one of the eighteenth-century firms survived to the mid-nineteenth century.'[30] The subsequent history of these charcoal iron concerns is considered later.

In Cheshire the foremost works was Vale Royal, which during the period 1688–1713 was an important link in the integrated concerns of the Foley family, supplying pig iron to neighbouring forges.[31] In the eighteenth century there were important connections between the forgemasters of the Furness area and the Kendall family who owned forges at Holmes Chapel and elsewhere in Cheshire. The few account books which survive of W. Latham of Sparke Forge[32] reveal that the three brothers, Henry, Edward and Jonathan Kendall, were not only partners in the Lancashire forges, but also had considerable amounts of ore and pig iron shipped to Frodsham, Liverpool (for Warrington) and Chester, whilst Jonathan Kendall acted as a middleman for the sale of Lancashire iron to the Midlands ironmasters via Chepstow. For instance, in 1773, Kendall's account was debited with over £1000 for nearly 200 tons of pig iron sent by this route.[33]

But despite this early industry the Lancashire and Cheshire sections of the iron industry became backwaters when the concentration of the industry on the coalfields of Shropshire, Staffordshire and South Wales had taken place. It will be remembered that Abraham Darby's offer to introduce his methods of coke-smelting

[30] Marshall, *op. cit.*, pp. 37–9. The rich haematite ore from Cumberland became an important staple of export from the district as early as the middle of the eighteenth century; it was a 'standby' to the modest incomes of the Furness forgemasters. The ores found markets as far afield as Scotland, Yorkshire, the Forest of Dean and South Wales. In the year 1807, 11,202 tons of haematite were sent to the port of Ulverston for export.

[31] For fuller details see B. L. C. Johnson, 'The iron industry of Cheshire and North Staffordshire, 1688–1712', *loc. cit.*

[32] *Latham MSS* (Lancashire Record Office, DDX/191/1–4).

[33] Ledger 1772–79 (DDX/192/4). The accounts also reveal the Scottish connections of the Furness forgemasters e.g. with the Argyll company.

had been turned down by the Backbarrow partners, but although there is a coalfield at hand in West Cumberland, the products of its mines were not altogether suitable for the ironmasters and Durham metallurgical coke had later to be sent into Cumberland by rail.

Manchester, the commercial headquarters of the cotton industry, also, of course, provided many of the ancillary services for the industry, such as the engineering works and foundries making steam-engines and textile machinery. It was through this connection that Manchester played an important part in the second half of the nineteenth century in developing the new product, mild steel.

It was a Manchester engineer, John Galloway, who built the early converting apparatus for Bessemer and who became a partner in the Sheffield steel works. (Nasmyth, too, of course, whose own patent had helped Bessemer arrive at his method of conversion, was the owner of the Bridgewater Foundry, Patricroft, near Manchester, and had been experimenting at the Bolton works of T. L. Rushton.) Experiments had, in fact, been carried out by Bessemer at Knott Mill[34] in Manchester, but they had been a failure. Galloway, it seems, being on very friendly terms with Bessemer, had taken out a licence before the Cheltenham meeting, but when Bessemer was forced to withdraw his concessions, the Manchester engineer and his brother entered into a partnership with Bessemer and Longsdon.[35]

The success of the process having eventually been demonstrated by the Sheffield venture of this partnership, the prominent Manchester engineers were quick to adopt the processs, as Lord points out. Whitworth, Beyer, Peacock, Sharp, Stewart, Fairbairn and Platt Bros.—all but one, locomotive makers—owned converters.

Again, in 1862 Lancashire capital in the hands of a syndicate headed by John Platt of Oldham was attracted by the process and paid Bessemer and Longsdon £50,000 for a quarter share in their patents relating to the pneumatic conversion principle. At the same time the Lancashire Steel Company was formed[36] with a capital of £150,000 with the object of making 200 tons of steel a week in the form of rails, engine tyres, axles and shafts, piston rods, boiler and ship plates, and weapons of war. In the company's prospectus the confidence in the Bessemer process, partly fostered by the triumph of the exhibits in the International Exhibition, is evident from the following quotation: 'It is now universally admitted by scientific men

[34] W. M. Lord, *loc. cit.* This section is largely based upon this paper.

[35] As a result, the monopoly of the manufacture of converter plants was shared with one other firm only.

[36] E. F. Lang, 'The Old Lancashire Steel Company', *Memoirs and Proc. of the Manchester Lit. and Phil. Soc.*, Vol. 82, 1937.

and engineers, that Bessemer's steel must ere long supersede the use of iron, when extra strength and lightness are essential, this fact will thus guarantee a large and continuous demand.' The steelworks began operations at Gorton in 1864, but in the words of Lang, 'met competition and hostility'. Competition came from Krupp in the armaments field and from the other Bessemer plants of Dowlais, Butterley and Ebbw Vale; the conservatism was met from the safety-first minded locomotive makers. The hostility was that of the press against Bessemer. Moreover, there was, as Lang points out, a shortage of skilled steelworkers able to operate the converter. All these difficulties discouraged the shareholders from finding the promised capital and the concern was bought by Bolckow and Vaughan as a stop-gap whilst erecting their Eston steel works. Finally, they were sold again in 1876 to the Standard Iron and Steel Company.

The history of this concern was therefore not a success—the reason probably being that the concern had no engineering shops of its own to establish the use of the new material. But elsewhere in the country matters were different.

Lord gives details of the Mersey Steel and Iron Co., of Park, Liverpool, which went over to Bessemer's process from making puddled steel; and the Bolton Steelworks of Rushton and Eckersley, which were finally acquired by Sir Henry Bessemer and Co. in 1906. Both these, however, were inevitably at a disadvantage with the converters who received their metal hot from the nearby blast furnaces, hence it is not surprising to find that the first closed down in 1898[37] and the Bolton works was dismantled in 1924.

Before looking at this development of steel-making as it concerned the Barrow-in-Furness district, however, it will be as well to trace the beginnings of large-scale production of iron in the 1860s, both in Cumberland and North Lancashire. Of the charcoal furnaces, that at Duddon Bridge remained active until 1867; there was Backbarrow, of course, and the Newlands furnace, all operated by Harrison, Ainslie and Co. In Cumberland, eighteenth-century smelting activity had ceased when the blast furnaces of the Seaton ironworks (founded in 1762) had gone out of blast in 1813. From 1830 onwards, however, it had belonged to Messrs. Tulk and Ley and the furnaces were in blast between 1839 and 1857. This, evidently, belonged to the small burst of activity in the 1840s which saw the erection of the Cleator Moor and Harrington furnaces. The first-named works belonged to the Whitehaven Haematite Company formed in 1841 by seventeen local men, although Thomas Ainsworth was largely

[37] Lord remarks that an injunction requiring the silencing of a 15-ton steam hammer was the death-blow to this works.

responsible for the company's formation.[38] A third furnace was built during the railway boom of 1845. This concern depended, according to Wood, for 75 per cent of its coke from Newcastle and the rest was obtained locally but had to be washed to remove its impurities.

Thus by 1856 there were only 5 blast furnaces in the district, although another works was begun in the following year when only 56,511 tons of ore was consumed locally.[39] Yet between 1847 and 1857 the export of ore from the Barrow district had increased from 100,000 to 600,000 tons.[40] The export of ore to the other smelting districts both from Whitehaven and Barrow, of course, quickly mounted to exceedingly great proportions, so much so that before long local supplies had to be supplemented by Spanish ore. As a result of this pressure upon the ore reserves, to which was added the demand of the booming development of the Cumberland iron industry discussed here, the prosperity of the district was rather short-lived.[41] However, this is anticipating the conclusion of this account.

The establishment and rapid development of Barrow as the 'Sheffield of the North' from 1859 onwards is one which could be told within a very simple framework, that of the demand for haematite pig iron by the Sheffield makers of Bessemer steel and the initiative and skill of three men, H. W. Schneider, James Ramsden and J. T. Smith.[42]

H. W. Schneider, the owner of the rich Park Mine, as well as of mines in Somerset, had been prospecting in the district for about twenty years previously.[43] He and his partner Robert Hannay were persuaded by the directors of the Furness Railway, or rather its genral manager, Ramsden, to develop the coastal site of Hindpool,

[38] Oliver Wood, 'The development of the West Cumberland coal, iron and shipbuilding industries, 1750–1914', London Ph.D. thesis, 1952, pp. 172–3.

[39] *V.C.H., Cumberland*, Vol. II, p. 386.

[40] Report, *S.C. on the Rating of Mines*, 1857, evidence of H. W. Schneider, M.P., Q. 2127.

[41] Partly on account of the burden of high royalty charges incurred during the period of demand and partly upon the over-concentration of the district upon acid-steel making for the supply of the overseas railway market. (See Andrews, *op. cit.*, p. 114.)

[42] J. D. Marshall, *op. cit.*, *passim.* In fact, the dream was to make Barrow more than this—to eclipse the cities of Dundee, Glasgow, Sheffield and Liverpool together! In the 1870s, as a result of the development of the basic Bessemer process, Barrow's advantages and growth were curtailed.

[43] *S.C. on the Rating of Mines*, 1857, 0.2110. Schneider had also helped to promote the south Durham and Lancashire Union railway lines—the fuel artery for this district. (Marshall, *loc. cit.*) The Duke of Devonshire, judging by the extract from his diary which Marshall quotes: 'Schneider has nearly agreed to build a furnace' (7 December 1858), was not the *deus ex machina* behind this industrial promotion.

in circumstances similar to those prompting Bolckow and Vaughan to establish Middlesbrough as an industrial city. The technical direction of the first three blast furnaces was in charge of Smith, a manager with much British and foreign experience, especially of fuel economy, which, of course, was a very important consideration at Barrow. Nevertheless, despite being at a 100 per cent disadvantage in buying their coke as compared with the ironmasters of Cleveland, the richness of the haematite and the skill of the technicians enabled the Barrow-in-Furness blast furnaces to produce this valuable pig iron at a cost nearly equal to the price of the North-eastern product.[44]

Upon this sound basis and in face of a growing demand for Bessemer pig iron, the concern was all set for expansion. So that by 1867 there were 11 blast furnaces with 5 more under construction making about 4000 tons of iron a week.[45] Even five years earlier, Jordan, the French metallurgist, was able to report an output of over 60 tons every day[46]—sixty years before, this was considered a good weekly figure.

By 1867 another decisive course had been taken to integrate the flow of production from the mines which produced one-third of the total Furness output of iron ore to the finished steel rails, which the Sheffield rollers had previously been left to make. The Barrow steel works was a joint stock company set up in 1864 with a capital of £150,000, in which Schneider and Hannay owned £40,000 of the shares. According to Marshall,[47] who had investigated the company's files at the Registrar of Joint Stock Companies, in November 1865 the steel company agreed to purchase the Hindpool furnaces, the Park Mine, and other mines operated by Schneider for £500,000. Thus it was an amalgamation of the Furness railway interests, headed by Ramsden and the ironmasters. As a result the capital of the Haematite Steel Company was raised to £1,000,000 of which Schneider and Hannay were to hold £200,000 shares and were to leave £200,000 of the purchase price as a loan to the company.

The consolidation of railway interests in the company (which now had a majority over Schneider and Hannay) was the signal for an attempt to capture the world market for rails. This mighty Barrow

[44] Marshall, quoting *Colliery Guardian*, 30 May 1863.

[45] F. Kohn, *Iron and Steel Manufacture* (London, 1869), pp. 17–19.

[46] Marshall quoting an article in *Colliery Guardian*, 23 May 1863.

[47] Marshall, *op. cit.*, p. 252. See also F. Leach, *Barrow-in-Furness, its Rise and Progress* (Barrow-in-Furness, 1872). Other original shareholders were the Devonshire family; a Furness Railway director, and Thomas Brassey, the railway engineer. When the capital was increased, Liverpool merchant capital and other railway directors were represented upon the list of shareholders.

enterprise was, from its inception, an attempt to exploit the new and growing market for steel railway lines. The company's shareholders included Thomas Brassey, the builder and financier of railways on an industrial scale. The Furness Railway was the first purchaser, in fact, of its steel rails. The steelworks which adjoined the furnaces were, of course, equipped with the latest machinery and directed by the most modern methods introduced by J. T. Smith, an expert in fuel economy and furnace design.[48] In 1874 the capacity of the works was 3500–4000 tons of rolled steel a week, from 18 converters—as time went on fewer converters of a larger capacity were used.[49] One interesting technical feature of the production here was a method of pig mixing so as to ensure a more regular quality for conversion.[50] There were evidently good grounds for thinking that the works of the Barrow Haematite Iron and Steel Company commanded the most favourable combination of assets in location, capital and management, of any ironworks in the world. This was the largest and the most brilliant example of the effect of the new demand for Bessemer steel, but the North-western district as a whole was able to overcome its disadvantages of unsuitable coal on account of the premium attached to the special value of its ore.

In the immediate neighbourhood of Barrow there was the Carnforth Haematite Iron Co., Ltd., promoted by the Manchester firm of Walduck and Co.[51]—so that Carnforth was optimistically thought to be another Barrow in the making. On the coast north of Dalton, speculators joined with a Middlesbrough capitalist to form what was to be the Furness Iron and Steel Company of Askam. However, this concern was not a complete success and the steel manufacture projected was never taken up.

In Cumberland itself there was a period of renewed activity from the 1850s. There were new ironworks at Seaton, Harrington, Workington. In 1860 the West Cumberland Haematite Company was formed with a £400,000 capital by some colliery owners, and after building its first blast furnace in 1862, had a few years of moderate prosperity. Between 1863 and 1866 there were dividends of 10 per cent, 20 per cent, 8 per cent, and 5 per cent, after which the company embarked upon wrought iron manufacture.[52] Later, development took place at Maryport, Whitehaven and Workington; this, was along the coast and one factor here, as Wood points out, was that

[48] Marshall, *op. cit.*, p. 253. [49] Lord, *loc. cit.* [50] *Ibid.*

[51] Marshall, *op. cit.* pp. 254–5. See A. Harris, 'Carnforth, 1840–1900', *Trans. Hist. Soc. of Lancs. and Ches.*, Vol. 112, 1960, and A. Harris, 'Millom: a Victorian new town', *Trans. Cumb. and Westm. Antiq. and Arch. Soc.*, Vol. 66, 1966.

[52] Wood, *loc. cit.*, pp. 273–4.

the beaches were useful for tipping as well as facilitating the export of the iron.[53]

Thus this district was a thriving area for iron smelting and steel production for almost all of the period.[54] Yet for pig iron, the peak of this district's output was to be reached as soon as 1882 when it constituted 12 per cent of the total British output. This large output had been reached with the help of improved rail communications which reduced the cost of transport for coke, and a healthy regard for efficiency in fuel consumption.[55] In the production of steel, the accessibility of suitable pig iron was still an important factor, as was the coastal situation. In this connection, the case of the move of Charles Cammell and Co. to Workington is a famous example of industrial mobility. The burden of railway charges was a source of much complaint among ironmasters of this time, as will be seen in the following section, and the Sheffield firm is said to have saved £60,000 a year in railway charges, although this involved transporting all their workmen.[56] The basic Thomas process had evidently not yet made it clear that it would soon deprive haematite iron of its monopoly in the manufacture of steel. Still less was it foreseeable that in the next century steel production would only be retained at Workington because of the existing owners' feeling of social responsibility for keeping the town alive.

3 NORTHAMPTONSHIRE AND LINCOLNSHIRE

The development of iron and steel smelting in these two counties in the East Midlands, as we have seen, was the result of the latest adjustment of the geological factors determing the location of the blast furnaces. This is an inland area based on lean home ore. The

[53] *Ibid.*, p. 275. See also A. E. Smailes, *North England* (London, 1961). The map on p. 142 shows most clearly the move of the steelworks to the sea.

[54] The figure of profits and losses of the Solway Iron Company which Wood produces (p. 287) show that from 1870–75, £331,720 was paid in dividends, between 1875–9 a loss of £60,969 was incurred. From 1880–88 the balance swung the other way again and a profit of £19,857. Nevertheless by 1894 there was a debit balance of £67,492.

[55] Wood, *loc. cit.*, p. 276.

[56] They had taken over the Derwent Ironworks (founded in 1870) (P. W. S. Andrews, *op. cit.*, p. 113). The United Steel concern at Workington is the result of several amalgamations which first took in the Solway, Derwent and Lowther works, later the Workington Iron Co., the Moss Bay Haematite Iron and Steel Co., and the Harrington Iron and Coal Co., Ltd. As a result of this combine, as Andrews points out, only the Distington ironworks stood outside United Steel and eventually this joined the group.

extent of this recent trend in the production of British steel can be judged by the fact that in 1942 the ironstone output of the Northamptonshire Sands formation (which extend under the adjoining counties of Oxfordshire and South Lincolnshire) amounted to 10½ million tons, which would yield about 43 per cent of the national output of pig iron.[57]

It was the Northamptonshire ores which were discovered and exploited first. In 1851 S. H. Blackwell, a mining engineer and Midlands ironmaster, drew attention to their existence in a lecture devoted to the state of the iron industry, one of a series promoted by the Commissioners of the Great Exhibition. These ores had, in fact, been known to Blackwell for eight or nine years, but when specimens had been exhibited at the London Exhibition the note in the catalogue put them down as of no commercial value at that time.[58] Blackwell had realised their value to the handicapped South Staffordshire furnaces and that they could be supplied with cheap ore to 'enable it (the district) to hold its place amongst the competing districts'.[59] Indeed, news of the discovery seriously perturbed the Welsh ironmasters, for Lady Charlotte Guest noted in her diary on 5 February 1852: '(John Evans) came to give Merthyr an account of the Mines in Northamptonshire which appear to be most wonderful in extent and richness, and by their ready transport to Staffordshire to be likely to seriously affect our trade.'[60] However, these fears were groundless because for some years the ore was not well received by the conservative Black Country smelters.[61] The ore found more favour with the Derbyshire and South Wales ironmasters.

There were, on the other hand, local attempts to smelt the ore on the spot, particularly as there were good inducements to do this. To the ease of mining there was added the advantage of a good railway communication with the distant Durham coalfield for a supply of coke; the difficulty which restrained enterprise in this otherwise promising new field was the absence of a local metal using industry. It is true that some pig iron went to South Staffordshire, when it

[57] L. D. Stamp and S. W. Wooldridge (eds.), *London Essays in Geography* (London, 1951), art. 'The Development of the Northamptonshire iron industry 1851–1930', by S. H. Beaver, p. 33. See also the same author's 'The iron industry of Northamptonshire, Rutland and South Lincolnshire', *Geography*, 1933, pp. 102–14.

[58] S. H. Blackwell, *The Iron Making Resources of the United Kingdom* (London, 1852), p. 383.

[59] *Ibid.*, p. 385.

[60] *Diaries of Lady Charlotte Guest*, p. 290.

[61] *Mining Journal*, 12 June 1869, p. 415. A writer pointed out that the addition of a small proportion of Northants ore was sufficient to 'totally condemn the iron by the finished iron manufacturers'.

was used, according to Beaver, surreptitiously by the puddlers,[62] and as we shall see later, there were some unfortunate attempts to set up secondary establishments in Northamptonshire, but these did not create the industrial hinterland which was essential to large-scale development.

The first ironworks to smelt ore upon the spot was that of William Butlin, who erected a cold-blast furnace in Wellingborough in 1853; a second was built in 1855 at Heyford near Weedon and both these concerns attest to the importance of the railway, as Beaver points out, by their proximity to the Midland and London and North Western Railways.[63] The development of the ore quarries followed a similar pattern of development.

But although, as Beaver says, 'by about 1857, then, the industry was really on its feet and both ore extraction and smelting were proceeding apace', subsequent progress was slow. Even if Butlin had solved most of the problems of smelting—the ores were rich in silicon and sulphur, as well as phosphorus—others were chary of the difficulties and not until 1866 was a third furnace built. There were other considerations besides that of fuel economy, but in this respect the three furnaces were very efficient, having a consumption of under 30 cwt. a ton of pig iron.[64] Thus, as to price (45s. as against 64s. for Staffordshire pig iron), this district could compete very strongly with that of other districts.

However, 1864–1873 saw a more rapid expansion of ore output than of pig iron. Whilst nearly 1½ million tons of ore were dug in 1873, less than 60,000 tons of pig iron were made in the district. One curious instance, in this connection, of the use of technical facilities available elsewhere is the acquisition of the Golynos furnaces in South Wales by a Northampton ironmaster to smelt ore from the East Midlands county![65] Moreover, it might have been expected that the export of ore from the district might have reduced the cost of bringing in supplies of coke and so encouraged further smelting, but this was not the case. It is interesting to note that the Glendon Iron Company, which erected furnaces in the years 1866–9,

[62] This evidently should be the wrought iron manufacturers, as the puddlers were not responsible for providing their own supply of pig iron.

[63] Beaver, *op. cit.*, p. 35.

[64] *Ibid.*, citing *Mining Journal*, July and November 1859. Roepke, *op. cit.*, p. 70, comments that the competitiveness of the price of Northamptonshire iron in the 1860s—£2 5s. as against £2 9s. 6d. (Cleveland), £2 17s. 6d. (Scottish) pig iron—is 'an excellent illustration of the extent to which the cost structure had come to favour location on ore-fields and particularly on those which, as in Northamptonshire, could be worked by open-pit methods'.

[65] Beaver, *op. cit.*, p. 38.

soon afterwards bought a colliery in Derbyshire to ensure an adequate supply of coke.[66] However, integration was usually in the other direction, for between 1871 and 1879 three Derbyshire ironworks acquired large mining concessions in Northamptonshire.[67]

The years 1874–80 were a period at last in which the local consumption of the Northampton ores increased, although still as much as 70 per cent was sent to other districts. This time it was a Derbyshire coal-owner, C. H. Pelvins, who began another small burst of activity in building furnaces, with the erection of the ironworks of the Islip Iron Co. (In 1873 the first blast furnace was blown in.) Other works were built in the years 1873–8 particularly near Kettering.[68] Nevertheless, by 1880 the output of Northampton only formed 2 per cent of the country's pig iron.

The development of the iron industry into the manufacture of wrought iron was even less successful than the limited growth of the blast furnace section of the industry. Beaver lists four forge establishments in the county projected or begun in the late 1860s and only in the case of one of them, which in fact imported scrap iron from elsewhere, was the enterprise carried through to a satisfactory conclusion.

Finally, it is interesting that William Siemens, the exponent of the open-hearth process of steel making in England, came to Towcester in Northamptonshire to perfect his 'direct process' of making wrought iron from the ore. Experiments to supersede the blast furnace were conducted here for several years from 1873 onwards, but, although some iron could be made which was considered to be of the same quality as Swedish bar, the project was abandoned in 1878. The rotatory furnaces, however, continued to make 'sponge iron' for filtering purposes until 1882.[69]

The Northamptonshire iron industry stagnated from 1880 until the end of the century; schemes for developing the district came under the cloud of depression in the iron industry partly brought on by the competition of steel.

In Lincolnshire, however, conditions were more favourable to the continuance of the exploitation of local ores which began from 1859 onwards.[70] In the first place, Scunthorpe in North Lindsey, Lincolnshire, was conveniently situated to river and sea communication, being eight miles south of the River Humber and four miles east of

[66] *Ibid.*, p. 42.

[67] *Ibid.*. p. 43. The Lincolnshire works, at Scunthorpe, also imported Northamptonshire ore.

[68] *Ibid.*, pp. 43–4.

[69] *Ibid.*, p. 45.

[70] See G. R. Walshaw and C. A. J. Behrendt, *The History of Appleby-Frodingham* (privately published 1950).

the River Trent.[71] The first exploiters of the rediscovered ore (made, it is said, through an investigation of clinkers of burnt limestone unsuitable for the land and used for the roads) were W. H. and G. Dawes of Elsecar who sent samples to Barnsley by river and canal transport for trial at their furnaces.

As in Northamptonshire (the district in fact has the strata of the Northampton Sands underlying it) the ore was easily available for open-cast mining in the outcrop which extended to a length of 7 miles and a width of 1½ miles. Not all the ore, however, is suitable for smelting, for the content of sulphur in the iron ore increased to an excessive amount as it extends eastwards. It was, incidentally, the intention of the Dawes to mix the Scunthorpe ore with the clayband iron ores to be found in the South Yorkshire coalfield, yet in 1862 the Barnsley firm of ironmasters decided to erect blast furnaces. Apparently the cheapness of the Lincolnshire ore, with its economy in lime (it was self-fluxing), outweighed the disadvantages of overcoming the difficulties of adjusting smelting to the local ore. This, according to the authors of the Appleby-Frodingham history, provoked one operator in the district to describe his ore supply as 'a mixture of monoliths and mud'.[72] In all the ore-fields where the deposits were not associated with strata of coal there was necessary a period of experimentation with methods of smelting to discover the most productive charges and the rate of driving (that is, the pressure of blast necessary to maintain an even flow of the materials down into the hearth); with the clayband ironstones this need for adjustment to the local ores does not seem to have caused so much trouble to the ironmasters. With the Scunthorpe liassic ores it was found advantageous to mix them with the Northampton ores containing a large amount of silica and in the case of the Frodingham Company these were obtained from the vicinity of Grantham.

As regards the establishment of the Appleby-Frodingham concern which today is the principal steel works in this area, the historians of the company have been unable to discover the reasons why Joseph Cliff of Wortley, near Leeds, and William Hirst, his son-in-law, were drawn in 1863 to the idea of setting up furnaces at Frodingham rather than in the Clevelands or in Northampton.[73] Cliff was a brickmaker and, it seems, had no previous connection with or interest in Lincolnshire, nor had he experience of iron-making. The works were built and managed by a Stockton-on-Tees man.

[71] Scunthorpe's access to the importation of ore and pig iron is through Immingham, which is 19 miles away.

[72] Walshaw and Behrendt, *op. cit.*, p. 7.

[73] Walshaw and Behrendt, *op. cit.*, pp. 36–50.

Difficulties with the raw materials were also experienced by this concern. One of the furnaces exploded and, in 1865, Hirst's partner left the business. But after seeking the help of the Cleveland expert John Gjers of Middlesbrough, who reorganised the works and heightened the furnaces, consistent results were obtained. Incidentally, the history of the company reveals the primitive methods of working in some respects even at this late date. For, in consequence of the gobbing up of the slag-hole, it was necessary to take down and rebuild the brickwork in order to dig out the obstruction with a heavy iron shovel slung by a chain from the top of the furnace. All this meant that the furnace would be out of blast for four hours. Eventually a brick-lined door was contrived which could be quickly replaced.

Nevertheless, in respect of the consumption of coal required for the smelting of a ton of pig iron (in this case Durham coke) Lincolnshire compared favourably with other districts. According to the statistics compiled in connection with the enquiry into the country's coal reserves in 1871 and 1873 Lincolnshire furnaces required 55 cwt. The lowest figure was that of Durham, where only 2 tons of coal were needed to produce 1 ton of pig iron, but Northampton furnaces required 58 cwt. on the average, as did those of South Staffordshire.[74] By 1880, the consumption of Lincolnshire had fallen to 46 cwt., and that of Northumberland and Durham to 38 cwt.

Lincolnshire at that time possessed 21 furnaces; however, as their average yearly output was rather low—13,847 tons as against the 23,445 tons of the North-eastern furnaces—the total production was something less than half a million tons. These furnaces belonged to the Trent Iron Company (7); Frodingham Iron Co., and the North Lincolnshire Iron Co. (4 each); and two each owned by the Lindsey Smelting Co., the Redbourne Hill Iron Co., and the Appleby Iron Co.[75] The exploitation of the district was thus by no means a Klondyke rush to newly discovered wealth, for apart from the pioneer concern—the Trent Iron Company—the establishment of the rest was spread over the ten years 1866–76.

The industrialisation of this part of Lincolnshire was concentrated upon the townships which eventually were united to form the town of Scunthorpe. This sparsely populated area of sandy heathland and commons, where the chief economic activity was the sale of rabbit skins at the Brigg fur market,[76] had a population of only

[74] Walshaw and Behrendt, *op. cit.*, pp. 45–54. One would not, of course, accept these figures as being anything more than indication of the relative fuel efficiency of these districts, bearing in mind the deficiencies of the iron and coal statistics.

[75] *Ibid.*, p. 45. [76] *Ibid.*, p. 1.

1235 in 1851. It cannot be claimed, even so, that the coming of the iron industry made the district grow at a pace comparable to that which transformed Merthyr Tydfil. In fact, by the turn of the century the total population had only reached 11,167.[77]

At Frodingham (today part of Scunthorpe), although the village suffered from its intersection by the railway, Charles Winn, the lord of the manor and a farsighted encloser of common land in 1832, planned a small community of a half a dozen streets. He also made provision for a town hall, National school, shops, allotments and a cricket ground for Frodingham's miner and furnacemen inhabitants. However, owing to the social class distinctions between these workers and the clerks, skilled workers and tradesmen, who were unable to buy plots of land in order to assert their respectability, Scunthorpe received the influx of these upper classes of ironworks' society and rapidly outpaced Frodingham.

The importance of the Lincolnshire district depended, of course, upon the change-over of the ferrous industry to steel-making and the eventual utilisation of the phosphoric iron ores of this geological formation in the East Midlands. The beginning of this development was the adoption of the basic open-hearth process in 1888 on the advice of Gilchrist.[78] The co-inventor of the basic processes recommended for this purpose the employment of an Austrian, one M. Mannaberg, who had assisted in introducing the new technique to the Wishaw works of the Glasgow Iron and Steel Company. Although the history of this new venture falls strictly outside our period, the details of this project in terms of the capital involved, the plant laid down and the output expected is of interest here as showing the extent of technical progress in the 1880s. The history of Appleby-Frodingham reveals that a capital outlay of £60,000 made possible the construction of two open-hearth furnaces each of 15 tons capacity and a cogging mill and rack, a billet stand (with rolls 7 feet long and 28 inches in diameter), hot shears and a hot saw.[79] It is apparent these steelworks were in the vanguard of technical progress and Mannaberg, who eventually succeeded to the managing directorship, although the Cliff family maintained its connection with the firm until its amalgamation, kept Frodingham abreast of modern ad-

[77] *Ibid.*, p. 2. The last population figure is an estimate for 1948.

[78] Walshaw and Behrendt, *op. cit.*, p. 63. The authors point out that Brymbo Iron and Steel Co. was not unique in pursuing the basic open-hearth process at this time, for *Engineering*, 13 June 1890, stated that during the year previous the number of open-hearth furnaces was 233 acid and 14 basic furnaces and another 43 new acid furnaces and 3 basic furnaces were under construction.

[79] *Ibid.*, p. 64. The building, admittedly bought second-hand, was of wrought-iron construction.

vances. He also secured the ore reserves of this expanding concern, thus providing the necessary foundation for the important role this company has come to play in the United Steel Company.[80]

The history of this development brings out very clearly the great advantage of this steel-making district in that it was conveniently near to the Sheffield market. This fact, not the negative one of suppressing competition, led to the technical and financial integration of the Sheffield and Lincolnshire steelworks.

[80] The account of the history of Frodingham leading up to its joint purchase of the neighbouring ironworks of the Appleby Iron Company (established ten years later than its neighbour), and this amalgamation's eventual merger into the Sheffield steel concerns of Samuel Fox and Co., and Steel, Peech and Tozer Ltd., will be found in P. W. S. Andrews and E. Brunner, *Capital Development in Steel* (Oxford, 1951), 109–13.

CHAPTER XVI

The Triumph of Steel over Wrought Iron

'The Revolution we are about to announce is not in Sicily against King Bourbon, nor in France against the Emperor, but here in England, in science, in the manufacture of iron and in arts in which all the world is interested. What but a complete revolution in all our ideas is produced by the fact that a blast of cold compressed air directed on a molten metal, without any fuel . . . lashes it into most furious heat, makes it boil and bubble and throw up scum like a soup kettle under which a great fire is blazing? What but a complete revolution has begun in the whole manufacture of iron, when it is shown that we can draw off melted iron from the blast furnaces . . . into another furnace and from it . . . run off in half an hour, into any shape, cast steel or perfectly pure malleable iron, fit for all the many purposes to which steel and iron are put? The invention of the railroad, with its adjunct to telegraph, was not the beginning of a more important revolution than is, we know, now commenced in the iron trade . . . '[1]

'Its history may be suitably divided into three epochs—the first, the experimental and struggling period, which extended from 1856 . . . to 1865, when it had been taken up at several of the leading works in this country, and had been adopted in Germany, France and the United States; the second, the development period, which extended from 1865 to 1884, during which the process was developed rapidly and largely, not only in this, but in other countries as well, and the third, the stationary or retrogressive period which extended from 1884 to the present time (1895) and during which, in this country at least, the make of Bessemer steel has greatly declined . . . what is to become of the Bessemer steel industry? Is it destined to dwindle and decay, like the industry which it superseded for a time, but which now threatens to outlive it?'[2]

[1] *The Economist*, August 1856, pp. 952–3, quoted in Gibson, 'The establishment of the Scottish steel industry' (*Scottish Journal of Political Economy*, Vol. 5, 1958).

[2] 'Before Bessemer and after', *Iron and Coal Trades Review*, 26 July 1895, pp. 113–14.

This chapter is not a history of that decline of the Bessemer steel industry, which seemed so imminent in the 1890s. Indeed, a full account of that period of the relative decline of the British industry and the frightening and meteoric rise of the steel industry of the Austro-German empire would take this study far beyond its intended limits. However, it is necessary to see the first and second phases of the Bessemer steel industry in this country in perspective. The achievement of Bessemer in creating 'mild steel' was, as we know, overshadowed in Britain by Siemens' open-hearth process. On the Continent, the availability of the iron ores of Alsace-Lorraine and certain metallurgical improvements made possible a rate of expansion quite outpacing the growth of both the Bessemer and open-hearth sections of the industry in Britain. The shock of this Continental prowess gave rise to what Armytage, ironically but aptly, terms 'intimations of industrial mortality'.[3] And D. L. Burn, of course, had previously interpreted the period of adjustment to the metallurgical revolution started by Bessemer as a study in competition. His first section is entitled 'Mid-Victorian Alarms'[4] and his starting-point is the Paris Exhibition of 1867, when British exhibits made such a bad showing, and the products of the hitherto unchallenged British iron and steel industry appeared as 'pieces of rusty iron' among 'slovenly intruded heaps of raw material'. This apprehension was to be confirmed, too, by the experience of the British iron and steel industry in the crisis, which heralded in the long period of uncertainty, pessimism and, as some saw it, stagnation—the Great Depression of 1873–96.[5] In essence, we can see now, the discussion turned on the question of the 'rate of growth' in various sectors of the economy. In the iron and steel industry, while there was expansion, particularly in the new steel works making open-hearth steel, particularly in Scotland,[6] and from the mid-1880s, in the North-east of England,[7] this contrasted with the plight of the malleable iron manufacturers deprived of their former markets for wrought iron rails. Finally, as Burn argues, there was a sensational fall in iron prices, but this was

[3] W. H. G. Armytage, 'Portents and Polytechnics: the efflorescence of Civic University Colleges in England 1867–1898', *The Universities' Review*, October 1952.

[4] D. L. Burn, *Economic History of Steelmaking* (London, 1940).

[5] Burn, *op. cit.* Chapter III, 'Competition and the crisis of the 1870s'. On the general question of the Great Depression the reader is referred to A. E. Musson's article, 'British Industrial Growth during the Great Depression 1873–96, *Econ. Hist. Rev.*, 2nd Series, Vol. XV, No. 3, April 1963, and the previous references given by him.

[6] Gibson, *loc. cit.*

[7] W. A. Sinclair, 'The growth of the British steel industry in the late nineteenth century', *Scottish Journal of Political Economy*, Vol. 6.

due not only to competition between iron and steel, but to the general economic situation of deflation.

However, the immediate task is to look at the development of the Bessemer steel itself, to trace its introduction as the raw material of modern heavy industry, railways, shipbuilding, engineering, etc.; and, if we can, to study its economics, the cost advantage of steel over wrought iron.

The fortunes of the Bessemer steel industry during our period were overwhelmingly tied up in the use of steel for rails. As Skelton, economist of the iron and steel industry, said: 'the iron rail trade was the first attacked'. Wrought iron rails wore out quickly; even at the outset, steel seemed a better proposition even if it was much dearer per ton. Bessemer and his supporters argued that the steel rail might be of lighter construction and economies made that way. The first test rails of authentic Bessemer steel to be laid down were at Crewe Station in 1861. A year later, a comparative test was made outside London, at Camden, when one section of steel rail outlasted seven wrought iron rails laid on to the line of track.[8] The price of Bessemer steel was then reckoned by John Brown of Sheffield to be £22 per ton. In 1865 the price fell to £17 per ton and the next year even further.[9] In that year, 1866, the first order for steel rails came from America. According to Dredge, it was sent by the Erie Railroad Company to Messrs. John Brown and Co.—an order for 1000 tons at £25 per ton. In England, a start had been made; 63 miles of steel rails were laid on the London and North Western Railway.[10] This company had its own converter at the railway works at Crewe. The victory of steel rails over the traditional iron was in sight. A conservative railway company like the Great Northern Railway Company might hedge—in January 1869 it called for tenders for 200 tons of steel rails and 5000 tons of iron; but the difference in price between iron and steel rails narrowed, reflecting the greater economies possible with the converter, and in the next decade the defeat of the iron rail was accomplished. In 1869 the price of steel rails was then £9, and of iron rails £6 10s., and there was a hope of a further reduction of the £1 Bessemer royalty which expired in 1870.[11] The effect of this

[8] In 1857, according to Mushet, who forged an ingot and had rails rolled from steel made at Ebbw Vale, a similar test was carried out at Derby Station. Cf. A. Birch, 'The first Bessemer steel rail', *Engineering*, 8 February 1952, pp. 181–2.

[9] According to Dredge, *loc. cit.* In 1866, with the price at £12 per ton, the royalty payable to Bessemer was £2 per ton, but there was a 'drawback' of £1 per ton if rails were rolled from the ingot.

[10] Dredge, *loc. cit.* pp. 944–5.

[11] Burn, *op. cit.*, p. 22. Burn however quotes a Barrow steel maker as thinking that a further reduction was not possible.

competition is to be seen in the following figures:

GREAT BRITAIN: PRODUCTION OF RAILS 1870–82 (1000 TONS)[12]

	Wrought iron		Steel	
1870		1350	—	—
1	1000	1370	200	—
2	975	1270	250	—
3	750	1005	300	—
4	700	1005	350	—
5	300	865	400	—
6	250	855	470	400
7	200	820	565	510
8	120	775	700	630
9	70	675	560	520
80	170	350	810	730
1	150	150	1080	1020
2	60	60	1280	1230

In the same period, according to Burn's figures, the exports of railroad iron fell from 945,000 tons in 1872 to 464,000 in 1879.[13] However, other figures which he cites make the decline of the wrought iron rail industry even more dramatic and disastrous than one might infer from these export totals. In the United States, where changes were apt to be great, the imports of iron rails from Britain, in 1871, amounting to more than half a million tons, completely fell away by 1876. As Burn says, 'by 1876 it was recognised in England that the iron-rail trade was dead'.[14] Then, only a few years later, there was to be another frantic spell of railroad building in the United States, when her own extravagant records in railway extensions were broken. Between 1880 and 1882, 27,000 miles of track were laid in three years. The steelmakers, making the most of their new opportunities, shipped 2·8 million tons of steel rails across the Atlantic during that boom.[15] The Bessemer steel industry was launched on a wave of expansion. The rail shops, 'triumphs of engineering organisation', were geared to the output of these immense quantities at a very low cost of production. We have quoted earlier the comparative

[12] This table is reproduced from Burnham and Hoskins, *op. cit.*, p. 158. The second series of figures refers to data published by Sir David Dale; source is not given for the first set of figures.

[13] Burn, *op. cit.*, p. 27.

[14] Burn, *op. cit.*, p. 28.

[15] *Ibid.*, p. 73. As early as 1869, according to Dredge, the three main English steel rail makers—Brown's, Cammell's and the Barrow works—supplied 102,000 tons of rails out of a total of 110,000 tons laid down. By then, however, the American manufacturers of Bessemer steel were beginning production.

costs of steel rails during the early years of this development; unfortunately we do not have any exact comparable figures for the cost of production by the late 1880s: but Skelton points out some of the comparative economies of steel. One cwt. of steel needed 10 cwt. of coal for its manufacture as against 30 cwt. for wrought iron: there was a saving of 4½ cwt. of pig iron. Thus were economies to be won in the use of the two raw materials. Then, there were economies in the size and cost of the plant needed—£25,000 against £50,000 to make the equivalent of 1,000 tons of puddled bars of iron or 1,000 tons of steel ingots. Only a quarter of the number of workmen were necessary to work the Bessemer steel converters. Skelton summed up his comparisons, arriving at a cost of £3 13s. a ton of wrought iron, and £3 7s. 6d. for one ton of Bessemer steel.[16]

Already by 1880 the output of Bessemer steel exceeded one million tons and this was nearly one-half of the production of puddled iron.[17] Now, this 'silent revolution' when 'iron rails (were) being gradually improved off the face of the earth'[18] was being effected when the wrought iron industry itself had been expanded to meet the mounting demand of the world for railways. There was a very real basis for this demand, as is evidenced by the statistics of miles of railway completed in the world:[19]

1840	1850	1860	1870	1880
5262	23,304	65,558	127,887	221,887 miles

The 1870s, as will be apparent, was a decade of large-scale extensions, and the year 1870 was in fact a year of 'mania' with lines projected in America, India, the Balkans and Central Europe. According to Burn, Russia alone ordered from the Cleveland rail rollers £3m. of railway equipment. In all, at the peak of the boom, in 1872, exports of iron amounted to more than three million tons, of which railway iron was about one-third. At that time the North of England, the chief district manufacturing iron rails,[20] was producing 325,000 tons.[21]

[16] H. J. Skelton, *Economics of Iron and Steel* (London, 1891), pp. 238–41. He assumed the cost of pig iron to be 40s. per ton.

[17] Siemens-Martin (o.h.) steel, in addition, amounted to another quarter million tons, whilst the basic Thomas steel was only just beginning to get into its stride with a total of 10,000 tons.

[18] *Circular of Messrs. Shaw and Thompson*, December 1877 reprinted in *Steel News*, December 1952, p. 11.

[19] W. J. Larke, 'The Iron and Steel Age', *Iron and Coal Trades Review*, 1927, p. 129.

[20] Scotland, the other important centre of the malleable iron district, was not so involved; the bulk of production here went into the local shipbuilding industry.

[21] Burnham and Hoskins, *op. cit.*, p. 158.

The measure of the changing fortunes of this industry, as a result of the competition of steel, can be seen from the fact that, ten years later, only 5000 tons was being made at Cleveland. Twenty out of the forty-four puddling firms in the district, according to Burn, became bankrupt. These had owned 821 out of a total of 2158 furnaces.[22] In fact, a number of firms had seen the danger signs and tried to break into the market for constructional iron and plates for shipbuilding; but all this was of no avail.

At Dowlais, where one of the pioneering licences had been taken out, the subsequent experience of the company was instructive. During the first epoch of the Bessemer innovation, up to 1865, it is clear that the trustees of Dowlais regarded their rash enthusiasm for the new process as 'an unfortunate transaction'. According to the first agreement they had paid £5000 cash down; the balance of £5000 was to be paid in two instalments of £2500 each after three and six months respectively. By December 1856, when it was evident the Bessemer experiment was going to be unsuccessful, the trustees, naturally were not anxious to pay the last instalment, even though they conceded they had entered into the agreement 'on the chances of failure or success'. They wrote to Bessemer asking 'whether some further efforts ought not to be made to render your Patent applicable to the mixtures of ores used at Dowlais before you insist on our fulfilment of the strict letter of the bargain'.[23] They further averred: 'While we believe that we have done all in our power to aid your experiments we do not think that these experiments have received from you all the attention which we expected you to bestow.' Finally, putting it to Bessemer that his reputation and interests were 'to a great extent bound up to the event', they asked for a postponement of the payment until 'you can declare to us that you have tried in vain to remove the difficulties which have hitherto thwarted us'. The outcome of this approach was that, after an interview with Bessemer, Dowlais paid the outstanding sum 'to get rid of what they considered an unfortunate transaction'; a new agreement was drawn up, but not signed.

There the matter lay until 1861, when Charles Attwood of the Weardale Iron Company apparently approached Dowlais about the possibility of taking over the defunct Bessemer licence. This company's mines produced some non-phosphoric iron ores, also ores

[22] Burn, *op. cit.*, p. 29. One of these was the firm of Hopkins Gilkes, makers of the ill-fated wrought iron Tay Bridge.

[23] *Dowlais MSS.* This and the following quotations are taken from a recital of the correspondence between D.I. Co. and Messrs. Bessemer and Longsdon in a brief prepared for counsel's opinion, 1861.

which could be smelted into spiegeleisen.[24] Both these were, as we know, essential ingredients of Bessemer steel. It is significant, too, that now, when counsel's opinion was being sought to discover if Dowlais had any legal powers to appoint Weardale its agent under the licence, the question was raised whether the original patent or licence also extended to the manufacture of steel. We know nothing of the actual negotiations between the two companies and of the probable Dowlais participation in the venture at Trealaw. However, in 1861 the Weardale Coal and Iron Company did have a 50 cwt. converter in operation.[25] It had been suggested by counsel that since the original licence did not specify the actual location of the place of manufacture it was open to Dowlais to sub-contract the manufacturing operations to Weardale. On the question of whether the pneumatic process, as originally licensed, applied to the manufacture of steel, the opinion of the counsel retained by Dowlais was that it did. However, among the Dowlais MSS there is the curious and damaging confession by Bessemer that the manufacture of steel was a new departure, quite distinct from the patents of 1854–6 for making wrought iron: 'I could not allow steel to be made under the iron licence.'[26]

Dowlais did not take up active operations in Bessemer steel making in South Wales until 1865. In the Report of the Trustees for the year ending 31 March 1866 it is categorically stated: 'During the year the manufacture of steel under the Patent of Bessemer and Co. was commenced upon, the first ingots being cast on the 5 June 1865 and the first steel rail rolled on 10 June 1865.' The sum of £33,227 had been expended on a new steel works,[27] of which £20,000 was being recovered from Bessemer.[28] Up to that time 2180 tons of steel ingots had been made and there appeared to the Dowlais steel makers 'every indication of a fair profit accruing for solid steel rails.'

[24] Erickson, *British Industrialists*, p. 154.

[25] Possibly this had been transferred from Dowlais.

[26] *Dowlais MSS*. Folder correspondence with Sir Henry Bessemer, 1896–7. A letter of 4 April 1897 describes the negotiations (in 1865 or 1866) when, according to Bessemer, 'on condition of their entering into a steel licence with me at the same terms as other people, I spontaneously offered to pay back the £10,000 I had received for the useless iron licence.' The outcome of this negotiation will be described later.

[27] *Dowlais MSS*, Report, 31 March 1866.

[28] Dowlais asked for £30,000 for the old licence. Then, in Bessemer's words: 'I slept at Dowlais House and after breakfast next morning while the carriage was waiting at the door to take us to the Railway Station, Mr. Clark said should I be satisfied to deduct £20,000 from their first steel royalties. This I cordially accepted.'

The subsequent history of steel making under the Bessemer process of Dowlais must be summarily related. To summarise one's impression of the results of the introduction of this innovation, at the outset it may be observed that, although Dowlais steel rails apparently did not enjoy the same high reputation as those from Sheffield and did not command the same competitive power, nevertheless, the profits and rate of return on the Bessemer investment were 'very handsome'. In 1867–8 these amounted to 52 per cent on an outlay of £41,469. The trustees of Dowlais could report: 'The profits already made have recouped the amount of capital invested.'[29]

The year by year reports of the Dowlais works enable us, however, to trace in more detail the working out of the economics of the new steel manufacture as against wrought iron, particularly in the rail trade up to 1879. At first, even as late as 1865–6, there were technical problems. The Dowlais reports mention 'the numerous experiments made with various kinds of Pig Iron in the casting of the steel and in the rolling and hammering of the ingots'. Indeed, 'this new business of steel making' was described as 'from the beginning one of great anxiety and labour.' Nevertheless it is clear that these technical problems were soon solved; costs were brought down continuously, and, as we shall see, the problem facing the Dowlais managers was the commercial one of overcoming sales resistance to the product. With a productive capacity of four times the outputs achieved in 1866–7, there was a rich potential to be tapped by increasing sales. In 1868 it was admitted: 'The manufacturing department has produced extraordinary results, and it is to be regretted that our powers of effecting sales of steel rails are not at all equal to our skill in making them.' Dowlais had sold 7125 tons of rails in the previous year.[30] But as there was a good margin of profit—of £2 12s. 6d. on a Cardiff selling price of £12 18s. 6d.—there was the possibility of price reductions to compete in the market. Since this profit margin, again, depended upon the cost of production, and since the Dowlais efficiency could yield a cut of 8s. 1d. per ton in the cost of manufacture of rails each year, this was the source of a 'most profitable trade'. Indeed, there was a great expansion in the manufacture of steel ingots, as the table on p. 360 shows.

By that time just over half of the total make of the Dowlais blast furnaces was being converted into steel. By this time, too, close on 20,000 tons of steel rails were being rolled and shipped to America. This trade yielded an average profit of £2 8s. $8\frac{2}{10}$d. per ton of steel rails as against a mere 18s. 3d. margin on wrought iron. This explains why the Dowlais steel works contributed more than £54,102 profit

[29] *Dowlais MSS*, Report for 1868. [30] *Dowlais MSS*, Report for 1868.

BESSEMER STEEL INGOTS MADE AT DOWLAIS[31]

	Long tons		£	s.	d.
1866	2,106	Cost	9	2	8
7	5,793		8	18	$1\frac{3}{10}$
8	8,551		8	6	$8\frac{6}{10}$
9	9,088		7	10	$0\frac{4}{10}$
70	19,767		7	7	$10\frac{1}{10}$
1	24,288		6	16	$7\frac{5}{10}$[a]

[a] The cost difference between 1870 and 1871 was accounted for by the reduction in the royalty to 2s. 6d. per ton.

out of a total £174,568—more than 30 per cent.[32] It is not surprising the Dowlais company installed another steel rail mill, a decision which was justified by mounting orders, a bigger profit margin, and increased sales in 1871–2.

This prosperity was not to last, at least on this scale. In 1876–7 the prices of both iron and steel rails had fallen. In March 1877 Dowlais had orders for 15,180 tons of wrought iron rails at £5 16s. 11d. per ton and 28,810 tons of steel rails at £7 per ton. By then, further economies were achieved by running the molten pig iron direct into the converters, at a rate of 600 tons a week; Dowlais was also by then making its own ganister for the linings and the spiegeleisen for the final processes. So, whilst the profit on finished steel was down to £1 8s. $4\frac{6}{10}$d. a ton, on the bigger turnovers the Bessemer steel works were able to make £76,335 net profit on its year's operations.[33]

By the end of the decade, here at Dowlais the triumph of steel had again been demonstrated. More than twice as many Bessemer ingots were being made as puddled bars; wrought iron was still the cheaper product, costing £3 2s. 4d. a ton and Bessemer steel 5s. 6d. more. However, even in the wildly fluctuating markets for rails in the post-depression period—prices of Dowlais sales of steel rails fluctuated between £4 2s. and £9 11s. 5d—these sales consistently yielded a

[31] *Dowlais MSS*, Report, year ending 1 April 1871.

[32] *Ibid.* In 1871, it is worth while noting, the Dowlais forges were producing 94,105 tons of finished iron—nearly five times as much wrought iron as Bessemer steel; but the cost differences were down to £1 18s. per ton in favour of wrought iron. In other words, taking into account the lower cost of manufacture and the lower rate of profit, wrought or finished iron had an advantage on balance of £1 8s. per ton.

[33] *Dowlais MSS.* Report for 1877. Dowlais had by this time also introduced Siemens o.h. furnaces and was making 18,070 tons of Siemens steel as against 54,660 tons of Bessemer ingots.

higher margin of profit than the now obsolescent wrought iron of the first Railway Age. It is easy to understand why, in those years of uncertainty, the Dowlais trustees were sufficiently confident in the competitive power of steel to expand production and yet to put over a third of the make into reserve.[34] The economies of Dowlais and of the iron and steel industry were geared to steel.

Elsewhere in Wales, 'the old established Welsh makers largely withdrew from business in 1877, before their resources were dissipated',[35] and this presaged, as we know, a reorganisation of the industry in that region. There was decline and withdrawal, too, in the Black Country, the traditional home of the wrought iron industry, where malleable iron was used for the many items of ironmongery—edge tools, locks, chains, nails, tubes—as well as for machinery and industrial equipment. Unfortunately we do not know much detail about this period of crisis and readjustment to the new forces. But, as W. K. V. Gale says: 'The technical developments which made cheap mild steel possible took away a great part of the wrought trade. In the last thirty years of the nineteenth century a great number of ironmakers closed their works. Some, like the Thorneycrofts of (the Shrubbery Works) Wolverhampton, sold up and retired while they were still in a good financial position, but many were less fortunate.'[36] However, the reliance of the old-established iron trades on the puddling furnace for wrought iron in the manufacture of such products as chains and anchors for the Admiralty helped surely to keep the industry alive in the Black Country. Just before the First World War there were still about 661 puddling furnaces still in existence: this was nearly one-half of the 1500 in the country.[37]

In Scotland, however, the situation was not so acute. There had been an annual average production of around 200,000 tons a year in the period up to 1880, and it is estimated by Gibson that the Clyde shipbuilders must have absorbed between 65 and 70 per cent of the total production.[38] The Scottish malleable iron makers met some competition from displaced English and Welsh manufacturers; however, as Gibson says: 'To them the rise of the new English steel firms was reflected only indirectly in (this) increased competition . . . so long as malleable iron retained its supremacy in ship- and

[34] *Report for 1880.* Make of puddled bars and slabs 42,187 tons; make of Bessemer ingots 83,460 tons.

[35] Burn, *op. cit.*, p. 28.

[36] W. K. V. Gale, 'Development of Industrial Technology in the Black Country 1700–1904', *Birmingham and its Regional Setting*, 1950, p. 18.

[37] Gale, *loc. cit.*, and Burnham and Hoskins, *op. cit.*, p. 161.

[38] Gibson, *loc. cit.*, p. 33, fn. 24.

bridge-building, there could be no large-scale change-over to steel production in Scotland.'[39] However, when the Siemens steel from the Steel Company of Scotland won orders from the Admiralty for steel plates for its ships in the late 1870s, then the Scottish malleable iron firms began to put in open-hearth furnaces. This movement accounted for the great spurt in O.H. steel production in Scotland up to 1885.

Bessemer steel, incidentally, had first demonstrated its suitability for the job, in 1863, when a stern-wheel barge had been built. In the following year, according to Dredge, a paddle boat of 377 tons was built for the Humber Steam Packet Company. Two sailing ships, each of about 1200 tons displacement, were also launched in that year. By 1863 there were another six small steel ships to be added to 'Bessemer's' fleet.[40] But, as one would expect, headway could not be made until steel as a constructional material had been tested and officially recognised by the Lloyd's Register of Shipping. The following table,[41] in fact, shows how little progress could be made until 1877, when the underwriters of Lloyd's first accepted steel-structured ships for insurance.

TONNAGES OF SHIPS UNDER CONSTRUCTION FOR REGISTRATION BY LLOYD'S UNDERWRITERS

	Iron	Steel
31 December 1877	333,770 tons	1,110 tons
8	362,537	4,096
9	438,432	37,669
80	750,941	82,827
1	1,081,785	183,818

Finally, the question of costs was inescapable. Here, as in the manufacture of rails, large-scale production enabled costs to be cut from £20 per ton in 1876 to £7 per ton in 1884. The greater strength of the mild open-hearth steel enabled further economies and steel soon became the cheaper material. From 10 per cent of the Clyde launchings in 1879 the proportion of steel ships grew to nearly 100 per cent, within the decade.[42] The impact of this change on the wrought iron industry is to be gauged by the facts that in 1883 the production of iron plates reached its maximum, some 732,000 tons; by 1891 this line of production had all but disappeared.[43]

[39] Gibson, *loc. cit.*, p. 33, fn. 24.

[40] Dredge, *loc. cit.*, p. 935.

[41] J. W. Hall, 'The Development of the Rolling Mill with the Advent of Steel', *Iron and Coal Trades Review*, 1927, p. 159.

[42] Gibson, *loc. cit.*, p. 34.

[43] Burnham and Hoskins, *op. cit.*, pp. 158–9.

In sum, the total production of puddled bar iron certainly underwent a major contraction in the 1880s. In 1870, when there were about 8000 furnaces, they were turning out some 2,600,000 tons of bars; in 1882, after an erratic period of production in the 1870s, a maximum was reached—2,841,000 tons. Ten years later, when only about a half of the former number of furnaces were at work, their make was down to roughly 1½ million tons in a drop of more than 50 per cent. The decline thereafter was to be progressive until the pre-war years, 1908–14.[44]

These figures mark the absolute decline of the industry, which has been the principal subject of this study of the iron and steel industry in the nineteenth century. This is indicated, too, in the continuous fall in the percentage of pig iron—the raw material of both wrought iron and steel conversion—actually made into bar iron. Again, according to Burnham and Hoskins, this branch of the industry accounted for getting on for a half of Britain's 6 million tons of pig iron in 1870. In the next ten years, the result as we have seen of the triumph of Bessemer steel, this proportion was again approximately halved.[45]

Finally, to complete the picture of this process of supersession, we have to give the figures for the rapidly expanding outputs of Bessemer steel (acid and basic); and Siemens open hearth steel.

IRON AND STEEL PRODUCTION 1868–80[46]

	Bessemer steel	**Basic Bessemer steel**	**Open hearth steel**	**Crucible steel**
1868	110,000			
1869	160,000			
1870	215,000		25,000	
1871	329,000		28,000	
1872	410,000		40,000	
1873	496,000		77,500	5900
1874	540,000		90,000	3300
1875	619,785		87,969	4000
1876	700,000		128,000	4150
1877	750,000		137,000	3900
1878	807,527	20	174,000	
1879	834,222	1150	174,939	
1880	1,044,020	10,000	250,913	

44 Burnham and Hoskins, *op. cit.*, Table III, p. 320.

45 *Ibid.*, p. 156. It is curious, however, that the actual figures of the production of 'puddled bars' do not confirm this trend. It is possible, however, the great drop in rail output explains this discrepancy.

46 Source: Figures supplied by the British Iron and Steel Federation.

Of course, the foundations of the modern steel industry were, in a very real sense, the blast furnaces reducing the ore to pig iron. Here there was no major innovation, such as the Bessemer converter or Siemens furnace to overthrow the furnace and render it obsolete. However, there was much scope for improved yields and greater outputs. This transformation occurred mainly, following an experimental period, during 1890–1900. It is a striking instance of the effect upon production of a more scientific practice of metallurgy that in this case no major change in the blast furnace and its hot blast stoves was involved. The terrific increase in its productivity initiated in the United States, it is worth noting, came from a further study of the furnace's interior to secure an efficient and high rate of driving, the substitution of constant *volume* for constant *pressure* blast, and independent blast for each furnace.[47] However, by 1880 another important feature of modern smelting practice had already been introduced—the economy of fuel. As a result of the innovations already mentioned—the Cowper, Siemens and Cochrane blast stoves (incorporating the regenerative principle exploited by Siemens); the Parry bell and hopper closed top—the average yearly output of the blast furnaces had increased to a great extent between 1870–80. Output leapt up from 8981 to 13,667 tons and, what is more, the consumption of coal had fallen by anything up to 30 per cent of the previous figure.[48] Nevertheless, in one respect, in the adherence of the ironmasters to the ancient beehive coke oven and in their reluctance to adopt the Simon-Carves by-product recovery oven, the British industry lagged behind Continental practice.[49] Upon this subject, Lowthian Bell, who, indeed, was responsible for introducing scientific analysis into nineteenth-century metallurgy and whose opinions carried perhaps too much weight, was not converted until the present century.

Nevertheless, at this stage, it is probable that the modern works, particularly those engaged in steel manufacture—Bolckow Vaughan's, the Steel Company of Scotland, the Barrow Haematite, Brymbo and Frodingham—were not far behind up-to-date practice. As we have seen, these firms employed experts who were aware of Continental experience and innovations. This, of course, is an indica-

[47] A. K. Reese, 'A Review of Blast Furnace Practice, 1867–1927', *Iron and Coal Trades Review, Diamond Jubilee Issue*, 1927, pp. 147–9. It is true that, as a consequence of these improvements, larger hearths were necessary in the blast furnace.

[48] W. W. Rostow, *British Economy of the Nineteenth Century* (Oxford, 1948), p. 212.

[49] R. A. Mott and G. J. Greenfield (eds.), *The History of Coke Making* (Cambridge, 1936), pp. 64–83.

tion that the intiative for making technical progress had passed to other countries. In the last stages of manufacture, the conversion to steel, cogging and rolling, the chief factor in fuel economy was the avoidance of heat losses between the different processes and above all between the blast furnace and the converter. It was here, as Burn points out, that the Americans especially were at an advantage, with their new works planned as complete units to include the processes of manufacture from ore to steel; moreover, the American genius for mechanised handling and scientific lay-out had soon manifested itself in the great steel works of the U.S.A.[50] Yet one important innovation, the 'soaking' pit, in which the ingots are put, after being teemed, to retain their heat, was introduced in this country by a North-eastern ironmaster, John Gjers, at the Darlington Steelworks in 1882.[51] One of the most fundamental innovations of steel manufacture was the 'direct metal' plan, which had been thought of by Bessemer and, indeed, in wrought iron manufacture and founding had been tried several times from the eighteenth century onwards. The objection always raised by this method of supplying the refinery or converter direct from the blast furnace was that it was impossible to know the quality of the molten pig iron. As Burn rightly says: 'The kernel of the problem was indeed still in blast-furnace management'; and this fuel-saving plan was eventually perfected in Sweden, Belgium and France before it was belatedly taken up here—Bolckow Vaughan's were once more first in this field.[52] With the other method of steelmaking, the open-hearth process, it was again Continental metallurgical chemists who improved the Siemens steel with the new concentrated ferro-manganese. In this respect, as Burn points out, the employment of metallurgical chemists in this country was a matter of unambitious routine analysis rather than research to develop the scientific control of the structure of the metal.

To sum up this period of technical change:[53] it was, as Bell maintained in his defence of the British skill in metallurgical operations, a time of cumulative change and improvement, not of dramatic inventions. Even though Gilchrist and Thomas's extension of the Bessemer process was to be revolutionary in its results, it was at this time con-

[50] Nevertheless, Andrew Carnegie, the first steel multi-millionaire, who was responsible for many of these developments in the U.S.A. steel industry, was of Scottish descent.

[51] H. W. Dickinson, *loc. cit.*, p. 10.

[52] D. L. Burn, *op. cit.*, p. 49.

[53] Mention, too, should be made of the efforts to mechanise the puddling furnace. Attempts were made in this country, but the most successful of any of these machines of moderate achievement was the American Danks's rotary puddling furnace.

sidered unperfected and Thomas steel in England, for a very long time, had a bad reputation. Many of the ideas which, as in the case of basic steel, the Continental makers took up were, as Burn admits, of British origin. 'The position in England,' Burn concludes, 'was indeed characterised not by a lack of ideas, but by some degree of failure on the part of the English iron and steelmakers either in the appreciation of or welcome given to novel principles, or in the mode of their development and application.'[54]

With the increasing mechanisation and furnace capacity of the modern steel industry, labour, of course, has become a less vital element and consequently there has been a conspicuous absence of labour troubles in the industry.

Nevertheless, another important reason for the relative stability of labour in the iron and steel industry was the early adoption of the machinery of conciliation and arbitration, first successfully advocated by Mundella in the hosiery industry. David Dale, a director of the Consett Works, the man responsible for bringing together the ironmasters and the men through their union leaders, was a disciple of Mundella in this respect. But what is more remarkable than that an ironmaster should be the instrument of the negotiations, is that the two protagonists should have agreed to establish a Board of Conciliation in 1869 so soon after the bitter struggles of the 1860s.

The detailed account of Dale's part in bringing about this almost revolutionary change of opinion among ironmasters upon the standing of organised labour has been given by A. J. Odber,[55] but what is of immediate interest here is his explanation of the new circumstances. The first reason for Dale's success, which Odber gives, is the high capital cost of the ironworks of that time, so that strikes and lock-outs were costly. Nevertheless, this argument seems to have had little force with the Welsh ironmasters, in 1875 and earlier, where the capital invested was possibly greater than anywhere else in the country. The second reason was the wish of the ironmasters to enter into a period of prosperity without the threat over their head of strikes upon a rising market. This is the most convincing of the reasons given; for the boom of 1870–3 was to make the iron trade very prosperous for a short time. What was the most fortunate circumstance of Dale's approaches to the North of England Iron Manufacturers' Association was the conviction of John Kane, the union's leader, in the

[54] Burn, *op. cit.*, p. 64.

[55] A. J. Odber, 'The Origins of Industrial Peace: the Manufactured Iron Trade of the North of England', *Oxford Economic Papers* (N.S.), Vol. III, No. 2, June 1951, pp. 202–20. See also a Memoir of *Sir David Dale* by Howard Pease (London, 1911), pp. 31–49.

value of arbitration. It is evident that the unionism of this period was not made of the same revolutionary fire as it had earlier manifested. Kane supported this policy of moderation because he regarded strikes and lock-outs as economically wasteful and as bringing unnecessary suffering to women and children.[56]

These are the rational reasons behind the establishment of the Board of Arbitration and Conciliation in March 1869–the first of its kind in the industry. There was, too, the element of chance which brought about the favourable concurrence of circumstances, but all the same, at first the Board had to fight for its existence. The Board, in order to establish its prestige in a dispute at a Stockton rail works, and faced with a strike, made its decision against this unconstitutional act and, moreover, got the order in dispute completed without the men's assistance, But afterwards the Board was asked to pay damages by the firm, which it agreed to pay. This was a hazardous way for the Board to establish its authority.[57]

As a result of the successful establishment of the Cleveland Board, its example was imitated elsewhere; first in the Midlands in 1872 and by the Cleveland and Cumberland Blastfurnacemen in 1879–80.[58] However, these bodies were not altogether successful: in the case of the Midlands organisation, this was because the men's union was represented upon the Board, and it was not strong enough to enforce its decisions.

In the operation of the sliding scale, too, the Board was not particularly successful. The first scheme of this new period of labour relations was that drawn up in 1871 for the North of England manufactured iron trade. If the selling price of wrought iron reached £7 a ton an advance of 2½ per cent in wages was granted; another 5s. rise in the price gave another 2½ per cent rise and conversely in the case of reductions.[59] This 'departure from competition as a regulator of wages' only lasted nine months; in 1874 the idea was revived and a 'Derby Scale' to regulate the wages of both the north and south was agreed upon. But in 1875 the Cleveland ironmasters withdrew because they said they could not compete with South Wales (which paid lower wages) and as the demand for rails was decreasing. In

[56] 'One of Them' (Sir Arthur Pugh), *Men of Steel* (London, 1951), p. 61.

[57] 'One of Them' (Sir Arthur Pugh) *Men of Steel* (London, 1951), p. 59. Pugh makes strong criticism of the subsequent operation of the Board (pp. 59–60) yet it is claimed by Pease that by 1900 it 'had settled some 1400 cases of dispute and 25 large general arbitration on wages questions affecting the whole trade'.

[58] Clapham, *op. cit.*, pp. 175–7. Joseph Chamberlain was the first president of the Midlands Board.

[59] *Transactions of the Manchester Statistical Society*, 1885, pp. 1–15. The above section on sliding scales is based upon this paper.

1879 the first sliding scale in the North of England to take in miners and blast furnacemen was drawn up and lasted until 1884—for every alteration in the price of Cleveland No. 3 iron by one shilling the wages of the ironworkers was to change by 1¼ per cent. At the same time the puddlers were paid according to another new scale. The longest period of successful operation was that of the South Staffordshire Mill and Forge Scales from 1880–3. However, in the rapidly changing situation of the iron trade no agreement could last for long. As was stated at the time: 'It was found . . . to be quite unsuitable for conducting the iron trade of Great Britain in its competition with the world.' We must now turn to consider how the ironmasters organised themselves to meet this threat.

An earlier chapter on the ironmasters' associations has already pointed out the tradition of association amongst the ironmasters and, what is more important, the insuperable difficulties in the way of a successful combination to maintain a profitable selling policy. Nevertheless, as we saw, there were a few instances of success in establishing a stable market by means of agreed prices and a restriction of output; it was shown, too, how these associations, despite their own precarious existence, had been working to achieve a national scheme of co-operation between the ironmasters of the various districts.

The history of the ironmasters' associations during the second half of the century up to 1880 is no more spectacular than the history of their earlier development. Indeed, there is great deal of truth in Clapham's statement: 'After 1850, for five and twenty years, with the markets of the world at their feet and prices generally good, Britain's primary producers had little need even to consider such extreme defensive policies (viz. restriction of output). . . .'[60] Indeed, evidence of the ironmasters' activities during the years 1850–70 is very slight. It is true, of course, that the associations had come out into the open by this time. The South Staffordshire association, the most active of the local bodies, had adopted a constitution in 1864 and another in 1871.[61] The latter's objects were stated to be: 'To promote interchange of information on all subjects of interest to the trade; to secure united action and mutual support for the general welfare of the Association; to watch Parliamentary Bills relating to railways, canals, and foreign tariffs affecting the interests of employers and employed; and to engage legal assistance when required.' Every firm had one vote for each blast furnace and one vote for every

[60] Clapham, Vol. II, p. 150.

[61] *Royal Commission on Trade Unions 1867–9*, Q. 9829. *Royal Commission on Labour* (C.6795-XII), Rules of Associations of Employers and Employed.

eight puddling or balling furnaces, but the maximum number of votes was eight.[62] No detailed provisions were made for conducting the business of the association and no mention was made of the thorny question of what was 'to be deemed current price'. But before the *Royal Commission on Trade Unions* it was revealed that the North Staffordshire makers had followed the example of the ironmasters of the south of the county and they 'endeavoured to get as near as possible to the Staffordshire price'. This was said to be 'really the price of a few firms in South Staffordshire'.[63]

But it was in the North-east that in the 1860s the most formal type of organisation was established. Again it was largely concerned with the question of labour, and the predecessor of conciliation and arbitration was the setting up of two associations, one for the blast furnaces and the other for the manufactured iron trade. The crucial part of the organisations, which shared a common secretary, was, as we have seen, the operation of a central fund to finance the resistance of the ironmasters against the strikes of their workers. There was no machinery for fixing prices in the North of England.[64]

It might be expected that the depression in the iron industry in the 1870s would have prompted a renewal of the negotiations to restrict output and maintain price. However, apart from any doubts which might have survived about the efficiency of such measures—the ironmasters, it seems, had short memories—there was, as we have seen, a considerable reduction in the output of puddled iron as a result of firms dropping out of the industry. In fact, it was not until 1881 that a measure of agreement between the Cleveland and Scottish districts to reduce output by 12 per cent was put into operation for six months. This, however, broke down upon its renewal in the way that these agreements had done in the past. This time it was the independent action of Bairds, the largest makers, which disrupted the combination of the ironmasters.[65] A few years later, in 1886, there was again talk of a national reduction, but this time a revival of trade put the idea of restriction out of the ironmasters' heads.[66]

[62] *Loc. cit.*, p. 100. [63] *Loc. cit.*, Q. 10,532.

[64] *Loc. cit.*, Q. 9551, a similar association among the West Cumberland ironmasters was formed in 1890 with this mutual indemnity fund against strikes. In this association votes were allotted according to the output of the members. (*Royal Commission on Labour*, 1892, p. 98.)

[65] Clapham, *op. cit.*, p. 160.

[66] See H. W. Macrosty, *The Trust Movement in British Industry* (London, 1907), pp. 57–60, and P. Fitzgerald, *Industrial Combination in England* (London, 1927), pp. 36–58, for the subsequent history of these associations. In the 1890s, says Skelton, *Economics of Iron and Steel*, there were three organisations based

The greatest effort of co-operation among the ironmasters was undoubtedly the formation of the Iron and Steel Institute to promote research into the technology of the industry in 1869. As Asa Briggs says: 'It was an important landmark in the history of scientific application.'[67] Moreover, the value of the exchange of experience between the men like Bell, Siemens and Bessemer, who were keenly interested in the scientific development of the industry and in the setting up of plants to perfect the newly-evolved different kinds of steel, is evident from the histories of the new concerns of this period.

The crisis of the 1870s must certainly have caused several of the ironmasters to wonder why they were not better informed about the movements of international trade. As a result of this desire for information, early in 1876, the ironmasters met 'to organise united action and the acquiring of such general information as was for the common good'. Nevertheless, once more regional interests obtruded themselves—the Cleveland representatives hoped to secure national agreement upon a wage policy in the industry, but were opposed by the Welsh, and the British Iron Trade Association was formed to collect statistics and to organise representation of the industry in its dealings with the Government. Even with these limited aims the Association did not work very well and Burn's summing up of this development sums up the progress of the commercial organisation of the industry at this time: 'Narrow and ill-defined in its objects,' he says, 'timid in its concessions to the claims of organisation, imperfect in operation within its limited sphere, it was the chief symbol of a recognition throughout the industry in the '70s of the changing face of world trade.'[68] This, itself, was a result of the transformation of the iron industry by the creation of the large-scale world-wide steel industry. It was not yet known that when these innovations of Bessemer, Siemens Martin and Gilchrist-Thomas had worked themselves out fully, the competitive position of the British steel industry

---

on the quarterly meetings—at Middlesbrough, Wolverhampton and Birmingham. He noted, 'It is seldom of late years that declared alterations of prices are reserved until the quarterly meetings.' The meetings were useful for the exchange of information and for settling accounts—the origin of quarterly meetings, according to this writer. The most remarkable instance of combination at this time was the formation of the International Rail Syndicate of 1883; under this agreement Great Britain was allocated 66 per cent of the total business in rails.

[67] Asa Briggs, 'A History of the British Steel Industry' *J.I.S. Inst.*, November 1962, Vol. 200, Pt. II: 'So too was the first publication of the *Iron and Coal Trades Review*. The industry was taking in hand the problems of diffusing knowledge of new techniques and their business applications.'

[68] Burn, *op. cit.*, p. 32.

would soon necessitate more drastic measures than were conceivable in 1880.

2 THE SIEMENS REGENERATIVE FURNACE AND THE OPEN-HEARTH PROCESS

The advances in the technique of steel-making associated with the name Siemens are two: the first concerns the principle of heat regeneration applicable to smelting and re-melting;[69] the second is linked with the work of the Martin brothers of Sireuil upon the open-hearth process.

In the field of fuel conservation, through the most effective use of the coal or coke burnt in the furnaces and the boilers of the iron and steel works, Britain lagged behind Continental practice. The question of the utilisation of the waste heat has been discussed earlier and the reason for our neglect of this now vital industrial economy—our apparently inexhaustible sources of cheap coal—has been made clear.

However, as early as 1816 there is evidence of an invention which on the face of it would appear to be a prototype of the regenerator. This was Stirling's economiser, a brickwork chamber to store surplus heat; yet, according to Dickinson, no contemporary application of Stirling's patent was made, and so little was it understood, that it was mistakenly thought to be an anticipation of Neilson's hot blast process.[70] But the later patentees of the regenerative principle were apparently unaware of their predecessor.

It was Ernst Werner von Siemens, the eldest of the family of a small landowner near the Harz mountains, who first set the Siemenses upon their industrial career. It was he who associated the name of Siemens with electrical power and, in the 1840s, established the partnership of Siemens and Halske at Berlin,[71] which was to be the great centre of that industry upon the Continent.

The brothers Carl Wilhelm[72] and Friedrich were the members of

[69] We have already seen that the Siemens' furnace had been adopted by the Sheffield crucible steel makers for melting the blister steel.

[70] H. W. Dickinson, 'The utilization of waste heat' (*Trans. Newcomen Society*, Vol. XXIV), p. 6. This was a Scots patent; it was never enrolled in England. In 1851 Fischer, the Swiss metallurgist, exhibited a model of a furnace which he had used for some time, with a honeycomb arrangement of bricks to use waste heat to pre-heat his crucibles. See also W. Pole, *The Life of Sir William Siemens, F.R.S.* (London, 1888), p. 74.

[71] *Ibid.* p. 10.

[72] In his youth he was called William, and later, when he became naturalised, his names were officially anglicised.

the family who played a decisive part in the development of the iron and steel industry. The younger son, Friedrich, patented the first 'improved arrangement of furnaces' in 1856 and this was the beginning of their joint research upon the regenerative principle of heat conservation.

William Siemens (1823–83) had come to England, after a technical education under the direction of brother Werner and an apprenticeship in an engineering factory at Magdeburg, in 1843. His account of his early career in England, given in later life, seems too dramatic to be true.[73] Landing in London, with the idea of invention in electroplating, and after discovering the curious English use of the word 'undertaker', he made a successful start to his career with the firm of Elkingtons, the spoon manufacturers, at Birmingham.

The work upon the regenerative furnace began some years later when his brother Friedrich had joined him in England. The first essay was with the steam-engine; then followed experiments with a regenerative engine and evaporator at Northwich in the early 1850s,[74] intended to revolutionise the salt industry. However, Siemens had received several disappointments with this proposed application of his theory—the storing of otherwise wasted heat by passing hot currents of air over suitable heat-conducting matter for later use. Thus, as Pole puts it: 'he simply turned his attention towards other possible modes of carrying them into effect'.[75]

But it was his brother who, in 1856, conceived the idea of applying the regenerative principle to ordinary furnaces where fuel was consumed.[76] In a paper read to the Institution of Mechanical Engineers in June 1857 by William, he claimed a saving of 70–80 per cent of the total quantity of fuel previously consumed. The first practical experiments were conducted at Scotland Yard, but, according to Pole, the furnaces were tried at a Sheffield steelworks and at Lloyd and Foster's ironworks at Wednesbury, Staffordshire, early in 1857. One very important consequence of the invention was the patenting of the Cowper hot-blast stove, which incorporated the regenerative principle.[77] This had many advantages over its predecessors—for one thing the temperature could be raised to 1500°F with no leakages of the hot-blast—but apparently its adoption by the British steelmakers was by no means universal. In 1882 only 112 British furnaces out of 968 had Cowper or Whitwell stoves.[78] Nevertheless, if

[73] Address of Sir William Siemens at Birmingham, October 1881 (quoted in Pole, p. 44).

[74] Pole, *op. cit.*, p. 79. His first patent concerning this principle was taken out in 1847.

[75] *Ibid.*, p. 99. [76] *Ibid.*, p. 100. [77] Patented in 1856.

[78] D. L. Burn, *op. cit.*, p. 45, n.6.

Pole's word can be accepted, the savings resulting from the regenerative stove amounted to £500,000 a year for the iron industry of the world.[79]

The Royal Commission on coal supplies in 1871 was, of course, interested in the progress of the Siemens regenerative furnace and listed the iron and steel works where they were to be found. In 1862 this was at only one works—Attwood's near Darlington, but by 1870 thirty-eight were in use in steelworks (fourteen at Vickers at Sheffield and eight at Landore). However, they were more extensively employed with reheating and puddling furnaces. The Barrow Haematite Steel Company alone had forty and the London and North Western Railway Company at Crewe possessed twenty.[80]

Improvements in the Siemens regenerator furnace, notably the use of gaseous fuel patented in January 1861, brought about by the joint work of the two brothers, were not of immediate direct consequence to the ironmaster. But William Siemens had from 1857 onwards, at Bolton, been investigating the applicability of the principle of regeneration to the puddling furnace. Although the first reaction of the industry to this proposal was first unpromising, considerable numbers of the furnaces had been brought into use by 1870. Siemens, however, during his analysis of the puddling process had been forcibly struck by the disadvantages of the puddling, as being 'extremely wasteful in iron and fuel, immensely laborious, and yielding a metal only imperfectly separated from its impurities'.[81] This was in 1868 when Bessemer steel had not yet completely overcome wrought iron. Yet already Siemens was connected with another process which today provides the greatest part of the world's steel supplies—in Britain today one acid Bessemer plant, the Workington steelworks, survives as a relic of that superseded method of steelmaking.

The idea of making steel by the immersion of wrought iron into a bath of liquid metal upon an open-hearth was by no means a new invention. The technique was current even before the publication of Réaumur's celebrated work on steel published in 1722.[82] In the

[79] Pole, *op. cit.*, p. 103.

[80] *Royal Commission on Coal*, 1871, Vol. I, pp. 113–50. According to Edwards, 'Sir Charles William Siemens, F.R.S.', *British Steel Maker*, August 1951, p. 415, the adoption of the furnace by the crucible steel trade was a decided landmark in the development of that industry. 'It was found that the aggregate economy even after allowing for the royalty of 5s. per ton, was not less than 35s. 6d. per ton of steel cast.'

[81] Pole, *op. cit.*, p. 141.

[82] It is mentioned in Biringuccio's *De la Pirotechnia* (1540, English translation, New York, 1942).

eighteenth and nineteenth centuries several metallurgists had tried to bring the technique to perfection upon a large scale, from Clouet in 1798 to Bessemer in 1855, and William Siemens with a French metallurgist, Louis le Chatelier, at Fourchambault, in 1863.[83] However, it was not possible to obtain metal of a regular quality.

It was left to the Frenchman, Pierre Martin of Sireuil, to obtain consistent and successful results by prolonged experiments along lines already patented. These had begun in 1862, but it had been impossible to obtain sufficient heat. In the following year, a Siemens furnace had therefore been installed.[84] The Siemens brothers had already in 1861 patented the use of an open furnace for melting and making steel.[85] This was with gaseous fuel so that the impurities of solid fuel were excluded from the process. The Darby brothers of Ebbw Vale had subsequently experimented at Sheffield with the open-hearth steel furnace, but their results did not lead to immediate success.

With the Siemens regenerator and a 1-ton furnace, pig iron smelted from phosphorus-free Elba ore, and Bessemer steel scraps, some success was achieved—sufficient to justify the patenting of the process in England upon 15 August 1864.[86] Nevertheless, the technique of operation was not perfected for a year, so that the definitive patent dates from July 1865. When the first trials had been made the wastage had been as much as 50 per cent of the charge, and although this had been much reduced the steel made in the open-hearth was for some time considered to be expensive to produce.[87]

Meanwhile in England Siemens had had trials made at Barrow and elsewhere, but the furnaces failed to stand up to the great heat evolved during operations. Finally, in 1866, Siemens had to set up the 'Sample Steel Works' at Birmingham. By August of that year he was claiming that the body of the furnace there would last for years and ". . . I can produce a ton of cast-steel for less money than Besse-

[83] Le Chatelier, 'Hommage à Pierre Martin' (*Mémoires, Revue de Métallurgie*, Vol. VII), pp. 573–5. See also 'Aux origines de l'invention de l'acier Martin: Fragments inédits des carnets d'Emile Martin, 1840–1870' by André Thuillier, *Actes du 86e Congrès National des Sociétés Savantes, Montpelier*, 1961, pp. 341–81 (Paris, 1962).

[84] *Ibid.*, p. 868. Pole, *op. cit.*, pp. 145–6. The terms were 2000 francs a year for a furnace to melt steel at the rate of 6 tons a day.

[85] Pole, *op. cit.*, p. 143.

[86] Thuillier, 'Fragments inédits . . . ', p. 348, notes that on 9 April 1864 Emile Martin wrote in his Journal: 'Pierre a obtenu le métal homogène qui a toutes les qualités de l'acier.'

[87] Cf. Thuillier, *loc. cit.*, pp. 348–9, the cost of a Martin furnace was 20,000 francs as against 200,000 for a converter.

mer, and of superior quality'.[88] It was Siemens' intention, as it had been Bessemer's, to use the steelworks as a demonstration centre to convert the British ironmasters. By 1867 he had perfected his technique of making steel and, in August, he patented two methods of making steel. The first was direct production from the ore, in which Siemens was more optimistic than successful; the second was the open-hearth process by immersion of iron and steel 'into a fluid bath of cast iron under the influence of very intense heat, where they are dissolved and converted into cast steel'.[89] How far this was radically different from the Martin process it is difficult to say, for, after a short legal fight between the two metallurgists, an agreement had been reached in November 1866 which recognised the contribution of each protagonist.[90] Certainly, the process as perfected was dependent upon the regeneration gas furnace and the technical management of the immersion process.

In England, the open-hearth process was taken up because it enabled old wrought iron rails to be cheaply converted into steel ones. In 1868 Siemens made a proposal to the Crewe establishment of the London and North Western Railway Company upon these lines: 'I have now fully succeeded to melt steel in from two to four ton charges upon the open-hearth of my Regenerative Gas furnace . . . the cost of melting a ton of scrap metal and its proportion (of 10 per cent) of pig metal does not exceed 30s. per ton including fuel, labour, wear and tear of furnace and ferro-manganese required. . . .'[91] With the G. W. R. he arranged a spectacular demonstration. Old rails were sent from Swansea to Birmingham, there they were converted into steel ingots, then after being rolled by John Brown and Co. at Sheffield, they were delivered as new rails to the railway company.[92] Thus by 1869 the process had been adopted by three steel works including the London and North Western Railway Company itself.

Nevertheless, the competition of Bessemer steel was acute, as Siemens' own experience at the Landore steelworks in South Wales must have showed him very forcibly. As mentioned before, costs were high and the price of Bessemer steel rails fell in the 1870s.[93] This steelworks had been set up by a private company, of which Siemens was the most important member, at Swansea and commen-

[88] Pole, *op. cit.*, p. 147, quoting a letter dated 29 March 1866 to C. A. Attwood of Tow Law.

[89] Pole, *op. cit.*, p. 149.

[90] Chatelier also made claims for his inventions.

[91] Pole, *op. cit.*, p. 152. [92] *Ibid.*, p. 153.

[93] See D. L. Burn, *op. cit.*, pp. 51–2.

ced operations in 1869.[94] At first the steel works prospered—despite being dangerously thought of by Siemens as the '*Cheval de Bataille*' for pursuing his direct method. Pole records that the output of steel per week had reached 1000 tons in 1873.[95] But as most of its production was in the form of steel rails, when the depression came, sharpening competition, the company had 'a lingering miserable existence'.[96] In 1888 it went into voluntary liquidation. South Wales, however, was a leader in the production of the new type of steel. Dowlais was among the early open-hearth steel-makers. At the end of the 1870s, nearly half of the open-hearth steel made in Britain came from this district.

As Burn points out,[97] it was only after the improvements brought about by the French metallurgists of Terre Noire—in particular, a ferro-manganese with a high concentration of this element—that a new field was opened up to the open-hearth steel manufacturers. When the special steels needed for the armour plate orders of the Admiralty (at long last converted to the use of steel in its warships) could be met by Siemens' steel, then a new field of prosperity was opened up to the company. It was true, indeed, that the open-hearth steel possessed a special advantage. Where the Bessemer product was converted in a short space of time—less than half an hour—without the possibility of much control being exercised over the composition of the final result,[98] the fourteen hours needed to complete the open-hearth process gave ample opportunity for chemical analysis so that a definite quality of steel could be made to order. Thus Landore received Navy orders[99] and again, as Burn remarks, when Siemens 'looked upon the steel which had latterly been brought forward so

[94] Pole, *op. cit.*, p. 154, and D. T. Williams, *The Economic Development of Swansea and of the Swansea District* (Cardiff, 1940). For a more detailed account of the company see Erickson, *British Industrialists*, pp. 156–7. By mid-1869 the firm was producing 75 tons of steel a week.

[95] Pole, *op. cit.*, p. 191.

[96] J. Riley, 'On recent improvements in the method of the manufacture of open-hearth steel', *Journal of the Iron and Steel Institute*, 1900.

[97] Burn, *op. cit.*, pp. 52–4.

[98] Edwards quotes the homely simile of a partner of Andrew Carnegie: 'After all, this Bessemer process is an ocular process only. The workman stands and looks at the colour of the sparks. If he is tired the whole batch of rails may be flawed, but open-hearth practice is slower and much more accurate. It is like a cook making soup and testing it every now and then until it is just right.' Another trouble with the Bessemer Process was to overcome the air bubbles in the ingot of steel. (Edwards, *British Steelmaker*, August 1951, p. 415.)

[99] Burn, *op. cit.*, p. 54. Pole, *op. cit.*, pp. 194–5, mentions that the company supplied, in the first instance, the steel for two armed despatch vessels, the *Iris* and the *Mercury*. Nevertheless, the company's existence came to an end in 1888 (Edwards, *loc. cit.*, p. 417).

much by the Terre Noire Company in France as a feather in the cap of the open-hearth process', he was presumably finding confirmation in his pocket.

As for Siemens' ambition to supersede the blast furnace, although he took out twenty-seven patents with that object in view and did achieve some positive results in small rotative furnaces at Towcester, in the new iron district of Northamptonshire, his hopes were never realised.[100] Bell, at this point, had remarked upon Siemens' researches: 'I have a kind of instinctive opinion that the Blast Furnace will be difficult to exterminate';[101] only recently, with the sponge-iron process, has a major modification been made in the reduction of the ore to the metal.

Siemens came near, in the course of his metallurgical researches, to being responsible for two other great developments in the steel industry. In the first place he had experimented with a basic lining for his furnace, but, unfortunately, he gave it up and reverted to the original silica bricks, on account of the additional cost to an already expensive process.[102] Secondly, in 1879, he brought out his electric arc furnace.[103] Again it was the economic factors—the small capacity and a heavy consumption of electricity—which delayed the exploitation of an otherwise metallurgically sound idea. Its merits were, of course, that it gave absolute control over the temperature in the furnace, and higher temperatures than achieved before at that. Moreover, there was no question of the contamination of the materials by the impurities of the coke. However, the only immediate application of the electric arc was to welding, and, according to Pole, the smelting of aluminium.

As already indicated, the adoption of the open-hearth process of steel-making made a slow beginning in Britain. In 1875 the British output was a mere 77,500 tons; by 1880, the conclusion of this survey of the industry, just over a quarter of a million tons was reached.[104] Not until 1894 did Bessemer steel yield its predominance to its rival. In that year 2,355,000 tons of Siemens-Martin steel exceeded the Bessemer production by half a million tons.[105]

[100] Pole, *op. cit.*, p. 196–200.

[101] *Ibid.*, p. 199, quoting Bell's letter to Siemens, 12 November 1878.

[102] Pole, *op. cit.*, pp. 311–13.

[103] Bessemer too experimented with different kinds of linings for his converter.

[104] Open-hearth steel was important to the steel-shipbuilding industry. Lloyd's issued its first set of rules for ship construction in 1877 (Gibson, *loc. cit.*, p. 34). Initially steel hulls were much dearer than iron; but the cost of o.h. steel plates rapidly fell from £20 per ton in 1876 to under £7 per ton in 1884. For a discussion of the economics of o.h. steel shipbuilding see Gibson, *loc. cit.*, p. 34, fn. 25.

[105] See W. A. Sinclair, 'The growth of the British steel industry in the late

Nevertheless, the open-hearth process, much improved and its productive capacity increased by larger furnace, fed sometimes directly from the blast furnace, is one of the two major factors determining the state of the British and world steel industry of today. The other factor, which again is technical, and which was responsible for that advance of the other countries which exposed Britain's steel industry to a period of relative decline, is the basic Bessemer process. It began as the key to the problem which prevented the British ironmasters from exploiting all the native resources of iron ore and promised new visions of prosperity. Instead, this discovery eventually unloosed the forces of competition and poverty upon the industry. There is, of course, a complex of qualifying circumstances around this radical change in the position of the British iron industry and these will have to be discussed so far as they concern us here. However at this point we should consider the technical development of the basic process by the two young metallurgists, Gilchrist and Thomas—the first major contribution of metallurgical science to the industry.

3 BASIC STEEL

The achievement of Sidney Gilchrist Thomas (1850–85) and of his cousin Percy Carlyle Gilchrist presents the most attractive aspects of industrial biography. Their success was the result of a single-minded devotion to metallurgical science and in their success they did not attempt to overlook the contribution of fellow workers in the same field of research.

The problem of dephosphorisation of the pig iron used in the Bessemer and the Siemens-Martin processes had occupied the attention of many metallurgists. Among these was that of Heaton, an Englishman, who must have come near to the final solution in patenting a converting apparatus to use limestone as a neutraliser of the phosphoric acid.[106] Thomas, himself, had been set upon the task,

nineteenth century', *Scottish Journal of Political Economy*, Vol. 6, 1959. Sinclair points out: 'The rapidity of technical progress in the open-hearth industry at this time requires some emphasis in that the achievement of the steelmasters has not been fully recognised in the past by historians who have been inclined to be preoccupied with Britain's competitive position *vis-à-vis* Germany and the United States.' By then, however, there was another technical factor at work. For soon after the introduction of the Gilchrist-Thomas basic process to the Bessemer converter, it was applied to the open-hearth furnace as well. Thus the basic open-hearth process is now the predominant form of steel production.

[106] Anon., *J. Heaton's process for the treatment of cast iron and the manufacture of steel* (London, 1869).

which was to become his short life's work, by Dr. George Chaloner, a lecturer in chemistry at the Birkbeck Institution. Thomas at this time, in 1870, was a junior clerk at a London police court, having been denied the opportunity of taking up medicine by the death of his father. The remark which Chaloner is reputed to have made was one worthy to inspire any hero of Samuel Smiles' philosophy of 'Self Help': indeed, young Thomas, fired with the prospect of making his fortune through eliminating phosphorus in the Bessemer converter, is said to have soon calculated the product of a royalty of sixpence a ton on 3 million tons of Cleveland pig iron.[107] However, as Burnie says, 'The bent of Thomas's mind would, in a society where money did not exist, have carried him quite as irresistibly towards discovery . . . as the desire of riches even for others.'[108]

The following years were truly years of preparation, in which he acquired for himself, with a zealous voraciousness, all the knowledge which the School of Mines could bring to bear upon the problem. Thus the first stage of the research was reached when Thomas gave the theoretical answer to the question—why was phosphorus retained in the Bessemer converter? Lowthian Bell, the so-called 'high-priest of metallurgy' of his day, had indeed come to an answer, which to him only pointed to a remote possibility of success. 'The phosphorus in the iron joined up with the oxygen blown through it to make phosphoric oxide, which could not be taken up by the siliceous slag, acid in character, of the silica-lined converters then solely in use . . . therefore it remained in the iron whilst it was refined into steel.'[109] The experiments which Thomas conducted in the Birkbeck laboratory, and even in a home-made improved converter in a fire-grate in his lodgings in London, were directed to discovering a suitable lining for the converter, basic in character.[110] The materials which suggested themselves to Thomas were calcium or magnesium limestones.

As early as 1872[111] George James Snelus, the metallurgist at Dowlais, whilst investigating the Danks mechanical puddling process, had, by experiment, demonstrated that in a normal ganister-lined

[107] H. W. Burnie, *Memoir of Sidney Gilchrist Thomas* (London, 1891), p. 31.

[108] *Ibid.*

[109] Reese Edwards, 'Percy Carlyle Gilchrist', *The British Steelmaker*, November 1951, quoting Bell.

[110] That is of a chemical compound which treated freely with the phosphoric acid and neutralised it.

[111] Indeed, according to Wedding's *Basic Bessemer Process* (New York, 1891), p. 30. Tunner, the German ironmaster, in 1860, recommended the use of burned magnesite (i.e. freed of its carbonic acid). However, no 'tolerably stable lining could be made'.

converter the phosphorus remained in the steel, whereas with one lined with limestone the phosphorus became absorbed in the slag. Accordingly, he patented the use of different sorts of limestone for 'the lining of all furnaces in which metals or oxides are melted or operated upon while fluid'.[112] Reese Edwards, the biographer of Thomas's cousin, Gilchrist, remarks: 'There is also clear evidence that Snelus conveyed the information to both Sir Henry Bessemer and to Dr. William Siemens that he had successfully tried a Bessemer converter with limestone for the removal of phosphorus. Yet neither of these two distinguished metallurgists followed up this gratuitous and exceedingly valuable piece of information.'[113] Ironically, Snelus did not pursue his work upon basic linings, since he took up the management of the West Cumberland Steel Company, and the haematite ores, of course, had a monopoly value to the producers of phosphorus-free pig iron. As Snelus rather ruefully admitted, his company's 'interests were opposed to the solution of the problem'.[114]

It was to Thomas's credit and advantage that, having elucidated by his own researches the theoretical answer to his problem, he persisted in practical trials until he met success. Again it is pleasant to record that Thomas received the co-operation not only of his cousin, who was the chemist at the Blaenavon ironworks, but also of Edward Riley of Dowlais, another worker in the same field, and of the technical experts of Bolckow Vaughan, Windsor Richards and Dr. Stead, who had a great interest in bringing the trials to fruition. The magnanimity of men who might have been embittered rivals, as Robert Mushet was of Bessemer, is striking evidence of the untainted ideals and ambitions of this young metallurgical chemist, Sydney Gilchrist Thomas. Snelus acknowledged that it was Thomas's persistence, which he himself lacked, which had brought him to his position as the head of the metallurgical world.

It was in 1876 that Thomas first sought the help of his cousin, then employed by the Cwmavon Copper Works. Only with reluctance did he overcome his incredulity of the possibilities of Thomas's approach to the problem, and then circumstances prevented any experimentation until the following years. The two cousins had, in the meantime, both applied for the job as analytical chemist at the Blaenavon Iron and Steel Works, and its manager, E. P. Martin, who was to play an important part in encouraging the research, chose Gilchrist because of his experience.

The experiments, with linings of limestone and silicate of soda and charges of the phosphoric ore of Northamptonshire, were conducted

[112] Patent No. 908. [113] Reese Edwards, *loc. cit.*

[114] *Journal of the Iron and Steel Institute*, Vol. 1, 1879, p. 139.

in a rough shed at Blaenavon by Gilchrist in his spare moments. Thomas kept his job as a magistrate's clerk—not until 1879, when he was sure of success, did he resign from his job. He supervised the research by week-end trips to South Wales.

In December 1877 the two experimenters consulted Martin to organise the experiments upon a large scale. He, having convinced himself of the potentialities of the Gilchrist-Thomas research, not only undertook to allow the works equipment to be used, but also promised to approach Dowlais for the same facilities and to take up a share in the patent rights. This situation, as we shall see later, was fraught with many difficulties when the time came for staking a claim. However, in the immediate future lay the abandonment of the Dowlais experiments (because the linings would not hold) and a partial success at Blaenavon.

It is true, of course, that Gilchrist had taken out his first patent, in November 1877, for linings made of limestone and sodium silicate. This was not a success and Thomas had been well advised to examine the patent situation very carefully. He was fully aware of the dangers of his position, for he wrote to his cousin: 'I am afraid to claim all basic linings for the simple reason that if you claim anything that is not both new and successful, you lose all.' There was the considerable number of prior patents to be considered and the need to cover all the stages of innovation, in the process of dephosphorisation: the manufacture of the bricks for lining the converters, and the treatment of the slag. Thus Thomas had to take out several patents in the course of the next two years, between 1877 and 1879. But, according to the metallurgical experts, it was his patent of March 1878,[115] for producing a hard lining material, in lumps or bricks, by heating certain kinds of limestone, that was the bedrock of his contribution to metallurgical success.

There is no need here to enter into the details of the scornful reception of the public announcement of the achievement of the two young metallurgists, aged 27 and 28. It seems, however, that it was fortunate for Gilchrist and Thomas, so far as their German patent rights were concerned, that their paper was held over and not read at the Iron and Steel Institute's autumn meeting in Paris in 1878.[116]

Nevertheless, it was through the Paris meeting that the first trial

[115] Patent No. 6080, dated 26 March 1878. Wedding, *op. cit.*, p. 32, says: 'It must be regarded as the most important step towards securing a durable dolomite lining.'

[116] Wedding, *op. cit.*, p. 30. The first half appeared in *Engineering* of 5 April 1878 by error and this instalment ended in the customary way: 'To be continued'. It is a nice example of the irony of circumstances.

of the basic Bessemer process was attempted. Windsor Richards, the manager of Bolckow Vaughan, a firm dependent upon Cleveland ores, naturally had a keen interest in the problem. He had discussed the matter there with Thomas, and after a demonstration at Blaenavon he determined to try the process at Middlesbrough. So two 30 cwt. converters were used; but, as with Bessemer's trials at Dowlais and elsewhere, they at first refused to confirm the experimental results. There was difficulty in making the refractory bricks to withstand the very great heat. But by 4 April 1879, according to Dr. Stead,[117] who was associated with the experiments at Middlesbrough, 'after much labour, many failures, disappointments and encouragements, we were able to show the leading gentlemen of Middlesbrough two successful operations'. But, once more, it is necessary to add that the goal of unqualified practical success was still far away. As Wedding succinctly puts it: 'For the success of the process three principal conditions are necessary, basic lining, basic flux, after-blowing. Of these after-blowing is the most important.'[118] Just when the process had been seized upon by the Continental ironmasters—the first being the Metz Compagnie of Luxembourg on April 21—failure came like Bessemer's 'bolt out of the blue'. Only after an agonising period of efforts at Middlesbrough was a remedy found. Again, this resulted from another piece of co-operation, for Dr. Stead suggested that over-blowing the charge (that is, releasing the blast of air into the mouth of the converter) would finally remove the stubborn phosphorus. As Reese Edwards explains: 'Stead thus showed that the secret of dephosphorisation was the prolongation of the blowing after the carbon had been removed and sufficient oxide of iron had been produced to oxidise the phosphorus.'[119] Gilchrist and Thomas patented this and another improvement suggested by the Bolckow Vaughan staff—the use of tar for binding the bricks—in 1879 and 1880.

Thus Gilchrist and Thomas slowly arrived at their hour of triumph, represented, fittingly enough, by the enthusiastic acclamations of the Iron and Steel Institute in May 1879. (Thomas made a typical comment over this vindication of their faith in their theories: 'We became *pro tem.* junior lions'.) There was still to be a struggle in front of them, but it was the legal battle to safeguard their patent rights.[120] This may have well been disastrous for them, but, as we

[117] Presidential address of J. E. Stead, *Journal of the Iron and Steel Institute*, 1920.

[118] Wedding, *op. cit.*, p. 26.

[119] Edwards, *op. cit.*, p. 585.

[120] It is true that the Blaenavon Company failed at this time and Gilchrist found himself unemployed. The inventors were therefore denied the opportunity

have seen, Thomas, by taking a reasonable attitude, did not founder upon the dangerous shoals of patent law. In so far as his process was to be the key to the unlocking of the great mineral wealth of the Continent, it was fortunate, nevertheless, that Thomas succeeded in defeating Alfred Krupp of Essen and the Bochum Steel Company. In 1877 both Krupp and, almost at the same time in England, Bell, had been working upon experiments with refining furnaces lined with material rich in oxide of iron. They also used fluxes rich in oxides. Krupp's establishment had succeeded in dephosphorising about 17,000 tons of pig iron.[121] A rotating furnace had been used; Bell's methods were different, but he succeeded in reducing the phosphorus content by 95 per cent. However, the practical advantages of Thomas's process, which enabled the removal of the unwanted substance after the extraction of the other elements, silicon and carbon, entirely superseded these two methods. But, more important to Thomas, the German courts, rather surprisingly, recognised his patent rights in preference to the claims of Krupp.[122]

Since Gilchrist decided not to take the risk of being involved in the dangers of the Continental patents, Thomas owned these by himself; unfortunately, he did not live long enough to reap the benefits from the royalties of the Continental producers.[123] He died in 1885, and it would not be over-dramatising his early death to say, as others have done, that he 'burnt himself out' with his efforts to bring his idea to fruition.

It would be beyond the scope of this chapter to examine carefully the rapid adoption of the Gilchrist-Thomas process abroad, and the disastrous effects of the refusal of the British steelmakers to change over from the acid to the basic Bessemer process. The striking difference between the British and Continental attitudes to the innovation is revealed by the fact that, where, in 1884 20 per cent and 10 per cent of the Continental and British steel output respectively was produced by the Thomas process, in 1890 the respective percentages were 64 per cent and 14 per cent.[124] Again, it is Krupp who epitomises

---

of securing a large-scale trial of the new process by the ironworks which had been the birthplace of their most important advances.

[121] Wedding, *op. cit.*, p. 8. Bell's patent preceded Krupp's by a few months.

[122] Bessemer's initial patents were rejected in Prussia on the grounds they were not original.

[123] In 1882 the Dephosphorizing and Basic Patents Company Limited was set up to administer the interests of the people concerned in perfecting the process. It consisted of Thomas, Gilchrist, Bolckow Vaughan, Ltd., Snelus, Edward Riley, John Inman, T. H. Richardson and W. H. Thorpe. (*Bolckow and Vaughan, 1879–1929* (privately printed), p. 12.)

[124] Edwards, *op. cit.*, p. 586. In 1890 the production of basic Bessemer steel on

the unhesitant determination of the foreign competitor to the British industry. In 1883 he wrote to Alfred Longsdon, who had been his representative in London and who was by this time a member of the Prokura (the Board of Proxies): 'I shall propose to you to do away with all the mines and vessels and high furnaces[125] and Bessemer works and to clear the substance never mind with a certain loss—and to dedicate all our power to the making of guns and projectiles and wheels, Siemens-Martin steel and *Fleiss-eisen*[126] . . . to do away with Bessemer steel and rails, which will be ruined by the number of competitors.'[127] Already ten years earlier, he had affirmed his faith in steel, the new metal, to the Emperor William I: '. . . We live today in the Steel Age. Railways, the greatness of Germany, the fall of France, belong to the Steel Age. . . .'[128]

In England, the first outside trial of the new process was at Sheffield, when the patent lining of Edward Riley, containing petroleum and powdered lime, was tried. This did not, however, lead to the adoption of the Gilchrist-Thomas process by the Sheffield works, the heart of the steel industry. Instead, the manager of Brown, Bayley and Dixon Ltd., where the Sheffield experiment was held, himself went to the North-east to establish the North-Eastern Steel Company at Middlesbrough. The first Thomas steel from this enterprise was made in 1883.[129]

Before going on, however, to examine very briefly the adoption of the basic process, it should be mentioned that a valuable by-product was created by this innovation. Bell, the ironmaster-scientist, had declared: 'The yearly make of iron from Cleveland stone alone contains about 3000 tons of phosphorus, worth for agricultural purposes, were it in manure as phosphorus acid, above a quarter of a million. . . .' Once again, this was a statement which fired the imagination of Thomas and when the North-Eastern Company, previously mentioned, in which he had some interest, were proposing to sell the slag as a by-product at a low price, he threatened to sell out his shares in the company.[130] He was convinced that this was a major product, he even wrote to Gilchrist: 'However laughable you may consider the notion, I am convinced that eventually, taking the cost of production into consideration, the steel will be the by-product

---

the Continent was 2,021,903 tons as against 1,140,000 tons made by other methods; in Britain, 503,400 tons of basic steel were outweighed by 3,075,643 tons of other steel.

[125] I.e. blast furnaces. [126] Mild steel.

[127] *Letters of Alfred Krupp* (letter dated 28 October 1883), p. 384.

[128] *Ibid.*, p. 266 (23 April 1871). [129] Stead, *loc. cit.*, pp. 54–61.

[130] Thompson, *op. cit* p. 256.

and the phosphorus the main product.' The adoption of this fertiliser by agriculturalists was delayed, as was the supersession of the acid process, by the conservatism of those it was intended to benefit, and Thomas himself did not live to see the development of the basic slag industry.

As for the firm of Bolckow Vaughan, Ltd., which was the pioneer ironworks to exploit the basic process, it had in the first place been forced into the making of steel by the decline in the market for wrought iron rails.[131] As a result it had acquired the Gorton steelworks of the unsuccessful Lancashire Steel Company. Whatever motive the firm may have had in acquiring these steel works it was pleased at all events in retrospect, to look upon this venture as the acquisition of experience in the process whilst the new Cleveland steel works were under construction. In 1875 the new plant was completed, but, of course, this depended upon the import of foreign haematite ores in large quantities.[132] The new works, according to the report of the company for 1876, were to possess six blast furnaces to smelt the foreign ores and a weekly capacity of 2000 tons of finished steel products was expected. However, it was not until 1877 that the first steel rails were rolled and not until 1879 was the steel works working at full capacity.[133]

When the validity of the Gilchrist-Thomas process had been established, a 15-ton converter was set up in October 1880. The following year, 'a second pair of 15-ton converters, suitable for the peculiarities of this new process', was erected.

Nevertheless the new process had certain disadvantages. As D. L. Burn says: 'The new process must be dearer than the old. Basic additions were a new expense; there must be more slag; relining would be more onerous, and working more interrupted.'[134] Thus, it was a question if the cheapness of the phosphoric pig iron (as compared with the haematite Bessemer pig iron) would compensate for the additional costs of the new process; in this country the advantage was not so marked;[135] therefore the Thomas process was more quickly seized upon by foreign competitors. It was they, in fact, who demonstrated not only that basic steel was a satisfactory substitute

[131] H. G. Reid, *Middlesbrough and its Jubilee* (1881), p. 91. In 1872, the district's 1000 puddling furnaces had an output of over 440,000 tons of manufactured iron. It is true that there was a change-over to the making of plate and angle iron for shipbuilding.

[132] *Bolckow and Vaughan, 1879–1929*, pp. 15–16. The company was a partner along with Dowlais and Krupp's in the Spanish Orcanera Mining Company.

[133] Reid, *op. cit.*, p. 95. [134] Burn, *op. cit.*, p. 76.

[135] That is, after imports of suitable foreign ore had deprived the Cumberland pig iron of its scarcity value.

for Bessemer steel, but also that for some purposes (where ductility was required) it was superior. Thus, as Burn points out, 'Thomas had "dealt a blow"—a fatal one—at the supremacy of English haematite steel', and even in the Cleveland area there was an unprecedented slump in the prices of pig iron in the 1880s.[136]

The history of this period falls outside our present study. Of course, it ought perhaps to be pointed out that the basic process was not confined to the Bessemer method of making steel; in fact in the 1880s the basic open-hearth process was perfected. Unfortunately, because of the mistaken policy of the steelmakers of this later period, the adoption of this now almost universal method was delayed until the 1890s.[137]

Finally, we may note, with Sinclair, that by then the Bessemer steel industry in Britain was thought to be in decline. Certainly the number of Bessemer converters, for instance, declined from 126 in 1886 to 101 four years later. As the *Iron and Coal Trades Review* in July 1895 reported, 'As regards to Bessemer steel industry it is apparently in a semi-moribund condition. Year after year, it has not only ceased to make progress, but it has absolutely declined.'[138] However, the immediate task here has been to describe the supersession of the wrought iron industry.

[136] Burn, *op. cit.*, p. 77.

[137] In the mid-nineties, when more than 2 million tons of o.h. steel were made annually by the acid process, less than one-tenth of that amount was made in the basic o.h. furnace. (Sinclair, *loc. cit.*, p. 43.)

[138] Sinclair, *loc. cit.*, p. 44.

APPENDIX I

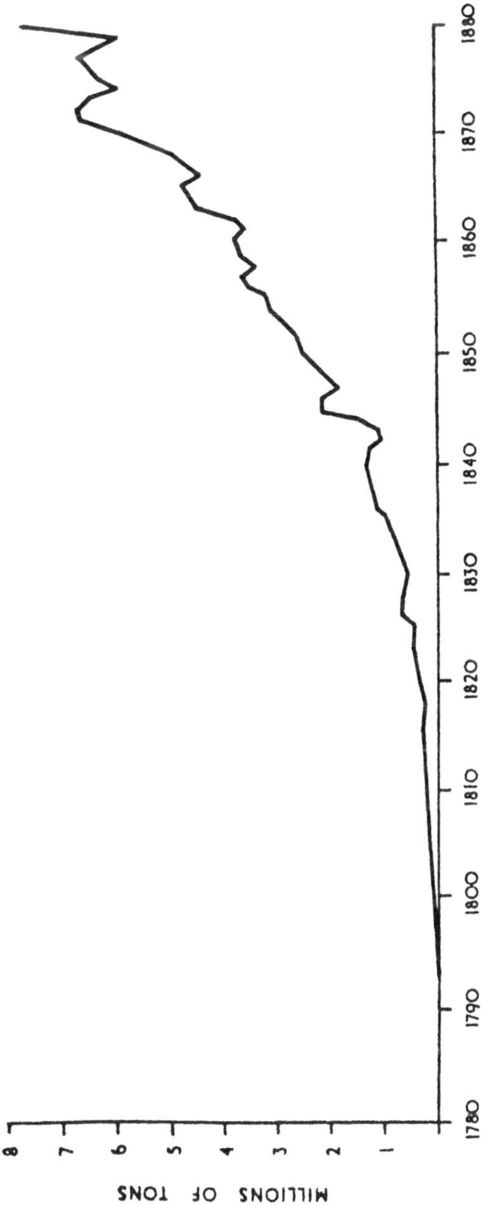

OUTPUT OF PIG IRON IN THE UNITED KINGDOM

1788—1880

APPENDIX II

List of Iron Furnaces in Blast, July 1849

	No. of furnaces
SCOTLAND	
Devon	1
Forth	5
Lochgelly	2
Kinneil	4
Carron	3
Govan	5
Clyde	7
Carnbrae	6
Monkland	3
Chapel	3
Calder	7
Omoa	4
Castle Hill	2
Kilbirnie	9
Blair	6
Kilwinning	3
Portland	4
Dalmellington	3
Muirkirk	3
Garscube	1
Lugar	4
Nithsdale	3
Shotts	4
Coltness	6
Langlean	6
Gartsherrie	16
Summerlie	6
Dundyvan	9
N.E. ENGLAND	
Ridsdale	3
Hareshaw	1
Bedlington	1
Wylam	1
Tyne	2
Walker	2
Birtley	2
Derwent	7
Crook Hall	7
Taw Law	2
Wilton Park	4
Stanhope	3
CUMBERLAND	
Cleator Moor	2

	No. of furnaces
S. YORKS & DERBYSHIRE	
Renishaw	2
Staveley	4
Wingerworth	3
Clay Cross	2
Alfreton	3
Codnor	3
Butterley	3
Morley Park	2
Stanton by Dale	1
SOUTH WALES & MON.	
Coelbrook	2
Trym Saren	3
Brym Anmar	2
Ynyscedwyn	7
Ystalyfera	11
Banwen	1
Onllwyn	2
Jarvis & Co.	2
Hirwain	5
Neath Abbey Co.	2
Cwm Avon	7
Maesteg	3
Garth	3
Llynvi	3
Tonddu	2
Cefn Crib	1
Pentyrch	2
Typwca	1
Blaindar	3
Pontypool	3
Bristol Iron Works	5
Varteg	2
Blaenavon	5
Nantyglo	7
Beaufort	7
Ebbw Vale	4
Victoria	4
Sirhowy	5
Tredegar	9
Rhymney	10
Dowlais	14
Pen-y-darren	7
Plymouth	8
Cyfarthfa	7
Aberdare	6

	No. of furnaces
YORKSHIRE	
Farnley	1
Leather & Co.	1
Low Moor	4
Bowling	3
Brierley	3
Shelf	3
Glasshouse	2
Elsecar	3
Masbro'	1
Milton	2
Thorncliffe	3
Chapeltown	2
Parkgate	1
Holmes	2
Unstone	1
Newbold	1
Duckmanton	2
SHROPSHIRE	
Donnington Wood	2
Lilleshall	6
Ketley	2
Oakengates	2
Wormbridge	2
Old Park	2
Langley	2
Dawley	2
Madeley Court	3
Madeley Wood	3
Lightmoor, Horsehay } Coalbrookdale Co.	5
SOUTH STAFFS.	
Oldbury	2
Union	3
Old Park	2
Crookhay	3
Goldshill	3
Wednesbury Oak	3
Willingsworth	2
Broadwaters	2
Old Birch Hills	1
Pelsall	2
Darlaston	1
New Darleston	2
Dudley Port	3
Coneygre	3
Horseley	3
Tipton	3
Priorfields	3
Deepfield	3
Caponfield	3
Ettingshall	2
Stonefields	1
Halefields	1
Bonvereux	2
Coseley	2
Bilston	2
Millfield	2
Parkfields	3
Priestfields	2
Stow Heath	3
Osier Bed	3
Chillington	4
Netherton	1
Dudley Wood	2
Bumble Hole	2
Corngreaves	2
Ketleys	3
Russell's Hall	3
Dixon's Green	1
Withymoor	1
Parkhead	1
Corbyn's Hall	6
Level Iron Works	3
Lays	3
Shutt End	4
Oak Farm	1
Woodside	2
Wolverhampton	2
Moseley	3
New Birch Hills	2
Old Hill	1
Tipton Green	1
NORTH WALES	
Leeswood	2
Coedtalow	2
Frwyd	1
Brymbo	2
Ruabon	1
FOREST OF DEAN	
Dean Forest	3
NORTH STAFFS.	
Shelton	3
Lane End	2
Latebrook	2
Kidsgrove	3
Apedale	3
Silverdale	3
Madeley	1
Goldendale	2
TOTAL	573

Based on *Report of Select Committee of House of Commons on Application of Iron to Railway Structures, 1849*.

Appendix III

Appendix III

The Coal and Iron Districts Showing Furnaces in Blast in July 1849

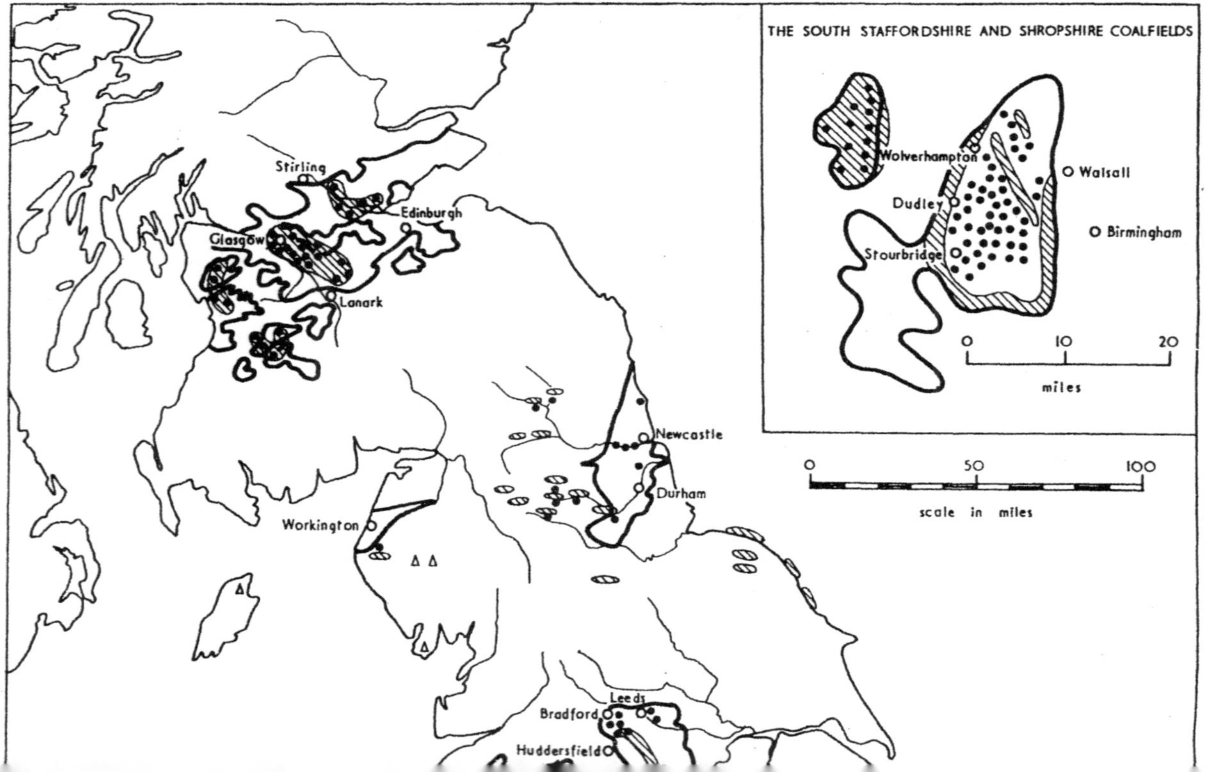

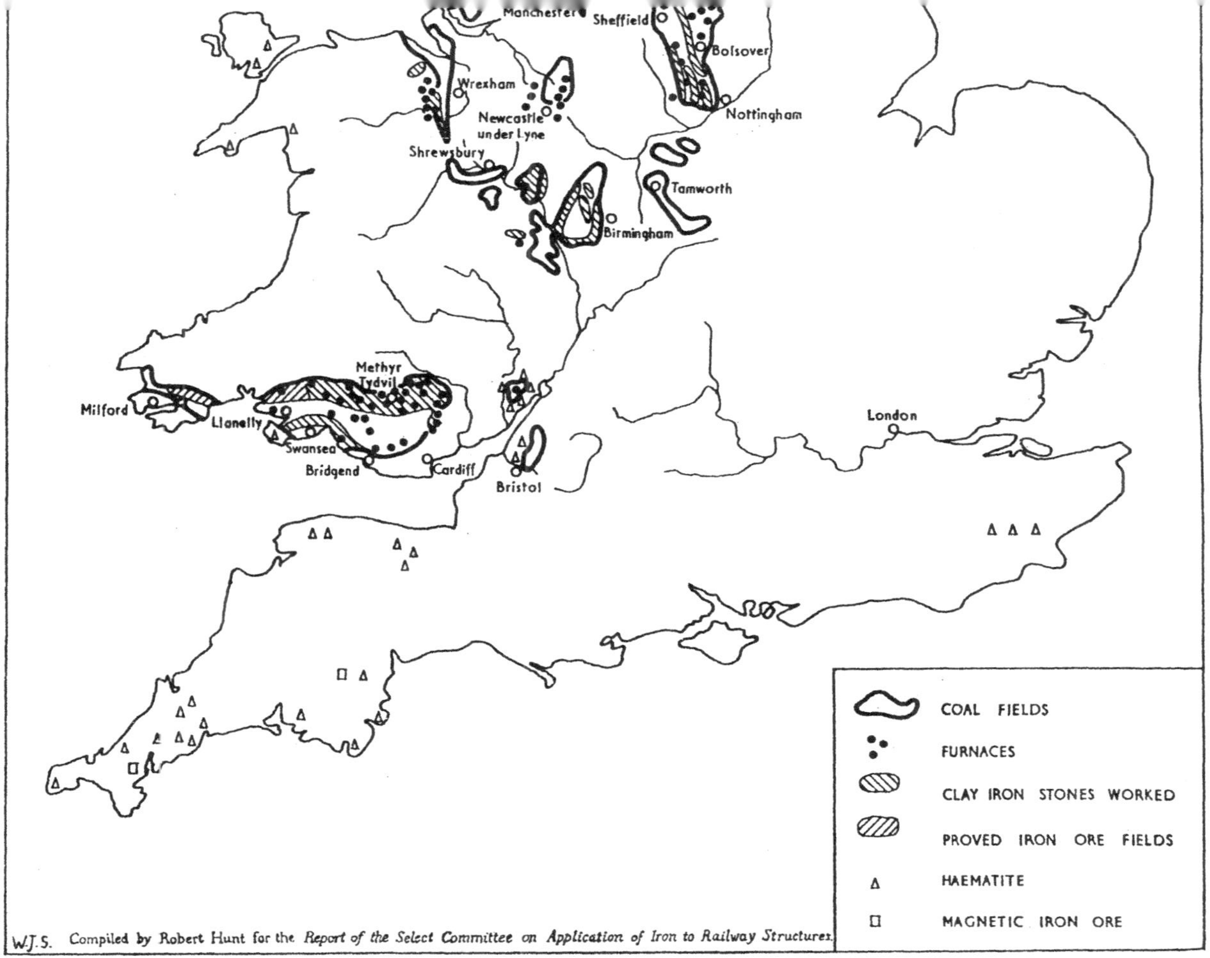

W.J.S. Compiled by Robert Hunt for the *Report of the Select Committee on Application of Iron to Railway Structures.*

Index

R

S

For Product Safety Concerns and Information please contact our EU representative GPSR@taylorandfrancis.com
Taylor & Francis Verlag GmbH, Kaufingerstraße 24, 80331 München, Germany

www.ingramcontent.com/pod-product-compliance
Lightning Source LLC
LaVergne TN
LVHW011106300926
849160LV00008B/292
* 9 7 8 1 1 3 8 8 6 5 3 2 7 *